39th Riso International Symposium on Materials Science 2018

Fatigue of Composite Materials: Microstructure, Mechanics and Methods

IOP Conference Series: Materials Science and Engineering
Volume 388

Roskilde, Denmark
3 – 6 September 2018

ISBN: 978-1-5108-6784-0
ISSN: 1757-8981

Printed from e-media with permission by:

Curran Associates, Inc.
57 Morehouse Lane
Red Hook, NY 12571

Some format issues inherent in the e-media version may also appear in this print version.

Printed by Curran Associates, Inc. (2018)

For permission requests, please contact the Institute of Physics
at the address below.

Institute of Physics
Dirac House, Temple Back
Bristol BS1 6BE UK

Phone: 44 1 17 929 7481
Fax: 44 1 17 920 0979

techtracking@iop.org

Additional copies of this publication are available from:

Curran Associates, Inc.
57 Morehouse Lane
Red Hook, NY 12571 USA
Phone: 845-758-0400
Fax: 845-758-2633
Email: curran@proceedings.com
Web: www.proceedings.com

Table of contents

Volume 388

39th Risø International Symposium on Materials Science 'Fatigue of Composite Materials: Microstructure, Mechanics and Methods'

3–6 September 2018, DTU Risø Campus, Roskilde, Denmark

Accepted papers received: 4 July 2018
Published online: 19 July 2018

Preface

Papers

Invited papers

Contributed papers

IOP Conf. Series: Materials Science and Engineering **388** (2018) 011001 doi:10.1088/1757-899X/388/1/011001

27 June 2018

Preface for the 39th Risø Symposium Proceedings, IOP publication

PREFACE

Proceedings of the 39th Risø International Symposium on Materials Science: *Fatigue of Composite Materials: Microstructure, Mechanics and Methods*

S. Fæster, S. Goutianos, H. Lilholt, B. Madsen, L.P. Mikkelsen, B.F. Sørensen, H.L. Toftegaard

The 39th Risoe symposium aims to address the current research activities and future trends on the fatigue of high performance composite materials for structural applications (e.g. wind turbine blades). It aims to provide a forum for discussion, which covers all aspects of fatigue of composites: from microstructure to composite structures.

The focus of the symposium is on polymer matrix composites for structural applications, i.e. composites with long, aligned glass, carbon, metal, natural or polymeric fibres. With such a strong focus, we aim to enhance interactions among scientists and engineers from both Academia, Research Institutions and Industries on the specific topic of fatigue of structural polymer fibre composites. Together with the high quality presentations on the basis of strict selection and long time for discussion, we hope to promote exchange of ideas between the participants.

The main topics of the symposium can be summarized as follows:

• Effect of properties of fibres, matrix and interface on fatigue properties • Fatigue of two-fibre hybrid composites • Improved fatigue test methods • Size and scaling laws for structural design • Delamination fatigue crack growth • Multiaxial fatigue • Microscale damage evolution • Micromechanical modelling of fatigue life / fatigue limit • Effect of residual stresses on fatigue • Effect of microstructural defects on fatigue • Monitoring fatigue damage in composite materials • High cycle fatigue

The Proceedings contain 8 invited papers, and 11 contributed papers.

The 39th Risø International Symposium is organised by the Section of Composites and Materials Mechanics, Department of Wind Energy, Technical University of Denmark (DTU), at the Risø Campus. We would like to thank all those at DTU Wind Energy who assisted in the preparation of the Symposium. We appreciate additionally the support from the international advisory committee consisting of: Povl Brøndsted, DK; Douglas Cairns, US; Kristofer Gamstedt, SE; Todd Griffen, US; Hiroyuki Kawada, JP; Theodore Philippidis, GR; Marino Quaresimin, IT; Ramesh Talreja, US; Janis Varna, SE and Anastasios Vassilopoulos, CH.

IOP Conf. Series: Materials Science and Engineering **388** (2018) 011002 doi:10.1088/1757-899X/388/1/011002

Peer review statement

All papers published in this volume of *IOP Conference Series: Materials Science and Engineering* have been peer reviewed through processes administered by the proceedings Editors. Reviews were conducted by expert referees to the professional and scientific standards expected of a proceedings journal published by IOP Publishing.

Improving the Performance of Glass Fiber Laminates

Georg Adolphs

Owens Corning, C/55 km 20, E-08295 Sant Vicenç de Castellet, Barcelona, Spain

Georg.Adolphs@Owenscorning.com

Abstract. The renewable energy sector and wind turbine industry in particular have improved their cost of energy to competitive levels [1]. For the materials used to produce wind turbine blades this means to look for ways to improve mechanical performance vs cost while keeping reliability, light weight and durability. For all this, knowing the limits of a material's reliable performance is key. In the past years more and more so called heavy fabrics are proposed. Heavy fabrics are understood as those fabrics with an area weight above commonly used 1200g/m2. Another point which has been of concern in the market in recent years is the evolution of mechanical properties during extended terms of storage and during transportation. This paper will address these 2 aspects.

1. Heavy fabrics

Heavy fabrics reduce the number of layers to be placed and allow a reduction in processing time and cost. Main applications are spar cap and the root joint area. Resin infusion properties have to be controlled and – if possible – improved. A possible solution is shown in figure 1.

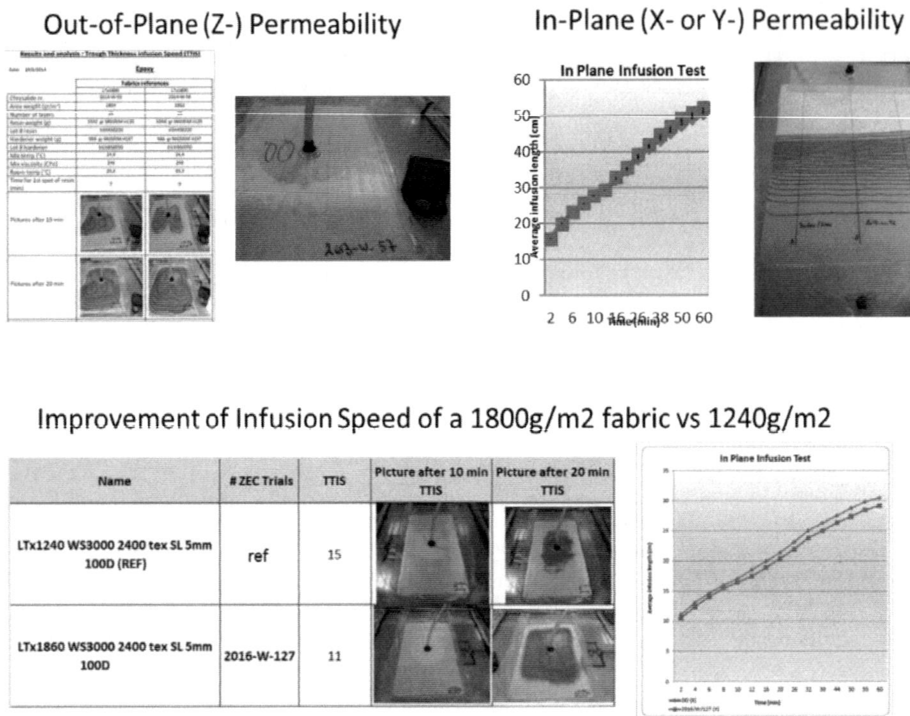

Figure 1: Infusion permeability tests carried out

While the infusion (flow) speed is critical for the processing time, the variation of fiber volume fraction is critical for the variation of stiffness. In a series of infusion tests, an example is shown in figure 2, this variation can be shown and its impact further studied.

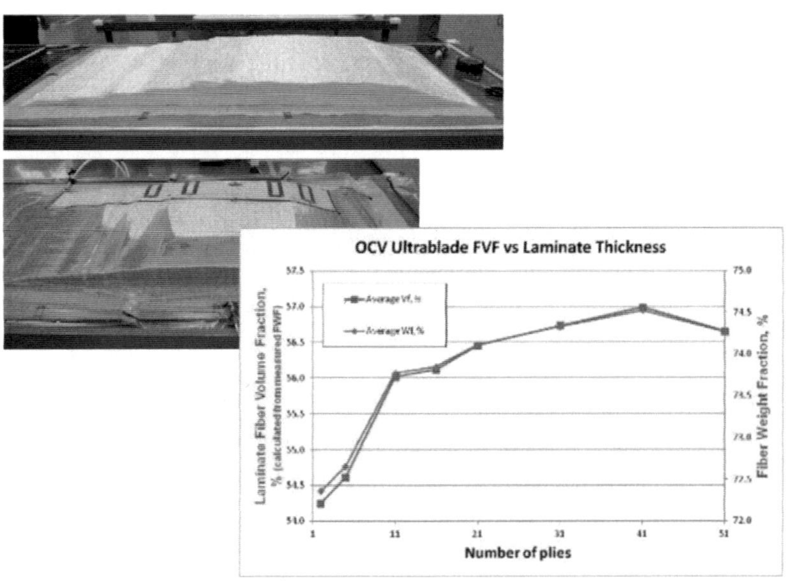

Figure 2: Effects of number of layers/plies on Fiber Volume Fraction (FVF)

As it can be observed, FVF values increase from values below 55% rapidly to 56% when +10 layers of UD fabrics are reached. Within this laminate thickness however further increase does not exceed another 1%. This means that within current process technology, use of membranes for example and proper permeability of the UD fabrics, the increase of FVF can be well managed below +2%.

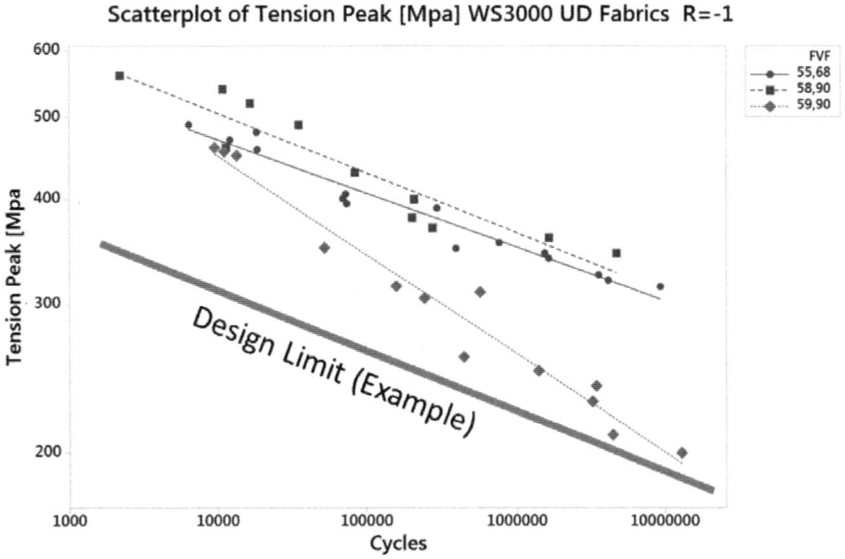

Figure 3: Fatigue results, R=-1, UDFA method, UD laminates vs FVF

In figure 3 fatigue curves for an identical fabric resin combination at different fiber volume ratios are shown. The 50% likelihood curves are indicated as well as a hypothetical fatigue limit considered for a blade design. The green curve at 59,9% FVF will most likely produce values which are below the design limit values considering usual spread of observations in the cycle range of 10^6 upwards, while the red curve at 58,9% is still uncritical over the whole shown frequency range.

The recommendation for this given requirement – performance combination is hence not to increase in laminate FVF values beyond 58,90%. Taking into account the before mentioned process tolerance of +/-1% for the fiber volume fraction, the target value of FVF in this laminate would be rounded 58%.

At this value the fatigue properties can be reached.

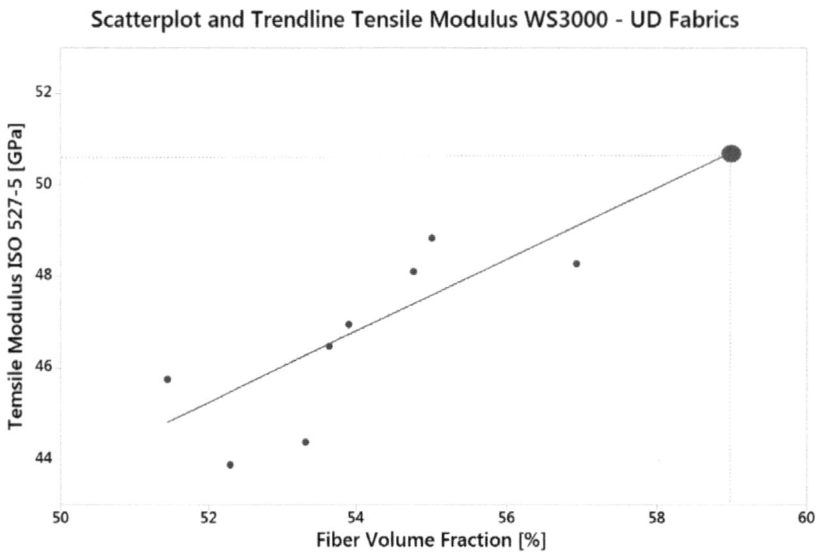

Figure 4: Modulus vs FVF correlation of Owens Corning WS3000 UD fabrics

With the information of several tensile modulus tests a statistical correlation allows to predict at this FVF of 58% a modulus of 49,6 GPa for a given group of fabrics with identical fiber modulus and cross fiber amount. Important here is to stress that this is the solution for a given fatigue requirement. If fatigue requirement can be reduced, at 59% FVF for example a modulus of 50,5GPa can be obtained.

With the information that lower FVF values are found at lower number of layers, or mainly in areas close to the blade tip, the reduced modulus will not affect strongly blade tip deflection for geometrical reasons.

Other applications of heavy fabrics are in root joint laminates where quickly thick laminate structures are build. High modulus glasses have shown their advantages in these applications. By selectively increasing the blade root stiffness better flanging designs can be build.

2. Long term storage effects on laminate properties

Storage can affect negatively the performance of commonly used composite materials. While this is widely accepted in the area of reactive resin systems, a possible reduction of glass fiber properties has not been considered for many applications.

Thee reduction of properties can be accounted for in different ways as shown in figure 5 and 6.

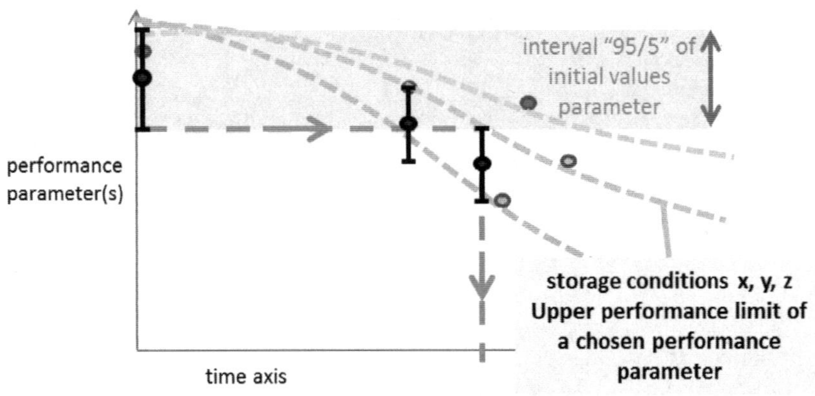

Figure 5: Variation of average properties

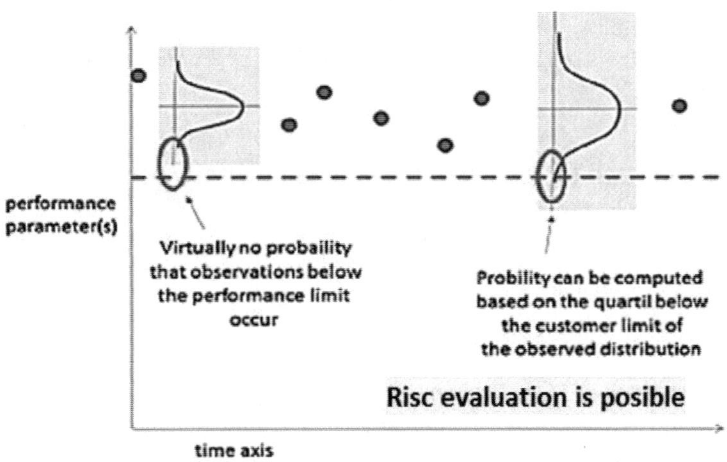

Figure 6: Variation of distribution of properties

In figure 5 a reduction of properties over time and under measurable conditions occurs. When the initial lower confidence limit is higher than the observed upper confidence limit after storage, the property of interest has sufficiently reduced to state at that moment the material has significantly lower properties than initially, so its storage lifetime under the indicated parameter has been reached. This value can be computed independently from customer requirements, by applying standard alpha and beta coefficients of a normal distribution for example. A good example here is the so called 5%/95% rule.

Another possibility is to use the observed distribution and compare this against a set criteria of minimum performance. This method is shown in figure 6. A possible variation of the average property is not required. The likelihood of an observation below a limit performance can be computed. A risk level according to the application can be established.

Owens Corning stores industrially produced material, glass fiber roving and fabrics, for several years now at different own warehouse to evaluate the impact of storage and to establish at regular time intervals the mechanical laminate properties.

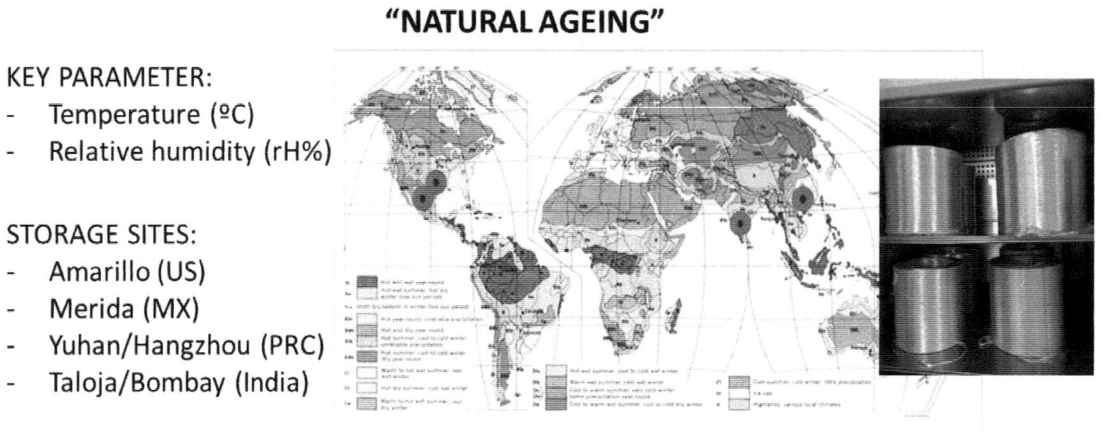

Figure 7: Storage and climate areas chosen by Owens Corning

As shown in figure 7 selected warehouses are used as "natural ageing" sites, but we test unpacked material as well and undergo specific ageing cycles in climate conditioned rooms. Chemical analysis is done on these materials too.

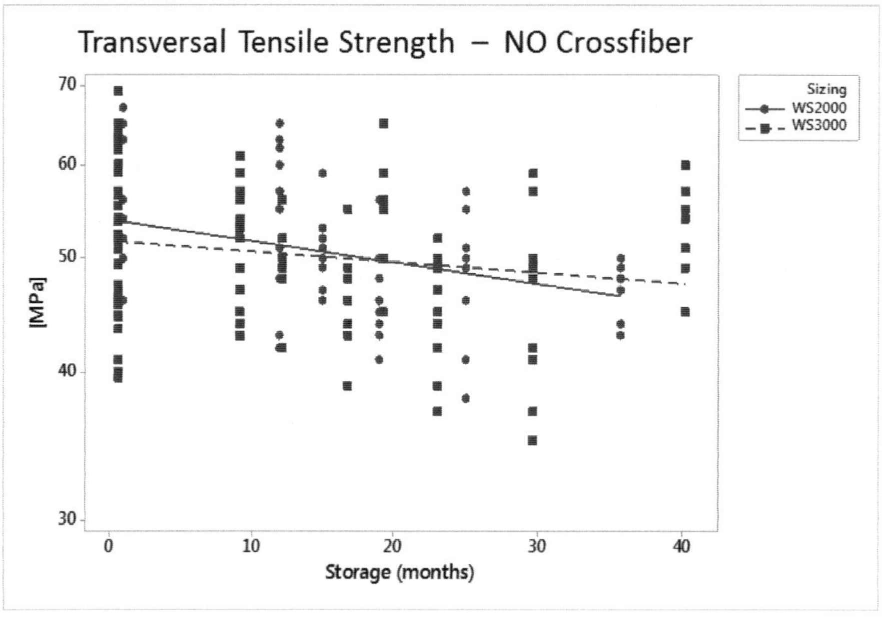

Figure 8: Findings on natural aged material, transversal tensile strength (wound, infused and cured roving panels)

Basis for these tests are roving and fabrics stored the indicated time under the local climate conditions. Temperature and humidity levels have been checked. The rovings as indicated in figure 8 have been wound into laminate panels, infused with a standard epoxy resin system, post cured, cut, and measured.

From these results both methods shown in slide 5 and 6 can be applied. The average degradation can be computed, or the likelihood to observe values below a certain limit value can be computed.

An average property degradation can be "seen" argued according to the regression lines, but a "significantly reduced performance" versus initial performance cannot be found yet after 40 months of test time.

Similar observations can be made on fabrics, stored under similar conditions and tested.

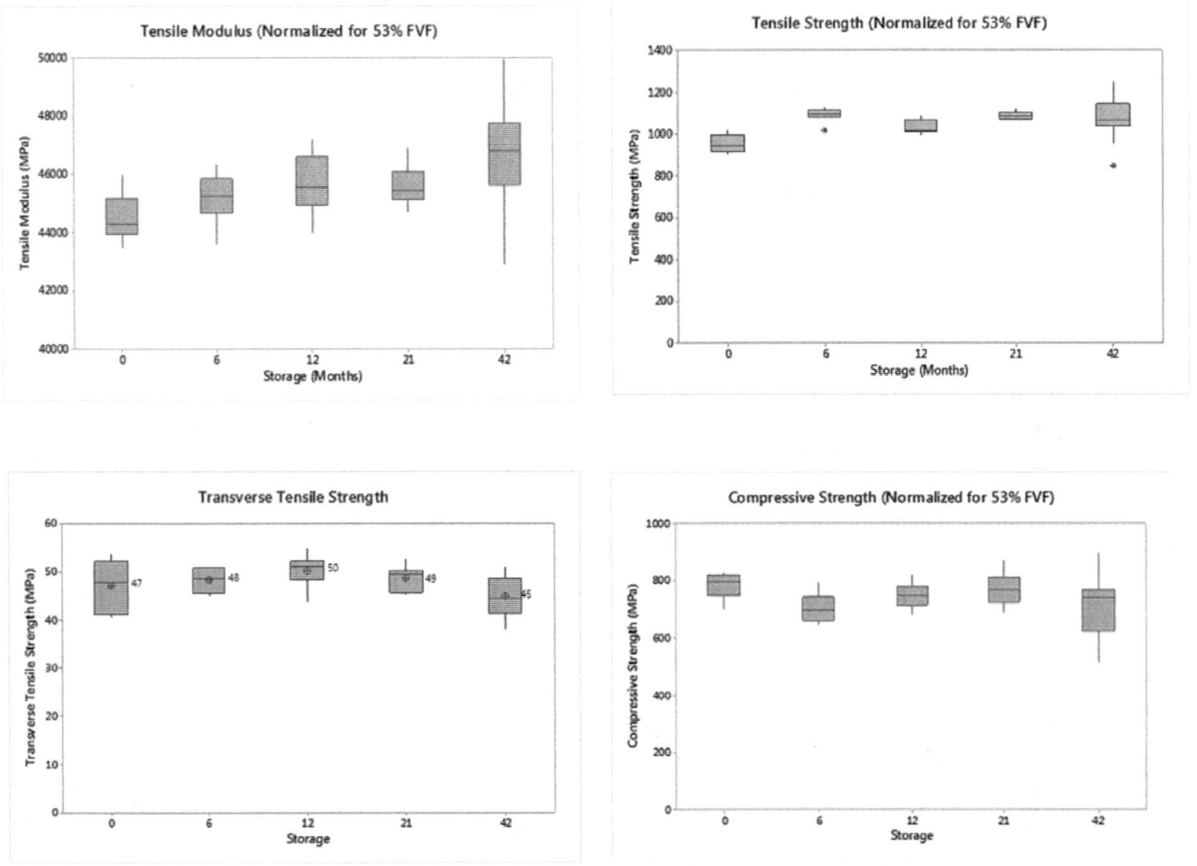

Figure 9: Key test results of stored fabrics after up to 42 months

References

[1] Make Consult (Wood Mackenzie), Aarhus, Denmark, Market Research Presentation "Global Wind Industry Trends – DWIA 2018", published 5 April 2018

Fatigue delamination of composite materials – approach to exclude large scale fibre bridging

René Alderliesten

Department of Aerospace Structures and Materials, Faculty of Aerospace Engineering, Delft University of Technology
Kluyverweg 1, 2629 HS, Delft, The Netherlands

r.c.alderliesten@tudelft.nl

Abstract. This paper proposes a straightforward methodology to generate fatigue delamination curves without the influence of fibre bridging using test data obtained on unidirectional double cantilever beam specimens with fibre bridging.

1. Introduction

Ply delamination growth in laminated fibre reinforced polymer composites has received ample attention in the past decades, as it constitutes a dominant failure mechanism for structures made of these materials. To assess the delamination fracture toughness, subject in quasi-static strength justifications, specimen geometries and test procedures have been standardized for the various opening modes. Take for example for mode I, the double cantilever beam (DCB) test [1]-[2], for mode II the end-notched flexure (ENF) test [3] for mixed-mode I/II, the mixed-mode bending (MMB) test [4].

For fatigue driven delamination growth, however, there is no agreement yet on a standard to correctly obtain fatigue delamination data. Although most researchers adopt standard specimen geometries from quasi-static test standards, no consensus exists on the correct method to test, nor on the proper similitude parameters to adopt [5][6]. Despite that several aspects have been studied through round-robin mode I fatigue fracture tests by both the European Structural Integrity Society Technical Committee 4 and the American Society for Testing and Materials, International, Subcommittee D30.06, a standard has not yet been proposed [7][8][9][10].

Among various factors causing scatter in the data, like e.g. accuracy in load cell and (initial) crack length measurements, it seems that the observed fibre bridging, shielding the crack tip, forms the dominant phenomenon hindering standardization.

Many researchers have studied this phenomenon of fibre bridging. Most studies however, focus on quasi-static loading conditions using critical strain energy release rate (SERR) or stress intensity factors to determine fracture toughness. In that case, large scale fibre bridging contributes to the fracture process zone, which complicates the sue of linear elastic fracture mechanics (LEFM). In fatigue delamination growth, these fibres shield the crack tip and store a portion of the applied cyclic load to be released upon unloading. Hence, the application of LEFM may remain more straightforward, as is discussed in this paper.

Fibre bridging appears to be mostly an artefact of testing unidirectional laminates. It occurs in for example DCB laboratory tests, but it is rarely observed and reported for naturally occurring delaminations in operational in-service conditions. This means that fatigue delamination data required in designing a structure for a given operational life should be conservative in that it excludes fibre bridging.

Hence, most work reported in literature aims to predict the fibre bridging contribution, by quantifying indirectly the magnitude of crack shielding imposed by bridging fibres. It seems though that these indirect methods do not allow to standardize fatigue delamination experiments; to come to delamination resistance curves, data is evaluated through theoretical models, which prohibits treating the data as material data.

This paper therefore discusses a method or approaches to unambiguously account for fibre bridging in mode I fatigue delamination experiments, and to come to conservative delamination resistance curves. First, a brief discussion is given on fibre bridging assessment, data processing and test specification. The paper then proposes a methodology compliant to such specification, demonstrating that material data in the form of delamination resistance curves equivalent to 'zero bridging' can be obtained from DCB experiments through a standard test procedure and standard data processing.

2. Assessment of fibre bridging

Literature reports various approaches to evaluate the fibre bridging contribution in mode I quasi-static and fatigue delamination tests. One approach aims at physically removing the bridging fibres [11], while most other approaches attempt to quantify the bridging contribution in order to subtract that from the total work applied. This category can be divided into studies that numerically or analytically model fibre bridging (loads or stresses) [12][13] and studies that attempt to (phenomenologically) correct loading parameters using experimental measurements [9][14][15]. This latter correction is either done with measurements within the same test, or measurements obtained from other tests and specimens.

What all these methods have in common is that the data taken from the experiments is evaluated using theoretical models describing the fibre bridging, which implies that the reported experimental model depends on the appropriateness of the theoretical model adopted. This could be acceptable in the circumstance where everyone agrees to adopt the exact same model, both in processing experimental data and in strength calculations using that data. However, literature evidently has illustrated that many models have been proposed and adopted, each resulting in different delamination growth data. This lack of consensus seems the prime reason for the lack of a standardized fatigue delamination growth test procedure.

3. Data processing and test specification

3.1. Data requirements

Following the discussion on proper similitude [5][6], the fatigue data should contain at least the two parameters that define the entire load cycle, i.e. G_{max} and G_{min}, or G_{max} and ΔG, or G_{max} and $\Delta\sqrt{G} = (\sqrt{G_{max}} - \sqrt{G_{min}})^2$. Here, one should consider that $\Delta\sqrt{G}$ constitutes the best analogy with the stress intensity factor range ΔK proposed by Paris and others [16][17].

From a physical perspective, one should not only consider a 'driving force' often described by $\Delta\sqrt{G}$ or ΔG, but also an inherent delamination resistance. Here one could consider the analogy with moving a box over the floor: the movement is described by the applied force minus the effective frictional force between the box and floor. For adhesive fatigue disbond growth, this resistance was observed to correlate well with the maximum work applied, equivalent to G_{max} [18], while the driving force then was described by $\Delta\sqrt{G}$. Hence the two-parameters necessary to describe fatigue delamination growth are not only required to uniquely describe the fatigue load cycle, but also to relate to both a 'driving force' and a 'material resistance' to the delamination growth rate.

In general, there are two approaches to monitor the crack growth, either visually measuring the length of the crack, or calculating the crack length based on compliance calibration. For generating fatigue resistance data for design purposes, either approach suffices. However one should keep in mind that the compliance calibration approach is deemed inadequate for proper material interface characterisation, because any discrepancy between crack growth and strain energy dissipation is not captured with an a priori assumed relationship between the two parameters. This is even more the case for mode II fatigue delamination growth where energy dissipates in process zone in front of the crack [19]. Hence, understanding observations on fracture surfaces in SEM investigations, require the visual measurements of the crack lengths, to determine growth rates, rather than calculating them based on compliance. Only then effective delamination resistance expressed as dU/dA can be established.

3.2. *Data scatter factors*

The crack length a represents the 1-dimensional parameter describing the growth in fracture surface area A. Although for specimen were the thickness or width is constant along the direction of crack growth, theoretically implies a linear relation between a and A, one should keep in mind that change in crack front curvature, or angled crack front in multidirectional laminates might add scatter to crack growth observations.

In that respect, one has to acknowledge the contribution to scatter through fibre bridging in the definition of the pre-crack length [20]. Defining the initial pre-crack length with a tolerance of ±1 mm, could impose significant scatter in the obtained crack resistance curve. This influence could be reduced following the procedure explained in the next section.

4. Methodology for obtaining 'zero bridging' resistance curves

This section outlines a procedure to obtain delamination resistance curves from DCB specimens prescribed in for example [1][2] while eliminating the unavoidable fibre bridging. Although it is recommended here to use $\Delta\sqrt{G}$ as the similitude parameter, one may adopt one of the other earlier mentioned parameters as well.

The main concept underlying the proposed procedure is that eliminating the influence of fibre bridging in a test should require data from that same test specimen only. Hence, the procedure consists of performing multiple fatigue delamination growth tests on the same specimen, which generates multiple crack resistance curves. In the data processing, the data from these curves are evaluated against the effective crack length at which each data point was obtained, and subsequently all data are translated to an effective zero pre-crack length using the relationship against crack length obtained through regression. This procedure can be performed for each specimen individually.

4.1. *Experimental procedure*

The experimental procedure initially corresponds to what is deemed common practice; the DCB specimen geometry is taken from for example [2]. One should note that the length of the DCB specimen should be sufficiently long to extract at least six to seven curves from that specimen. This could be equivalent to about 70 mm crack extension range [21][22]. Here, it is recommended to adopt the specimen geometry described in [21][22], in which the test procedure has been explained before.

4.1.1. Pre-cracking. Upon manufacturing the laminates for the specimens one has to determine the appropriate thickness, because the thickness has been observed to influence the experimental results [23]. An approximately 12-13 μm thin Teflon or FEP foil should be applied to create the starter crack. This crack is quasi-statically extended several millimetres to create the sharp crack tip. Note that this pre-cracking introduces already a first amount of fibre bridging over the generated pre-crack length.

4.1.2. Fatigue loading. The tests are performed in displacement controlled conditions to achieve stable delamination growth. With the initial maximum displacement equivalent to about 80-90% of the ultimate displacement for quasi-static fracture, the extension in delamination growth reduces the loads and the corresponding SERRs. The test is terminated after the delamination growth retarded, which may be between 10^{-8} and 10^{-9} m/cycle. Afterwards the fatigue delamination tip is quasi-statically extended another few mm, similar to the pre-cracking procedure. Whether or not this quasi-static pre-cracking is necessary or not, still has to be determined.

The procedure is repeated multiple times, until at least six to seven curves have been obtained covering about 60-70 mm delamination lengths. At these lengths fibre bridging has fully developed as also illustrated by Holmes et al. [24]. Limiting factors for generating multiple curves may be the specimen length, of the maximum machine actuator displacement; the longer the delamination length, the greater the (initial) opening displacement required to achieve the desired G_{max}. Examples of a set of such curves generated with a single specimen are illustrated in Figure 1.

4.2. Data Evaluation

Paris type power law relationships can be fitted through the curves illustrated in Figure 1 following

$$\frac{da}{dN} = c(\Delta\sqrt{G})^n = c\left[\left(\sqrt{G_{max}} - \sqrt{G_{min}}\right)^2\right]^n \quad (1)$$

These relationships can be evaluated similar to what is standard practice for *S-N* fatigue life curves. There, the slope of S-N curves is often described by the Basquin relationship [25], which also constitutes a power law relationship. Hence for each curve in Figure 1, the data points are translated along the slope of equation (1), described by the exponent *n*, to a selected average value of *da/dN*. This is illustrated in Figure 2 for the first three curves of Figure 1.

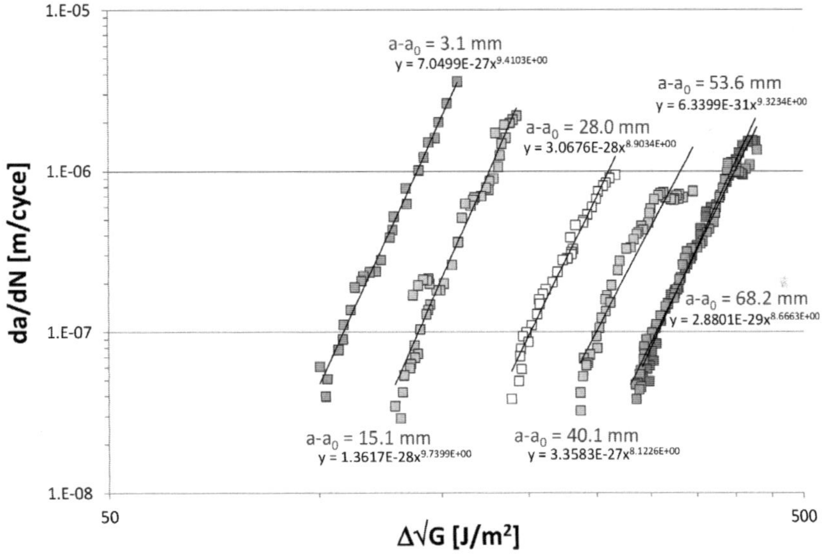

Figure 1. Six successive delamination resistance curves obtained from a single specimen, labelled specimen 7 in [26]

The equation to translate all data points illustrated in Figure 2 is given by

$$log_{10}(\Delta\sqrt{G_T}) = \frac{1}{n}\left[log_{10}\left(\frac{da}{dN_T}\right) - log_{10}\left(\frac{da}{dN}\right)\right] + log_{10}(\Delta\sqrt{G}) \quad (2)$$

where $\Delta\sqrt{G_T}$ is the value of the SERR after translation, and da/dN_T is the corresponding translated da/dN value. Note that the average value of $\Delta\sqrt{G_T}$ at an arbitrary value of $da/dN = 4.5\cdot10^{-7}$ m/cycle plotted against the corresponding crack increment a-a_0 yields a nonlinear relationship that either can be approximated with a bilinear presentation [27], or with a second order polynomial function of a-a_0. The reason for the curve to approximately become horizontal is attributed to full saturation or full development of fibre bridging [21]. The increase of fibre bridging in newly created crack lengths is compensated by the pull-out and failure of bridging fibres at the pre-crack side.

Regression analysis. Each data point in Figure 1 corresponds to a specific value of the crack length or $a\text{-}a_0$, hence one may fit a surface by regression through $\Delta\sqrt{G}$, $a\text{-}a_0$ and da/dN. To enable proper regression, one should first linearize the scales for $\Delta\sqrt{G}$ and da/dN. As mentioned before, the relationship between the crack length $a\text{-}a_0$ and $\Delta\sqrt{G}$ appears to be non-linear, as illustrated in Figure 3 for the data in Figure 1.

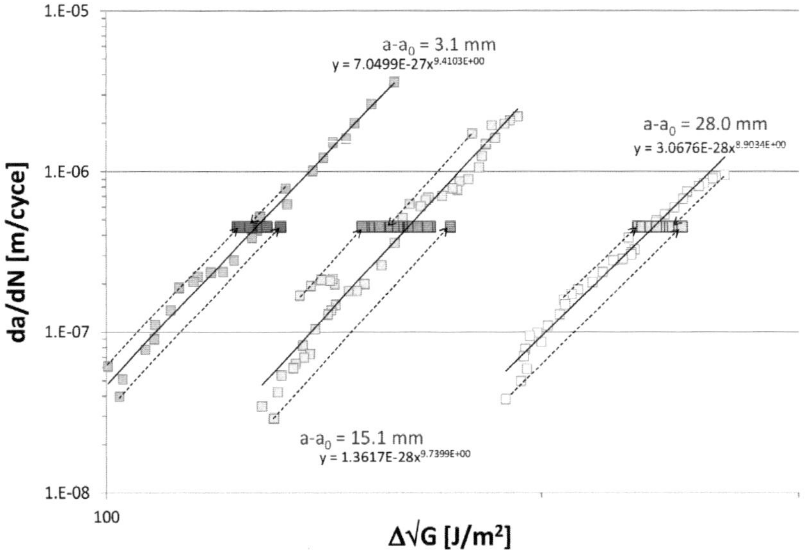

Figure 2. Translating all data along their respective curves to (an arbitrary value of) da/dN = $4.5 \cdot 10^{-7}$ m/cycle for the first three curves in Figure 1

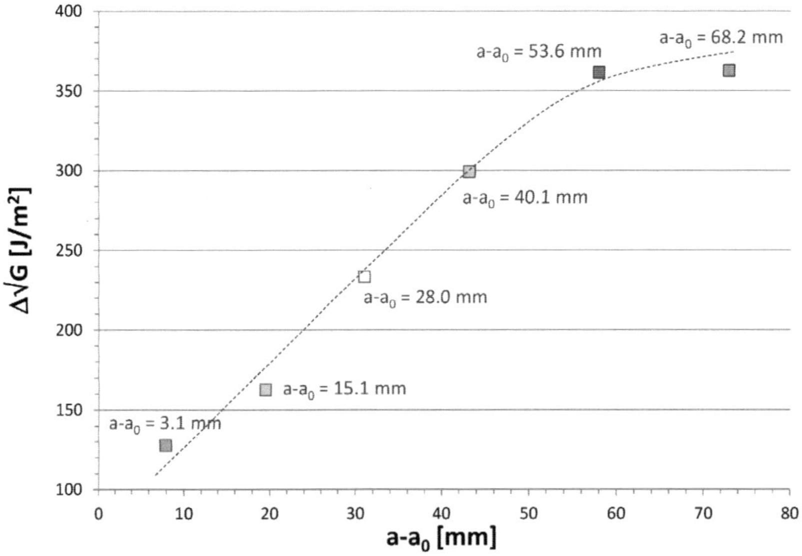

Figure 3. Average crack length at (an arbitrary value of) da/dN = $4.5 \cdot 10^{-7}$ for the curves in Figure 1

To capture the non-linearity in regression one therefore should verify that sufficient curves have been generated up to the full development of fibre bridging. If that is the case, an appropriate equation for regression is

$$Log(\Delta\sqrt{G}) = C_0 + C_1(a - a_0) + C_2 Log\left(\frac{da}{dN}\right) + C_3(a - a_0)^2 + C_4\left[Log\left(\frac{da}{dN}\right)\right]^2 \tag{3}$$

where C_i are the constants obtained through regression of all data in Figure 1.

4.2.1. Data quality assessment. The individual curves can be evaluated through comparison of slopes, while data scatter can be analysed for each curve by determining the distribution at the average level of crack growth rate as illustrated in Figure 2. These distributions can be compared, and if necessary (part of) curves may be omitted for having too much scatter.

A straightforward method to obtain a statistical distribution is by ranking the data points obtained after translation to a selected level of *da/dN* (see Figure 2) from low to high values. Afterwards, a probability P_f (in [%]) can be assigned to each value following [28]

$$P_f = \frac{100n}{n_{tot} + 1} \tag{4}$$

where n is the rank number and n_{tot} is the total number of data points per curve. This allows to plot cumulative density functions for each curve in Figure 1, as illustrated in Figure 4.

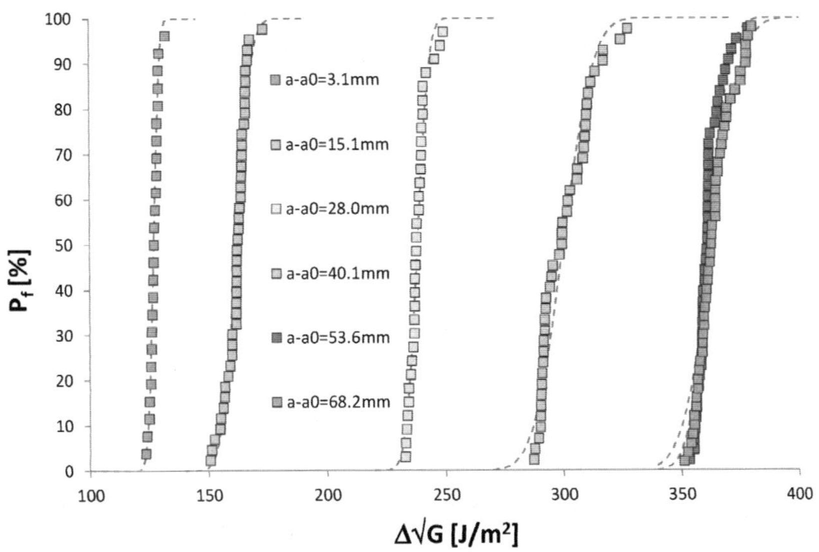

Figure 4. Cumulative density functions at da/dN = 4.5·10^{-7} m/cycle for the curves in Figure 1

Theoretically, the zero bridging curve corresponds to a delamination length of a-a_0 = 0 mm. Hence, once a surface fit has been obtained through regression using equation (3), the average zero-bridging curve can be described by

$$Log(\Delta\sqrt{G})_{av,th} = C_0 + C_2 Log\left(\frac{da}{dN}\right) + C_4\left[Log\left(\frac{da}{dN}\right)\right]^2 \tag{5}$$

Obviously, this average curve has limited value in practice, because an upper bound must be established including all material scatter, not only the scatter associated with fibre bridging. Hence, all individual data points must be translated to the zero-bridging curve of equation (5) while maintaining their relative offset from equation (3). For the curves illustrated in Figure 1, this yields a curve left of all curves as illustrated in Figure 5.

4.3. Reproducibility

To test the reproducibility, first the results of Figure 5 are correlated to the results of another specimen tested under the same test conditions, denoted as specimen 11 in [26]. This specimen has the same material lay-up, specimen configuration and test conditions, including the generation of multiple resistance curves. Few others specimen in [26] have been tested under the same conditions, but only yielding a single resistance curve, or maximum to curves. These will be discussed hereafter.

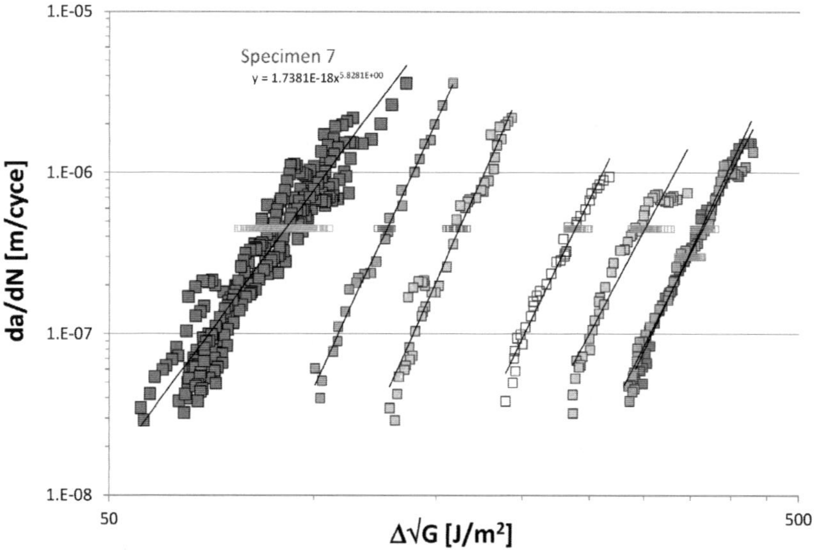

Figure 5. Comparison between the translated curve and the original six curves from specimen 7 in [26]

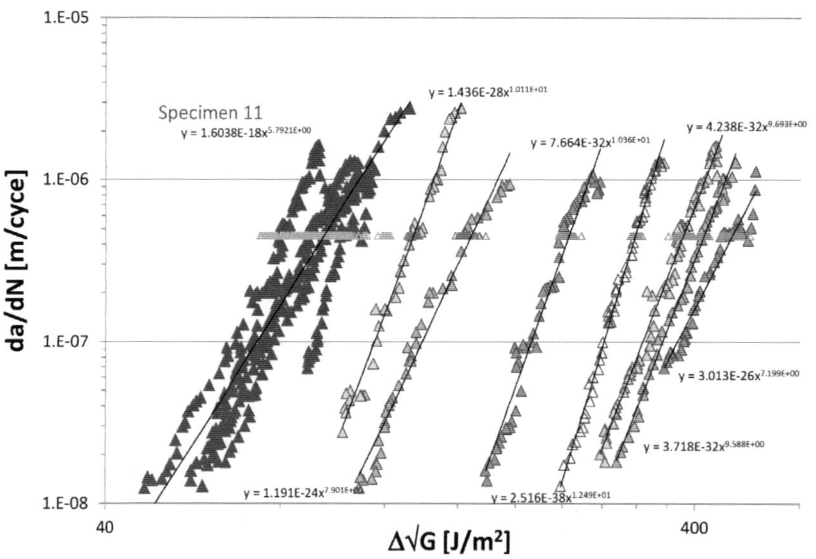

Figure 6. Comparison between the translated curve and seven original curves from specimen 11 in [26]

The obtained resistance curves, with their fit to equation (1), and the curve translated to zero-bridging are presented in Figure 6. To clearly illustrate the excellent correlation between the translated curves from Figure 5 and Figure 6, both are compared in Figure 7. This is further demonstrated by comparing both cumulative density functions for these translated curves in Figure 8. Except for the lower values of $\Delta\sqrt{G}$ both functions correlate very well.

Although a single specimen following the above described procedure yields substantially more data points than traditional single resistance curves, more specimens must be tested to generate so-called A-basis of B-basis allowables [29][30]. For that one must include at least specimens obtained from different material and manufacturing batches.

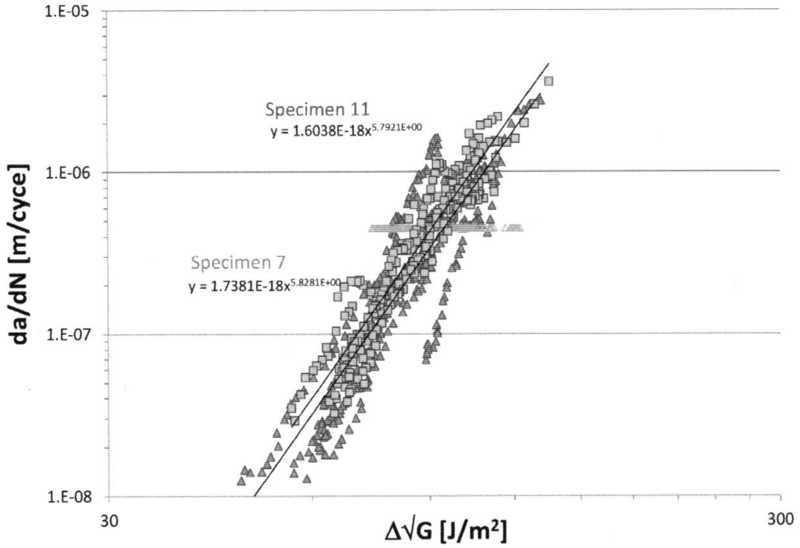

Figure 7. Comparison between the translated curves from Figure 5 and Figure 6

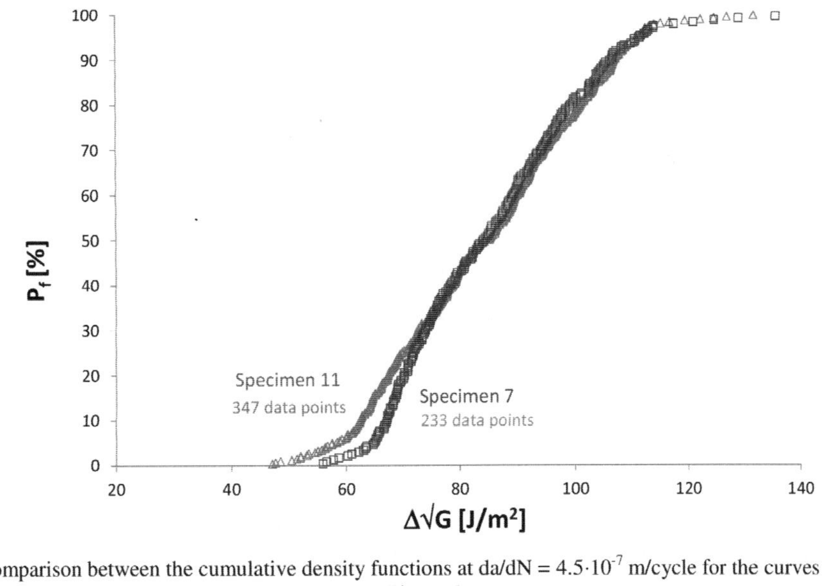

Figure 8. Comparison between the cumulative density functions at da/dN = $4.5 \cdot 10^{-7}$ m/cycle for the curves in Figure 5 and Figure 6

5. Discussion

5.1. Proper characterisation of behaviour and material variability

The results presented in the previous section show an interesting aspect related to the fatigue delamination resistance curves. All slopes obtained by repeating the test procedure, as illustrated in Figure 5 and Figure 6 exhibit similar steep slopes. However, the translation to zero-bridging curves demonstrate that these slopes are an artefact of fibre bridging in the test. As the test is executed in displacement controlled conditions, the high *da/dN* data is obtained at the start of the test, while the near threshold data is obtained at the end of the test. As result of crack growth throughout the test, the crack length at the end is greater than the crack length at the start, and thus fibre bridging is also more dominant at the end than at the start. Since bridging moves data (or curves) to higher values of $\Delta\sqrt{G}$, as illustrated in Figure 1, this implies that the slope of the curves end up steeper than they should be without fibre bridging: the upper portion of the curve smoves less to the right than the lower part, effect9vely changing the slope. Without modelling fibre bridging, the above described procedure incorporates this effect, and yields a zero-bridging resistance curve less steep than all the original curves.

Another observation one can make is that the scatter or variability in the original individual curves are smaller than the scatter in the translated zero-bridging curve. Here, one has to keep in mind that technically with each repetition of the test procedure to obtain multiple cures from a single specimen conditions change. Material variability will appear in the variation of fibre bridging and thus the magnitude of crack tip shielding effectively, which is included when translating all curves to $a\text{-}a_0 = 0$.

Here, one could argue that the material variability one seeks to obtain with testing multiple specimens, to some extent is also obtained with testing the same specimen multiple times. In any case, because of the relatively large amount of data points in the zero-bridging resistance curve, statistical methods can be adopted to establish upper bounds with sufficient confidence. Nonetheless, as mentioned before, this procedure does not eliminate the need to test multiple specimens for defining design allowables.

5.2. Limiting conditions to the proposed procedure

This procedure explained above constitutes a straightforward methodology to derive zero-bridging delaminations resistance curves without requiring theoretical models to describe the fibre bridging. Here one should note that recently Yao et al. [23] demonstrated specimen thickness has no significant effect on the zero-bridging curve. This may seem counter intuitive considering the thickness effects reported for quasi-static loading [12][31].

Nonetheless, one has verify whether certain conditions are met before applying the procedure. Take for example the earlier mentioned tests reported in [26], where only one or two delamination resistance curves were generated. For both specimens the two original resistance curves are plotted against the translated zero-bridging curves in Figure 9.

Evidently, the zero-bridging curves for specimens 1 and 2 do not coincide, nor do they correlate well with the zero bridging curves for specimens 7 and 11 illustrated in Figure 7. This can be best explained with looking at the regression equation, equation (3) and the trend illustrated in Figure 3. Since both specimen 1 and 2 comprise two curves, they effectively form each two data points through which a non-linear regression is performed. Hence, two justify the regression of equation (3), one needs at least a certain amount of resistance curves ranging over sufficient crack lengths $a\text{-}a_0$, to capture a non-linear relationship as illustrated in Figure 3. Simply one or two resistance curves will not work.

As a solution to the inapplicability of the methodology to data comprising too few curves, one could consider combining all data and applying the procedure to the combined data set. The reasoning then may be that the original resistance curves in Figure 9 all correlate to specific crack length (for specimen 2 even up to a-a_0 = 89.6 mm), which altogether contain sufficient range for the regression with equation (3).

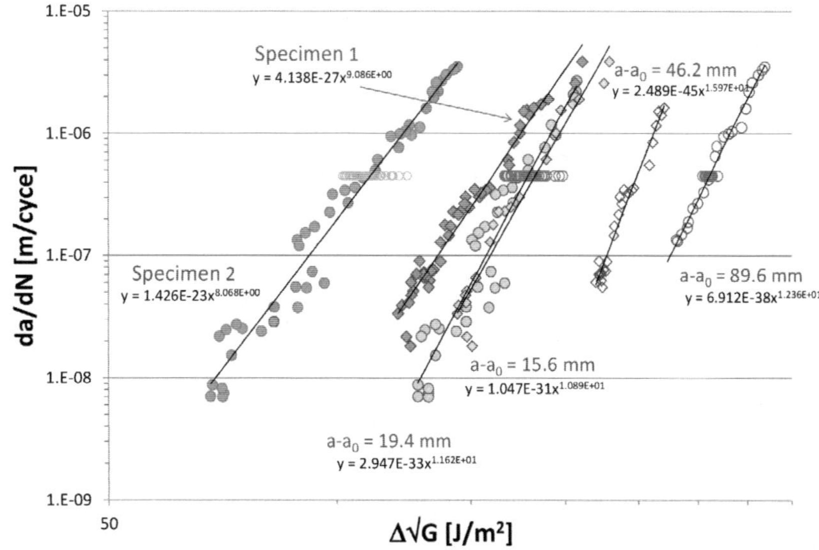

Figure 9. Comparison between the translated curves and two original curves from specimens 1 and 2 in [26]

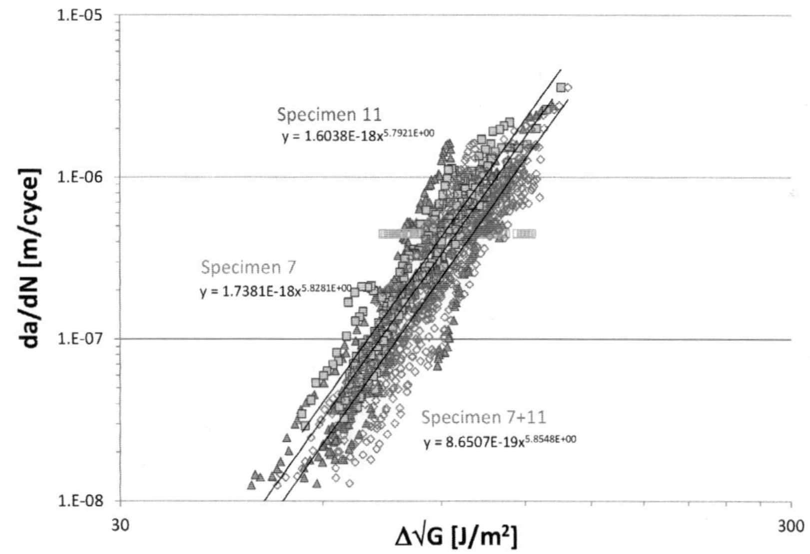

Figure 10. Comparison between the translated curve obtained after regression through the combined data from specimen 7 and 11 with the individual translated curves from Figure 7

To test whether such combined approach is valid, the data from specimens 7 and 11 in [26] can be used. Rather than performing regression per specimen, resulting in the delamination resistance curves shown in Figure 7, all data of both specimens could be combined after which the regression is performed on the total data set. The zero-bridging delamination resistance curves obtained after regression on both datasets combined is compared with the curves from Figure 7.

What is apparent from Figure 10 that the data points obtained with regression through the combined dataset does not overlay the data points obtained with regression through the individual data sets. It appears that the curve has moved slightly to the right compared to the curves from Figure 7. This can be further illustrated by comparing the cumulative density function of the curve with the individual curves shown in Figure 8.

It is evident that performing the regression on the combined dataset rather than the individual datasets, can yield a delamination resistance curve of which the upper bound is moved to the right. Compared to the original curves in Figure 8 this implies an unconservative curve. Whether this interaction between the two datasets relates to the crack growth rate *da/dN*, the SERR $\Delta\sqrt{G}$, or the crack length a-a_0 has yet to be further investigated.

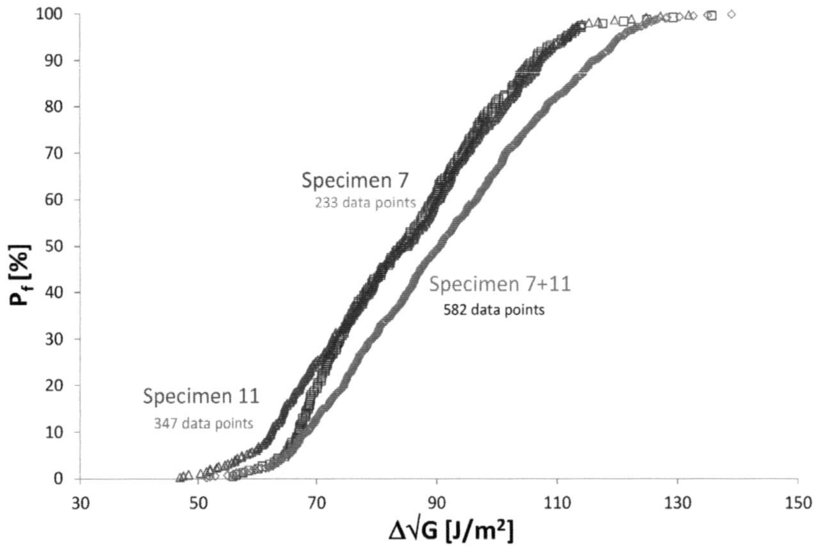

Figure 11. Comparison between the cumulative density functions at da/dN = $4.5 \cdot 10^{-7}$ m/cycle for the curve obtained with regression to the combined dataset and the curves from Figure 8

5.3. Ways forward to standardization

What the above explained procedure and the corresponding data and results have demonstrated is that it is very well possible to generate fatigue delamination resistance curves without the influence of fibre bridging while testing unidirectional laminated composites. Simply taking the DCB specimen and repeating the tests multiple times yields data that can be evaluated without requiring a theoretical model for fibre bridging.

However, the presented results also illustrate that care has to be taken that certain conditions are met before one can claim that the data is valid. The advantage of the presented approach, however, is that straightforward verification is available. One can verify whether the multiple curves obtained exhibit the non-linear relation illustrated in Figure 3 to capture fibre bridging development up to full saturation. One can also assess the individual slopes of the curves or distributions by comparing all distributions based on their individual cumulative density functions as demonstrated in Figure 4.

Translating all curves per specimen to a zero-bridging curve allows comparison between different specimens, to verify whether indeed all specimens exhibit the same delamination resistance curves, like shown in Figure 7 and Figure 8. Combining all individual zero-bridging curves will then allow the application for statistical methods to establish A-basis or B-basis upper bounds for strength justification.

The question left to answer is how to exploit all existing data that in most cases represent single delamination resistance curves per specimen. Following the approach proposed here, one needs to take the specimens and continue testing to generate the other resistance curves. However, if the specimens are no longer available, either because specimens have been taken apart for post-mortem investigations, or because specimens have been thrown away, data might still be used.

This is demonstrated recently in [32] using the same dataset from [26] but evaluated with the Hartman-Schijve relationship [33] proposed previously by Jones et al. [34][35] [36]. All data obtained on different specimens, tested under different loading conditions, collapse together when evaluated with the Hartman-Schijve relationship, yielding a similar curve as the zero-bridging curve presented here. The position and slope of the curve in [32] seems to agree rather well with the curves presented here, which indicates there are more ways to verify the validity of delamination resistance curves.

This seems further illustrated by Yao et al. [27] when proposing an alternative effective SERR parameter after [37], defined as

$$\Delta G_{eff} = \left(\sqrt{G_{tip}^{max}} - \sqrt{G_{tip}^{min}}\right)^2 \tag{6}$$

with

$$G_{tip} = \frac{G_0}{G_{IC}(a - a_0)} G_{applied} \tag{7}$$

where G_0 is derived from extrapolating the trend illustrated in Figure 3 back to a-a_0= 0, while the denominator relates to any point on the trend line of Figure 3. This approach is very simple, and technically performs an analogous or similar procedure as proposed above. It also appears that the zero-bridging curves obtained with this method correlate rather well with the zero-bridging curves presented for example in Figure 7.

The main difference at this moment is in the formulation of the trend in for example Figure 3. Here, it is assumed to represent a polynomial relationship gradually developing to a horizontal asymptote imposed by full fibre bridging development, while in [27] a bilinear relationship is assumed. Further scrutiny of the results is required to decide which type of relationship is preferred before fatigue delamination experiment can be standardized following these procedures.

Here the dataset from [26] might allow a first step into such study, because both 0//0 and 45//45 interface configurations have been adopted. As examined in [38], the amount of fibre bridging associated with these two interface types is substantially different. Hence, both methods of extrapolating or translating data to zero-bridging curves can be applied to these interfaces to investigate whether both interfaces yield similar zero-bridging curves. Nonetheless, one has to consider the possibility that still different curves are obtained, which may then relate to different fracture morphology or crack branching.

6. Conclusions

This paper presents an approach to determine upper bound fatigue delamination resistance curves obtained with tests on unidirectional laminated DCB specimens. The fibre bridging generally observed to occur in these specimen configurations is eliminated with the proposed methodology. The paper discusses some conditions that must be verified before the methodology yields valid results, and it demonstrates how the approach correlates with other approaches proposed in literature.

Acknowledgement

The author wishes to express his gratitude to Liaojun Yao and John-Alan Pascoe for the many fruitful and stimulating discussions during and after their PhD research performed under the author's supervision.

References

[1] ASTM D5528-01. *Standard test method for mode I interlaminar fracture toughness of unidirectional fiber-reinforced polymer matrix composites.*

[2] ASTM D6115-97. *Standard test method for mode I fatigue delamination growth onset of unidirectional fiber-reinforced polymer matrix composites.*

[3] ASTM D7905/D7905M-14. *Standard Test Method for Determination of the Mode II Interlaminar Fracture Toughness of Unidirectional Fiber-Reinforced Polymer Matrix Composites.*

[4] ASTM D6671/6671M-06. *Standard test method for mixed mode I-Mode II interlaminar fracture toughness of unidirectional fiber reinforced polymer matrix composites.*

[5] Rans CD, Alderliesten RC, Benedictus R 2011, *Misinterpreting the results: How similitude can improve our understanding of fatigue delamination growth*, Compos Sci Technol 71: 230–238.

[6] Alderliesten RC 2015, *How proper similitude principles could have improved our understanding about fatigue damage growth*, Proceedings of the 28th ICAF Symposium – Helsinki, 3–5 June 2015.

[7] Brunner AJ, Murphy N, Pinter G 2009 *Development of a standardized procedure for the characterization of interlaminar delamination propagation in advanced composites under fatigue mode I loading conditions*, Eng Fract Mech 76: 2678–2689

[8] Stelzer S, Brunner AJ, Argüelles A, Murphy N, Pinter G 2012, *Mode I delamination fatigue crack growth in unidirectional fiber reinforced composites: Development of a standardized test procedure*, Compos Sci Technol 72: 1102-1107

[9] Murri GB 2014, *Effect of data reduction and fiber-bridging on Mode I delamination characterization of unidirectional composites*. J Compos Mater 48: 2413-2424

[10] Stelzer S, Brunner AJ, Argüelles A., Murphy N, Cano GM, Pinter G 2014, *Mode I delamination fatigue crack growth in unidirectional fiber reinforced composites: Results from ESIS TC4 round-robins*, Eng Fract Mech 116: 92-107

[11] Khan R, Alderliesten RC, Yao L, Benedictus R 2014, *Crack closure and fibre bridging during delamination growth in carbon fibre/epoxy laminates under mode I fatigue loading*, Comp: Part A 67: 201–211.

[12] Spearing SM, Evans AG 1992, *The role of fibe rbridging in the delamination resistance of fiber-reinforced composites*, Acta metall, mater. 40(9): 2191-2199

[13] Daneshjoo Z, Shokrieh MM, Fakoor M 2018, *A micromechanical model for prediction of mixed mode I/II delamination of laminated composites considering fiber bridging effects*, Theoretical and Applied Fracture Mechanics 94: 46–56.

[14] Jones R, Pitt S, Brunner AJ, Hui D 2012, *Application of the Hartman-Schijve equation to represent Mode I and Mode II fatigue delamination growth in composites.* Compos Struct 94: 1343-1351.

[15] Murri GB 2013, *Evaluation of delamination onset and growth characterization methods under mode I fatigue loading*, Langley Research Center, Hampton, Virginia. NASA/TM-2013-217966.

[16] Paris PC, Gomez RE, Anderson WE 1961, *A rational analytic theory of fatigue*, The Trend in Engineering, 13(1): 9-14.

[17] Paris PC, Erdogan F, 1963, *Critical analysis of crack growth propagation laws.* ASME Transactions Journal Basic Engineering, 85D: 528-534.

[18] Pascoe JA, Alderliesten RC, Benedictus R 2017, *On the physical interpretation of the R-ratio effect and the LEFM parameters used for fatigue crack growth in adhesive bonds*, Int J Fatigue 97: 162–176.

[19] Amaral L, Zarouchas D, Alderliesten RC, Benedictus R 2017, *Energy dissipation in mode II fatigue crack growth*, Eng Fract Mech 173: 41–54

[20] Alderliesten RC, Brunner AJ, Pascoe JA (2018), *Cyclic fatigue fracture of composites: What*

has testing revealed about the physics of the processes so far?, Eng Fract Mech, submitted

[21] Yao L, Alderliesten RC, Zhao M, Benedictus R 2014, *Discussion on the use of the strain energy release rate for fatigue delamination characterization,* Composites: Part A 66: 65–72

[22] Yao L, Alderliesten RC, Benedictus R 2016, *The effect of fibre bridging on the Paris relation for mode I fatigue delamination growth in composites*, Compos Struct 140: 125-135.

[23] Yao L, Cuib H, Alderliesten RC, Suna Y, Guoaao L 2018, *Thickness effects on fibre-bridged fatigue delamination growth in composites*, Composites Part A 110: 21–28

[24] Holmes JW, Liu L, Sørensen BF, Wahlgren S 2014, *Experimental approach for mixed-mode fatigue delamination crack growth with large-scale bridging in polymer composites*, Journal of Composite Materials. 48: 3111–3128

[25] Schijve J 2009, *Fatigue of Structures and Materials*, 2nd edition, Springer Science + Business Media.

[26] Yao L, Alderliesten RC 2014, *Mode I fatigue delamination growth in composite laminates with fibre bridging– datasets, collection of datasets*; Available at: http://dx.doi.org/10.4121/uuid:6da548f6-f801-41b4-8d88-db9ae81f6913.

[27] Yao L, Sun Y, Guo L, Zhao M, Jia L, Alderliesten RC, Benedictus 2017, *A modified Paris relation for fatigue delamination with fibre bridging in composite laminates*, Compos Struct 176: 556–564

[28] Van Lipzig H 2016, *Spreiding bij Vermoeiing en Levensduur*, Hoofdstuk 12 PAO cursus Vermoeiing van Constructies, PAO Techniek en Management, Delft www.paotm.nl (in Dutch).

[29] Niu MCY 1992, *Airframe structural design, Practical Design Information and Data on Aicraft Structures*, Conmilit Press Ltd, Hong Kong

[30] Composite Materials Handbook CMH-17, Volume 1. Polymer Matrix Composites: Guidelines for Characterization of Structural Materals, 2012

[31] Suo Z, Bao G, Fan B 1992, Delamination R-curve phenomena due to damage, J. Mech. Phys. Solids. 40: 1-16.

[32] Yao L, Alderliesten RC, Jones R, Kinloch AJ 2018, *Delamination Fatigue Growth in Polymer-Matrix Fibre Composites: A Methodology for Determining the Design and Lifing Allowables*, Compos Struct 196: 8-20.

[33] Hartman A, Schijve J 1970, The effects of environment and load frequency on the crack *propagation law for macro FCG in aluminum alloys*, Eng Fract Mech 1:615-631.

[34] Jones R, Pitt S, Brunner AJ, Hui D 2012, *Application of the Hartman-Schijve equation to represent Mode I and Mode II fatigue delamination growth in composites*. Comp Struct 94: 1343-1351.

[35] Jones R, Hu W, Kinloch AJ 2015, *A convenient way to represent FCG in structural adhesives*, Fatigue Fract Eng M 38, 2015: 379-391.

[36] Jones R, Kinloch AJ, Michopoulos JG, Brunner AJ, Phan N 2017, *Delamination growth in polymer-matrix fibre composites and the use of fracture mechanics data for material characterisation and life prediction*, Compos Struct 180: 316-333

[37] Donough MJ, Gunnion AJ, Orifici AC, Wang CH 2015, *Scaling parameter for fatigue delamination growth in composites under vary loading ratios*, Compos Sci Technol 120: 39-48.

[38] Yao L, Alderliesten RC, Zhao M, Benedictus R 2014, *Bridging effect on mode I fatigue delamination behavior in composite laminates*, Composites: Part A 63: 103–109

Scatter, Scope and Structures: What fatigue fracture testing of fiber polymer composites is all about

Andreas J Brunner

Empa, Swiss Federal Laboratories for Materials Science and Technology, Laboratory for Mechanical Systems Engineering, Überlandstrasse 129, CH-8600 Dübendorf, Switzerland

andreas.brunner@empa.ch

Abstract. The approach to cyclic fatigue fracture testing of metals and alloys with the data presentation as "Paris-Graph" developed by Paul Paris and co-workers has proven immensely useful for structural engineering. It is, therefore, understandable that the later development of fatigue fracture test methods for fiber-reinforced polymer (FRP) composites followed this route also. However, recently questions have been raised by several researchers that require looking at fatigue fracture of FRP composites in more detail in attempts at using the full potential of this class of materials in structural applications. One of these questions is how the apparent analogy in the shape of the curves between Paris-Graphs of metals, alloys and FRP composites relates to the material specific, physical damage mechanisms. For FRP composites, the investigation of these mechanisms has a long history, but only recently, there seems to be a real advance using sophisticated pattern recognition of acoustic emission waveforms with complementary methods such as multi-physics simulations or high-resolution X-ray computed tomography. Comparing, e.g., crack sizes on the order of tens of micrometers occurring on time-scales of microseconds or less with visually observed delamination propagation, it becomes clear that the data analysis for the Paris-graph is averaging over orders of magnitudes in both, length and time-scales and this inevitably involves some scatter. Another question raised concerns the apparently larger scatter in Paris-Graphs obtained for FRP composites compared to those of metals or alloys. Again, a full understanding of the sources of this scatter has not been reached yet, but likely, the complex morphology and microscopic damage mechanisms of FRP composites play a role. There are further effects from mesoscopic or large-scale fiber bridging in fatigue fracture due to special fiber lay-up used in testing. For structural design with FRP composites, it is not sufficient to select the material performing "best" in the fatigue fracture tests, but it is essential to have safe, reliable, but on the other hand realistic design limits. Scatter in the data and how much of that has to be taken into account first requires understanding the extrinsic (test related) and intrinsic (material related) scatter sources. The concept of a fatigue threshold is also important for fracture-based structural design, since there are indications that threshold values for FRP composites may be quite low.

1. Introduction

A major approach to understanding fatigue fracture in material test coupons goes back to investigations by Paris and coworkers [1] on metals and alloys that finally yielded the so-called "Paris-Graph" of da/dN, the average crack advance per fatigue cycle versus the applied stress intensity K, or the range of applied stress intensity ΔK in a double-logarithmic plot showing essentially a power law relation with three regions. These are the transition to quasi-static and finally unstable fracture (at "high" values of da/dN and K or ΔK), a roughly linear region in the center (at "intermediate" values of da/dN and K or ΔK), and a so-called threshold region (at "low" values of da/dN and K or ΔK) described in more detail in [2]. Even though the first graph with data published in [1] had not yet adopted the exact form of the plot as described above (the X-axis showing the stress intensity K was plotted on a linear rather than a logarithmic scale), the authors concluded (cite).

> *"On the basis of the experimental data given, it is evident that rates of crack growth-for example, those in 2024-T3 and 7075-T6 skins of aircraft structure-may be computed by the theory presented over a wide range of nominal stress levels and crack sizes. The ramifications of such broad correlation imply an analytic theory of fatigue based on a concept of growth from initial imperfections through which structural life may be predicted."*

Applying this concept to engineering structural design proved to be extremely successful and, therefore, it did not take long before the methodology was applied to other types of materials, and, starting in the 1980ies also to fiber-reinforced polymer-matrix (FRP) composites, see, e.g., [3-8].

The successful application of Paris-plots with the empirical data correlation, and their later refinements, in structural engineering may have, at least in part, contributed to the fact that alternative descriptions of fatigue fracture, e.g., based on the Forman- or Hartman-Schijve equation [9] or similar equations used in the NASGRO® software with its database for fatigue crack growth analysis [10] were discussed much less in literature. Until recently, the application of NASGRO® was limited to metals or metal alloys, essentially with a focus on aerospace materials, e.g., [10-13]. Application of a modified Hartman-Schijve equation to fatigue fracture data of FRP composites was first discussed by [14-15]. This now looks promising for quantification of scatter and for determining conservative design limits FRP structures, at least for fatigue fracture under mode I tensile opening loads, as discussed in detail by [16,17].

Another drawback coming from the success of the empirical engineering approach to fatigue fracture characterization might also be the limited insight into the underlying physical micro-mechanisms obtained so far. For metals and alloys the possible effects of the microstructure on the fatigue fracture behavior are still debated; with a recent paper indicating that there no such effects [18]. Much less is known about the micro-and meso-structural effects on the physical damage mechanisms in fatigue fracture of FRP [19]. In the following sections, selected aspects of the characterization of fatigue fracture in FRP composite materials are discussed with a focus on scatter sources, the scope of respective test procedures and the implications for structural design.

2. Procedures for fatigue fracture testing of fiber-reinforced polymer composites and their scope

The only example of a standard fatigue fracture test method available so far is ASTM D6115 [20] for Mode I Fatigue Delamination Growth Onset of Unidirectional Fiber-Reinforced Polymer Matrix Composites. The scope defines (cite)

> *"This test method determines the number of cycles (N) for the onset of delamination growth based on the opening mode I cyclic strain energy release rate (G), using the Double Cantilever Beam (DCB) specimen ..... This test method applies to constant amplitude, tension-tension fatigue loading of continuous fiber-reinforced composite materials. When this test method is applied to multiple specimens at various G-levels, the results may be shown as a G–N curve ...".*

The procedure has been developed for a certain class of FRP composites referring to [21] for details:

> *"This test method is limited to use with composites consisting of unidirectional carbon fiber tape laminates with single-phase polymer matrices. This limited scope reflects the experience gained in round robin testing. This test method may prove useful for other types and classes of composite materials, however, certain interferences have been noted ....".*

These interferences mentioned are among others (cite):

"Linear elastic behavior is assumed in the calculation of G used in this test method. This assumption is valid when the zone of damage or nonlinear deformation at the delamination front, or both, is small relative to the smallest specimen dimension, which is typically the specimen thickness for the DCB test. .... Delamination growth may proceed in one of two ways: (1) by a slow stable extension or (2) a run-arrest extension in which the delamination front jumps ahead abruptly. Only the first type of growth is of interest in this test method."

Specifically with respect to application of the test method on other types of FPR composites, it is noted:

"Toughness values measured on unidirectional composites with multiple-phase matrices may vary depending upon the tendency for the delamination to wander between various matrix phases. Brittle matrix composites with tough adhesive interleaves between plies may be particularly sensitive to this phenomenon resulting in two apparent interlaminar fracture toughness values: one associated with a cohesive-type failure within the interleaf and one associated with an adhesive-type failure between the tough polymer film and the more brittle composite matrix. ... Nonunidirectional DCB configurations may experience branching of the delamination away from the midplane through matrix cracks in off-axis plies. If the delamination branches away from the midplane, a pure Mode I fracture may not be achieved as a result of the structural coupling that may exist in the asymmetric sublaminates formed as the delamination grows. In addition, nonunidirectional specimens may experience significant anticlastic bending effects that result in nonuniform delamination growth along the specimen width, particularly affecting the observed initiation values. ... Woven composites may yield significantly greater scatter and unique R curves associated with varying toughness within and away from interlaminar resin pockets as the delamination grows. Composites with significant strength or toughness through the laminate thickness, such as composites with metal matrices or 3D fiber reinforcement, may experience failures of the beam arms rather than the intended interlaminar failures."

The procedure described in [20] is mainly intended for quantification of the effects of materials or processing variables as well as effects of environmental variables, for quantitative comparison of different FRP composites, but may also serve (cite)

"To develop criteria for avoiding the onset of delamination growth under fatigue loading for composite damage tolerance and durability analyses."

These texts in "Scope" and "Significance and Use" of the ASTM Standard Test Methods for fracture of FRP composites are fairly general and, e.g., do not specify in detail how to develop the failure criteria for damage tolerance or durability analyses. The subject of structural design of composites is covered in more detail in [22]. In the earlier version [23] none of the ASTM fracture or fatigue fracture test methods are explicitly referenced, indicating that the direct use of the fracture mechanics test data in design may possibly require more research before becoming formalized in standards and guidelines.

2.1. Developments toward standard fatigue fracture test procedures for FRP composites

The development of test procedures for quasi-static of cyclic fatigue fracture of FRP composites has been reviewed in detail [24-27]. Currently, there are standard test procedures for quasi-static fracture of FRP composites under three types of loads, so-called Mode I with tensile opening load [21,28,29], Mode II in plane shear [30,31] and for various combinations of Mode I and Mode II [32] and one standard test [20] for fatigue delamination onset. Experimental research and activities in fracture mechanics test groups has explored the applicability of the respective quasi-static set-ups for cyclic fatigue facture testing under Mode I [33-34], Mode II [35,36], and Mixed Mode I/II [15,37,38].

Even though Paris-type graphs can be generated by all these draft procedures data analysis revealed certain limitations, e.g., relatively poor reproducibility [39-41] requiring large safety margins for design. There are also indications that the so-called threshold values that could be used to define design limits are fairly low, see Figure 1 and [34,42]. The straight line in Fig. 1 right indicates an estimated threshold at 10^{-7} mm/cycle (equivalent to 10^{-10} m/cycle defined as da/dN for an operational threshold in [2]) around 40 J/m^2, assuming that (1) linear extrapolation of the data is justified and (2) scatter in the visual data is not accounted for. If scatter is taken into account, e.g., by taking twice or three-times the standard deviation, the conservative value (design threshold) would be even lower. It can be noted that the data plotted without smoothing seem to indicate an incipient threshold of roughly 60-70 J/m^2 around 10^{-6} mm/cycle. Possibly, this cloud of data points represents scatter due to limited resolution of the load and displacement measurements, respectively, used in the data analysis.

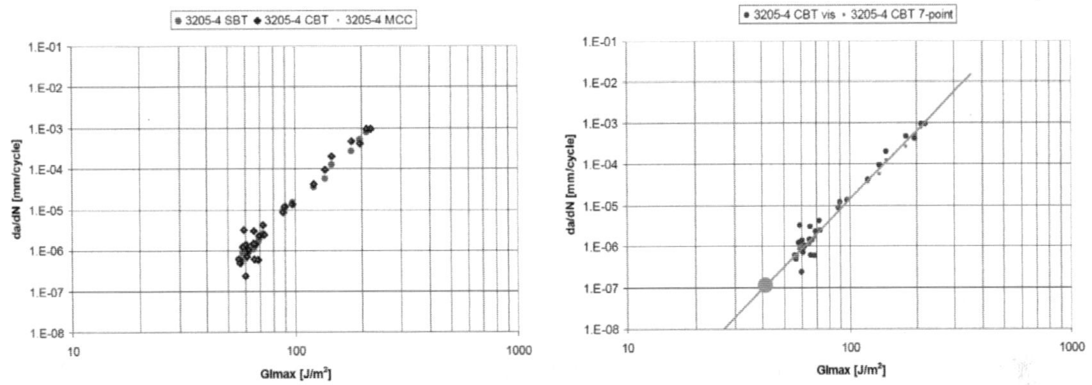

Figure 1. Paris-graph of Mode fatigue fracture data for one type of CFRP (carbon fiber G30-500 / R5259 type epoxy) showing (left) a comparison between different data analysis methods for visually observed delamination lengths (SBT = simple beam theory, CBT = corrected beam theory, MMC = modified compliance calibration) and (right) the data from the CBT-method smoothed with a 7-point polynomial according to [2], the straight green line drawn by eye estimates the da/dN for different applied G_{Imax} values, the green dot indicates the operational threshold $G_{Imax\text{-}thr}$ defined by [2] at 10^{-10} m/cycle.

However, there are recent investigations that point toward intrinsic problems with the cyclic fatigue fracture testing approaches based on the standard quasi-static tests and specifically the choice of the proper X-axis parameter for the data presentation. These issues are briefly summarized in the section below.

2.2. A brief discussion on data analysis and presentation

Since the data presentation for metals or metal alloys based on the double logarithmic da/dN versus K or ΔK graph by Paris et al. [1] was directly applied to FRP composites, the first publications showed da/dN versus ΔK, e.g., the graphs in [5]. Later publications changed the X-axis to either G_{max} or ΔG for the different fracture modes, e.g., the graphs in [39,40,43], since the stress intensity K is experimentally not well defined in FRP composites even though the theory yields (apparently) accurate expressions. As argued in recent publications, e.g., [44,45] replacing K or ΔK by G or ΔG in the Paris-graph does not properly present the fatigue fracture damage in order to predict the behavior of FRP composite structures and equations with "square-root-G" terms are proposed as proper similitude parameter. Rans et al. [44 hence conclude (cite)

> *"Finally, it may be argued that in adopting a Paris relation to fit experimental delamination growth data, the selection of the independent variable for fitting the data is arbitrary. It is merely an exercise in correlating experimental data. This, however, should not be the case. Applying a Paris relation is done not simply to curve fit experimental data, but to attempt to capture the material behaviour based on some basis for similitude. Indeed, numerous bases of similitude could be selected; however, for understanding fracture mechanisms, it is convenient to divide these mechanisms on the basis of cyclic (ΔK, $\Delta G = (\sqrt{G_{max}} - \sqrt{G_{min}})^2$) and monotonic ($K_{max}$, G_{max}) loading effects."*

Besides that, fatigue fracture data presented in literature often show a fairly large scatter in da/dN for a given value of the X-axis parameter (be it G_{max}, ΔG or a square-root G-type quantity). In the Paris-graph, scatter of one order of magnitude in da/dN is not uncommon (see, e.g., [39,40]) and this results in large safety margins that have to be included structural design noted above. Of course, this will limit the applicability of FRP composite materials in light-weight structures. The sources of scatter that have been identified so far are discussed in more detail later.

This brief summary of the current state suggests two main directions for future testing and design development: One is to improve the understanding of the microscopic damage mechanisms and their potential effects on scatter in the data in a rather long-term perspective (see, e.g., [19] for a first approach on that) and the other in drafting test procedures considering the identified sources of scatter and choosing the proper similitude parameter for data presentation. This can, e.g., be based on the Hartman-Schijve equation noted above or variations of that, essentially involving a dependence of the delamination rate da/dN on a "square-root-G" quantity as similitude parameter (see, e.g., [15,16] for more details). There is a perspective that the Hartman-Schijve equation will allow for obtaining a master curve for FRP structural design simultaneously with estimating safe design limits, at least for Mode I fatigue fracture [16]. Selected aspects of FRP structural design and of the scatter sources in fatigue fracture testing of FRP laminates are discussed next.

3. A note on fracture mechanics based structural design approaches for FRP composites

Even though in most designs of composite structures or elements to date interlaminar fracture or fatigue fracture data are not considered yet, the methodology for a fracture mechanics based design has been outlined almost twenty years ago [46]. In the abstract, the author notes (cite)

> *"In laminated composite structures, interlaminar failures or delaminations have continued to be a predominant life-limiting failure mode. The main concern for designers has focused on the damage tolerance of structures in the presence of delamination damage from impact events. The design approach has been to utilize existing stress/strain-based design criteria to design the part, then to impact the part consistent with the threat and to take a knock-down in design loads in the presence of damage. However, delaminations also occur during service in areas where high interlaminar stresses are present. These are generally at required discontinuities in the design, such as cut-outs, holes, joints or ply-drops. The application of interlaminar fracture mechanics as a design tool for optimizing these regions is still in its infancy and is adopted by very few original equipment manufacturers. This paper addresses this issue and presents the more recent developments in incorporating interlaminar fracture mechanics into design."*

The methodology described in the paper combines modelling with experimental data as follows (cite)

> *"The framework consists of identifying where cracks will initiate from structural discontinuities or high interlaminar stresses, or identifying likely damage threats. Using FE analysis to model the delamination, values of the strain energy release rate are compared with the interlaminar properties of the composite. An assessment is then made as to whether the delamination will initiate and, if so, how far it will grow until it causes failure in another part of the component. This failure may be another delamination or a structural collapse (e.g. fibre breakage or buckling) resulting in component failure."*

Both cyclic fatigue fracture test types, i.e., fatigue delamination onset [20] and a cyclic fatigue delamination propagation test [39,40] yielding Paris-type data plots can provide the experimental data for this design methodology.

Unfortunately, to date, there are very few published examples applying this concept to FRP composite structures or components. One example is the design and analysis of a rotorhub flexbeam for helicopters [47,48]. The conclusion in [48] states (cite)

> *"Nonlinear-tapered flexbeam laminates were cut from a full-size composite helicopter rotor hub flexbeam and were tested under combined constant axial tension and cyclic bending loads. Two different graphite/glass hybrid configurations were studied. The terminated plies*

> *were dropped in an overlapping stepwise pattern. All of the specimens failed by delaminations first starting as a crack between the tip of the ply drop group and the adjacent resin region, and growing at the interfaces around the dropped ply toward the thick region of the flexbeam. Delaminations grew with increasing number of cyclic loads, and new delaminations formed at other ply-drop locations. A 2-D finite element model was developed representative of the flexbeam geometry, boundary conditions, and loading. The model was analyzed using two commercially available finite element computer programs, ANSYS and ABAQUS. Delaminations of various lengths were simulated in the analytical model by releasing multipoint constraints (MPCs). Strain energy release rates (G) were calculated using the VCCT using both finite element programs. The calculated peak G values were used with material characterization data to calculate fatigue life curves, for comparison with test data. .... A fracture-mechanics-based methodology was used to relate maximum surface strain to the number of loading cycles at delamination onset for each hybrid specimen type. Good agreement was obtained between tests and analyses. The calculations represent the fatigue life from the opening of the crack at the ply-drop tip until delamination onset, but do not include the cycles necessary to open the crack initially."*

One important detail in the example is that the (cite)

> *"Initial delamination from the resin pocket at the plydrop growing toward the thick region of the flexbeam is dominated by mode II."*

Already more than 20 years ago, it had been pointed out that fracture test data from standard tests showed effects from fiber-bridging that do not occur, or at least to a much lesser extent, for FRP components or structures made from the same fiber-matrix system with different lay-up (fiber orientation). For mode I Cantwell and Kausch [49] noted (cite)

> *"Typical engineering composite structures are rarely unidirectional but contain fibres oriented at large angles to each other. The possibility of fibre nesting occurrence is therefore very limited. ... The effect of simply orientating the central plies by a few degrees relative to each other appears to have a significant influence upon the mode of crack propagation. The amount of fibre bridging observed on the fracture surfaces of the failed samples was considerably less than that in evidence on the standard DCB specimen. The influence of even relatively small amounts of fibre bridging on the mode of crack advance is clearly evident."*

A similar study for mode II [50] concluded (cite)

> *"Engineering structures based on composite materials are rarely unidirectional in nature but contain plies orientated at widely differing angles. Previous work ... has shown that the interlaminar fracture energy at such a ply interface may be significantly different to that associated with an interface between parallel plies. This difference in response is probably associated with fibre bridging, which frequently occurs in unidirectional composites. In this study a number of ENF tests were conducted on samples in which the two centermost plies were offset by four or five degrees. Although this offset angle is comparatively small, it is sufficient to effectively prevent fibre bridging occurring."*

This issue has been taken up in recent publications again, pointing out that in quasistatic and cyclic fatigue fracture tests under Mode I tensile opening loads, fiber bridging will inevitably occur and affect the resulting test data [16,17].

Fiber bridging is intrinsic to unidirectional laminate lay-up [51] which is specified as test material in the quasi-static fracture mechanics test standards [21,29]. Since most designs of FRP composites do not use unidirectional fiber lay-up, in the structural material the delamination propagation is much less affected by fiber bridging than in the standard test specimens. The reason that the fracture mechanics design methodology worked well for rotorhub flexbeams likely is that under Mode II in-plane shear loads fiber bridging is not as dominant as under Mode I, even though the Mode II fracture process zone can be of a quite complex shape and comparatively large size which makes the exact determination of the delamination length in the quasi-static and the cyclic fatigue fracture tests difficult [52].

Molent and Forrester [17] recently reviewed the available data on cyclic fatigue growth and residual strength for composite aircraft structures and concluded (cite)

> *"In this paper the growth of some (limited) composite laminate and bonded repair damage types was reviewed and found to broadly conform to the lead crack framework. Thus equations originally derived for metals may have utility for the assessment of the growth of defects in composite structures. One such equation is the Hartman-Schijve variant. With further research (and data) a damage-tolerance approach appears to be feasible and could be used to comply with the 2009 US Federal Aviation Administration (FAA) slow growth approach to certifying composite, adhesively bonded structures and bonded repairs."*

The approaches for testing and data analysis currently under discussion [19,41] likely provide a road-map toward this goal.

4. Extrinsic and intrinsic sources of scatter in fatigue fracture of FRP composites and their quantification

Identifying sources of scatter in fracture tests and quantifying the contributions from different sources is important for determining useful fracture mechanics test data [19]. Often, repeatability, i.e., the scatter from testing several, nominally identical specimens in one single laboratory, ideally by the same operator and the same test set-up is quoted as an average value with related standard deviation. Similarly, reproducibility, i.e., performing the same test on identical specimens at different laboratories, by different operators and possibly different test set-ups, yields again an average value with a standard deviation. The main purpose of a standard test procedure is to minimize the scatter in both, repeatability and reproducibility. For that, the sources of scatter have to be identified and the test procedure requirements defined sufficiently. Following [19], scatter sources are classified as extrinsic and intrinsic, respectively. Intrinsic scatter is essentially coming from the (unavoidable) variability in microscopic, mesoscopic and macroscopic properties of the FRP laminates. These typically reflect the manufacture and processing of the laminate and the test specimens prepared from that. Ideally, the specification of the specimen manufacture and processing should yield samples that show the full range of manufacturing and processing variability of the FRP composites structures or components. Extrinsic scatter is then attributed to, e.g., test setup, testing parameters, and possibly data analysis. This is discussed in more detail below.

4.1. Extrinsic sources of scatter

A first source of extrinsic scatter has been identified in the round robin tests reported by Stelzer et al. [39,40] coming from the resolution of the load cell. Typical loads measured in the fracture and fatigue fracture tests amount to 100-200 N and these values often drop to a few tens of N under displacement control. Therefore, the load cell measurement range shall be 250 N or less, or else a proper load signal calibration has to be performed in that range. Similarly, the displacement measurement may also be affected by the resolution of the measurement device. A recent study [53] illustrates this by providing estimates of the uncertainty in measurements of fracture toughness, however not on FRP composites, taking into account several experimental parameters. The so-called "human factor", i.e., the experience of the operator in setting-up and performing the tests [54] as well as acquiring the data, e.g., the visual measurement of the delamination length, also plays an important and often underestimated role. Specimen manufacturing, processing and preparation (often also done by humans, sometimes supported by robots or computer-controlled equipment that, however, also have limitations in their accuracy) also yields scatter, e.g., by variation in fiber orientation during lay-up resulting in fiber misalignment or waviness, variation of plate thickness and specimen width (some standards set limits on variation in specimen dimensions, e.g., [21,29], and quality of cutting from the plate. Variation in fiber orientation and specimen geometry or size may affect the compliance of the test specimens which in turn affects the test data [51]. Even if variations in specimen size are limited by specifying tolerances in the test standards or guidelines and unsuitable specimens are excluded; some *intrinsic* material variability and hence scatter will remain. Finally, the operator experience or human factor also comes into play in the analysis, e.g., when determining the non-linear or 5% change in compliance points for defining the initiation toughness as clearly shown by the example discussed by Davies [55]. This is one motivation for trying to automate the test performance and data analysis as much as possible and to eliminate operator dependent error and scatter sources [56].

The compliance of the test specimens, calculated from machine or test specimen displacement and applied load can be used to calculate the delamination length for quasi-.static or cyclic fatigue fracture under Mode I loading [39,40]. The compliance of the test set-up has to be determined separately and deducted from the compliance calculated from the test machine data. Fig. 2 left shows the specimen compliance as a function of load cycles in fatigue fracture of (carbon-fiber/epoxy type IM7/977-3) calculated in two different ways. The first calculation takes simply the maximum load P_{max} and the maximum displacement δ_{max} (the test was performed under displacement control), i.e., assumes that the respective load-displacement curve can be extrapolated back to zero load and displacement simultaneously (represented by the blue line in Fig. 2). However, if the specimen compliance is calculated as the slope (assumed to be a straight line) between the minimum and maximum values of displacement and load, respectively, the resulting compliance is higher for any given cycle number (red curve in Fig. 2). Therefore, the "effective" compliance of the specimen under cyclic, displacement-controlled fatigue between maximum and minimum displacements (defining the displacement-ratio) is higher than that evaluated from the peak displacements and loads and the same holds for the effective delamination lengths calculated from the compliance, i.e., longer delamination lengths for the analysis based on the larger effective compliance.

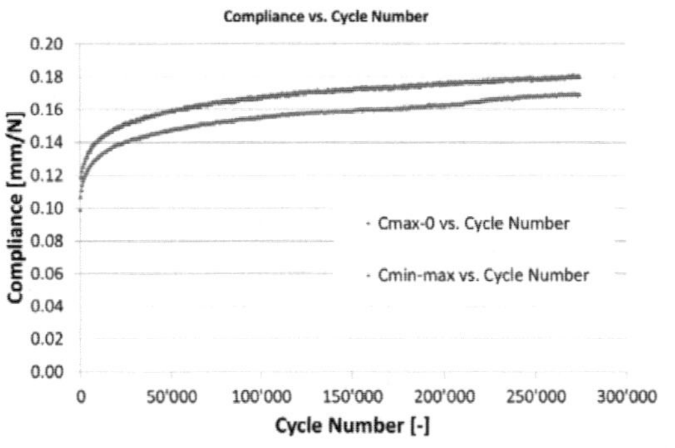

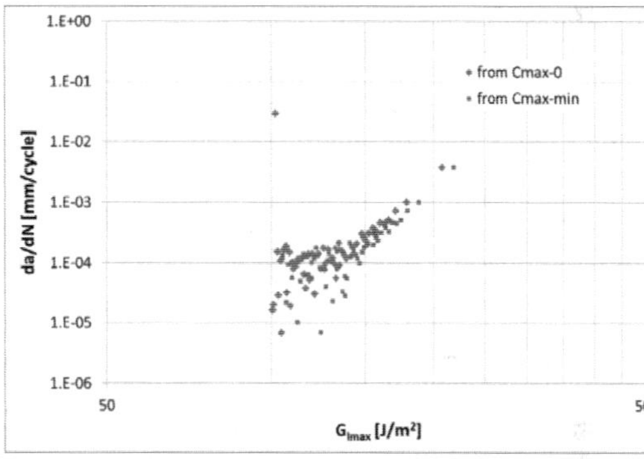

Figure 2: The compliance calculated via two different methods, once from maximum load and displacement (red dots) and once from the difference between maximum and minimum load and displacement (blue dots), respectively (left) and the comparison of the Paris-type graph resulting from delamination lengths calculated with the two compliance methods.

Fig. 2 right shows the resulting Paris-type graph (double-logarithmic da/dN versus G_{Imax}) for the two compliance measurements and the delamination lengths back-calculated from that, respectively. The delamination lengths for the two compliance measurements have been determined using the cubic relation between compliance and delamination length with the equations shown in Appendix 1 of [57] with estimated values of the tensile modulus $E_1$of 150 GPa and shear modulus G13 of 4.9 GPa. The G_{Imax} data have been calculated with the Modified Compliance Calibration Method without applying the load-block and large displacement corrections (Factors F and N according to [29]). It has to be noted that the data sets have been filtered to show data points for a minimum delamination length increase of about 0.1 mm only. Even though the delamination lengths differ due to the different compliance, the Paris-graph data for the two sets, at first sight, agree fairly well. There seems to be a trend to being less conservative and possibly showing slightly larger scatter for data calculated from the compliance based on maximum and minimum values of displacement and load, i.e., for the data yielding higher compliance, compared with those calculated from the maximum values. These trends will, in any case, have to be confirmed and quantified in further analysis comprising more test data.

4.2. Intrinsic sources of scatter

Intrinsic scatter sources have been defined as being material and/or processing related and these have to be taken into account in structural design, as discussed above. One important type of intrinsic scatter that has to be considered, if the data are being used in FRP structural design, i.e., the large-scale fiber bridging due to the unidirectional fiber alignment in the standard fracture test specimens, has already been discussed above. Possible approaches for the quantification of this are discussed in a separate section below.

Manufacturing and processing of FRP composites can yield intrinsic defects such as, e.g., porosity, inclusion of foreign particles, inhomogeneous micro- and meso-scale fiber volume fractions and fiber alignment, local fiber-matrix debonds, etc. [58]. Residual stresses also arise from processing, specifically the cure and post-cure conditions [59], and may interfere with the applied global stresses during testing and thus affect the delamination behavior. Scatter in fiber alignment for a carbon-fiber thermoplastic composite has been quantified by [60]. The fibers in this case lie within about ±3° from the nominal fiber direction with standard deviations between about 0.6° and 1.9°. Wavy fiber plies can also affect the mechanical properties of the laminate as discussed in detail for IM7/8552 [61]. The quantification of intrinsic defects and of their effects on fracture and fatigue fracture behavior requires further investigation for minimizing intrinsic scatter.

5. Accounting for large-scale fiber-bridging in Mode I fatigue fracture of unidirectional fiber-reinforced polymer composites

5.1. Materials approach

Since fiber-bridging is considered to be intrinsic to testing unidirectional FRP composite laminates, the question could be asked whether choosing another laminate lay-up with different fiber orientation might reduce or completely prevent fiber bridging. In early literature on fracture testing, it was recommended to use a lay-up with a slight offset angle (typically 3-5°) for the two plies between which the delamination was propagating in order to reduce fiber-bridging in Mode I and Mode II loading [49,50]. However, the effect of angle-ply lay-ups on fiber bridging in material coupon testing was usually not quantified. Most documented fiber bridging to date is essentially from visual observation or photography, see, e.g., [62-64].

Selected values of G_{IC} and G_{IIC}, respectively, determined from quasi-static fracture testing of angle-ply laminates reported in the literature are summarized in Fig. 3. The data (AS1/3502 [65], AS4/8552 [66], T800H/3631 [67], IM7-G/8552 [68], C12K/R6367 [69], AS4/3501-6 [70] and T300/914C [71]) are plotted as a function of the fiber orientation angle between the two plies between which the delamination initiates or propagates. This angle does not necessarily correspond to the angle between the main delamination direction and the fiber orientation of the adjacent plies, e.g., in ±15° or ±30° lay-ups with delamination propagation in 0° direction which would be shown under 30° and 60°, respectively. Nicholls and Gallagher [65] report average data for selected areas of delamination identified by fractography (described in their paper) and hence their data are labelled as "propagation" values rather than initiation. Sebaey et al. [66] distinguish initiation and propagation, the latter correspond to values from delaminations without jumps to other planes only. Chou et al. [67] also report propagation values for selected delamination areas in addition to initiation values. Some of the authors also provide information on scatter or standard deviation for their average values, but this is not considered in the graphs in Fig. 3.

There is scant data on "small" angles below 10° and it is not clear whether or how data for larger angles can be extrapolated to lower values. For Mode I, the data in Fig. 3 indicate that there is little effect of fiber orientation at delamination initiation (unless precracking has been performed). This is not surprising, since the fiber orientation likely does not affect the resin pocket around the starter film insert. However, delamination propagation after initiation or precracking results in clearly increasing R-curve behavior, i.e., increasing G_{IC} values with increasing delamination length, sometimes noted likely to be due to fiber bridging (from visual observation) in the references. G_{IC} initiation also seems to increase with increasing difference in angle between the fiber plies adjacent to the delamination. For Mode II, the data tend to show a slight decrease in the G_{IIC} initiation values, if the fiber angle difference is increased, but there are exceptions, e.g., [71]. High propagation values (high compared to G_{IC} initiation) have also been noted for cross-ply (0°/90°) stacked laminates [72,73] that can be considered the extreme case among angle-ply laminates. Multi-directional or quasi-isotropic FRP laminates often yield multiple cracking or crack branching, that, similar to cross-ply might involve mixed mode or inter-laminar and intra-laminar fracture, as discussed, e.g., by [74] and hence is not well suited for determining design values from coupon testing.

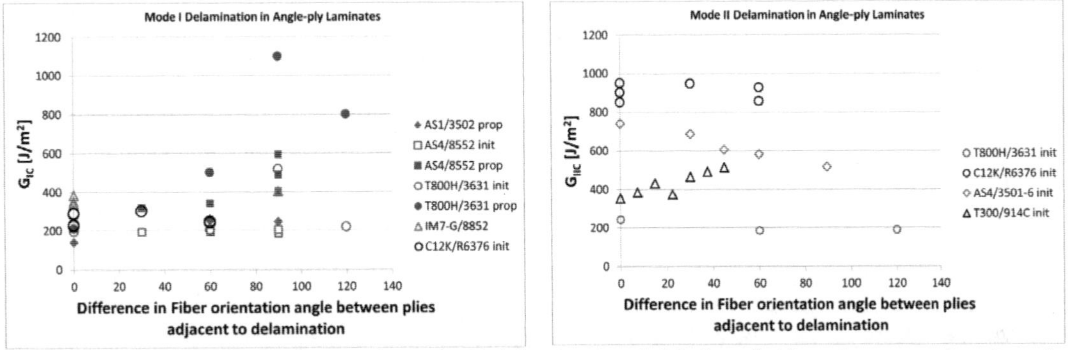

Figure 3. Quasi-static Mode I and Mode II delamination initiation (init) and propagation (prop) values for selected angle-ply CFRP laminates from literature (see text for references and discussion).

5.2. Experimental approaches

Recent publications by the Structural Integrity & Composites Group in Aerospace Engineering at TU Delft provide a perspective for an experimental approach for quantifying the effect of fiber bridging in unidirectional FRP composites subject to cyclic Mode I fatigue loading [41,75,76]. For details, the reader is referred to the referenced literature and the paper presented by Dr. Alderliesten at this symposium [77].

Another, independent approach for accounting for fiber bridging was published by Hojo et al. [78,79]. This method is based on fatigue fracture testing (again Mode I only was considered here) at constant level of G_{max} or ΔG. The data obtained from these experiments can then be extrapolated to short delamination lengths $\Delta a \sim 0$ mm yielding a fatigue fracture curve in the Paris-type graph which is not affected by fiber bridging. While the method proposed by the group at TU Delft likely is more easily implemented in a standard test procedure, the constant-G or constant-ΔG tests may serve as an independent method for validation of the other approach. So far, there is no data available performing both tests on the same material.

5.3. Modelling approaches

There have also been attempts at using modelling or simulation methods to quantify fiber bridging in quasi-static Mode I tests. One example of that is [80], where the author states (cite)

> *"This paper presents a beam model based on a linear traction–separation law that has been shown to be a good approximation to the early stages of fibre bridging. The present model enables the evaluation of accurate propagation GIc values through the fitting of experimental load–displacement curves."*

At this stage, it is not clear whether this or similar modelling approaches could be extended to cyclic Mode I fatigue fracture (private communication from Prof. A.B. de Morais), but it might be worthwhile to explore this in future research.

6. Microscopic damage mechanisms in quasi-static and fatigue fracture and their averaging on macroscopic scales

Due to the morphology of FRP composites, there are several different types of microscopic or mesoscopic damage mechanisms that can be activated by quasistatic or fatigue loads. These are usually noted to be: matrix cracks, fiber-matrix debonding or interface failure, and fiber breaks, see, e.g., [81]. Some authors include (interlaminar) delamination or large matrix cracks as an additional damage mechanism for FRP composites, but as discussed by [82,83], these consist again of a series of smaller size matrix cracks, adding up to larger damage between the ply levels. For more insight into the physical damage mechanisms in FRP, besides identifying the type, also the typical size and time scales are of interest [19]. These issues are briefly discussed here.

One nondestructive monitoring method that is quite sensitive to generation of small defects in many types of materials including in FRP composites is Acoustic Emission (AE), see, e.g., [81,84] for details. The combination of unsupervised AE signal pattern recognition and Multiphysics modelling of wave propagation from model signal sources embedded a FRP composite yielded three AE signal clusters and modelled AE signal parameters attributed to different model sources consistent with matrix cracks, fiber-matrix debonding and fiber breaks [85]. This methodology can be applied to large-scale FRP composite structures, such as wind-rotor blades and yields locations of microscopic damage sources in the structure as well as identification of the damage mechanism type, see, e.g., [86]. However, without suitable damage models, a direct prediction of the life-time of the component from a mapping of locations and type of accumulated microscopic damage, as argued in [87], is not feasible.

In spite of the success of the analysis of AE signals with pattern recognition and source mechanism modelling, this approach does not yield direct quantification of the size or extent of the microscopic damage created by a specific source mechanism. For this, imaging non-destructive test methods, such as X–ray radiography or computed tomography can provide some information on damage location and size. This has been explored in a preliminary analysis of carbon-fiber epoxy composite laminates after quasi-static and cyclic fatigue Mode I fracture testing. One example of a X-ray computed tomography slice taken from a Mode I specimen after quasi-static fracture testing is shown in Fig. 4. There are crack-like features that seem to be oriented parallel to the main delamination. The thickness of this slice (normal to the delamination) is about 3.5 mm, the length of the micro-cracks hence amounts to between about 350 and 1200 μm. The size of the micro-crack features in the thickness direction of the specimen is comparable to the voxel size of about 3 μm or less, and hence, they are difficult to clearly identify. Therefore, a 2-dimensinal Fourier Transform map of the X-ray image slice has been calculated and compared with the same algorithm applied to image slices from unloaded parts of the specimen. The latter do not show the features highlighted by red arrows in the Fourier Transform map in Fig. 4 and also did not yield indications of crack-like features in the X-ray images. This constitutes first evidence, but that will have to be verified by additional investigations, combining more measurements.

It has to be noted that the X-ray imaging by computed micro-tomography discussed above provides first evidence for the damage created in the so-called fracture process zone around the delamination created by the different fracture tests. Even though the quantification of the micro-crack sizes in the fracture process zone can still be debated, it is nevertheless clear that additional fracture surfaces are created in the FRP composites not detectable by the visual or technical observation (e.g., camera or video recording) of the delamination length on the edges of the specimens. On the other hand, this additional damage is reflected in the compliance of the specimens, and hence at least to some extent in the load and displacement measurements. If the fracture or delamination areas from visual observation together with load and displacement values are used to calculate the toughness or delamination resistance, i.e., the G-value, the effective facture surface might be underestimated, resulting in an overestimate for G and possibly also in less accurate estimates of the related scatter.

The currently available data on damage types and sizes are interpreted as follows: The evidence currently available from AE crack or damage sizing or from X-ray computed tomography indicates stochastic micro-crack formation on lengths scales between about 10 μm and about 200 μm. It is possible, that cracks sizes smaller than 10 μm do occur in FRP composite fracture and that the AE sizing described here reached its limit of sensitivity. X-ray computed micro-tomography with a voxel size around 3 μm poses a similar resolution limit. That the time scale of microscopic fracture events is on the order of microseconds or even less (see, e.g., [87] for details) indicates that the methods for determining delamination lengths currently used in the fracture tests (e.g., visual observation or photography or video recording) do average over several orders of magnitude in time, and, to a lesser extent, also on the length or volumetric scale (e.g., with the requirement to determine the delamination tip with an accuracy of 0.5 mm [21]. The implied averaging over microscopic and short-time events constitutes a potential source of (intrinsic) scatter that is difficult to quantify. The matrix crack sizes determined from the delaminations obtained by the fracture tests described above can be complemented by a study on another microscopic mechanism, namely fiber-matrix debonding discussed below.

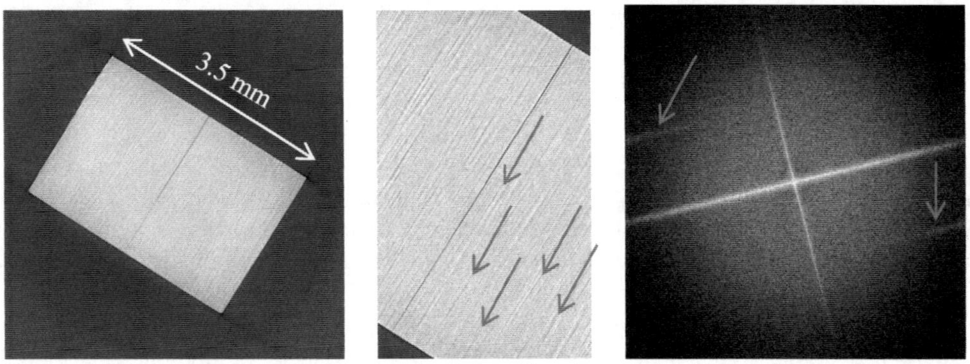

Figure 4: X-ray computed micro-tomography slice (voxel size about 3 μm) of a carbon-fiber epoxy specimen after quasi-static Mode I fracture test (left), a higher magnification of the same slice near the delamination running from the bottom left hand corner to the top right hand corner in the image, selected crack-like features are indicated by red arrows (center) and a 2D Fourier transform of the image (right) showing features (pointed out with red arrows) that are absent from untested specimen parts (see text for details and discussion).

For fiber-matrix debonding, estimates for crack size or debonded surface area can be obtained from micro-mechanical tests. If single fibers (glass or carbon) embedded in an epoxy matrix are subject to quasi-static tensile loading in the so-called single fiber fragmentation test, the fibers will break once a certain stress level is exceeded. Around the location of the fiber break, there occurs also some debonding between fiber and matrix that can be observed under the microscope (Fig. 5). The debond length of course depends on the fiber-matrix adhesion. Increasing the tensile load further after the first fiber break will induce additional fiber breaks until the fiber fragments are too short to be loaded to failure, i.e., reaching saturation. The observed fiber fragment lengths (between about 300 and 400 μm) can be statistically analyzed yielding an average and a width of the distribution [88]. An average debond length for each fiber break of around 80-120 μm can be estimated from the micrographs (Fig. 5), resulting in an average debond area between 1'500 and 2'300 μm^2 for the carbon fibers with a diameter of 6 μm. Similar values of debond length are obtained for glass-fiber epoxy SFFT specimens [88] where fiber diameters are between about 10 and 15 micrometer.

It is known that the SFFT tests do not represent the damage occurring under tensile loads in CFRP since using fiber bundles instead of single fibers indicates that fiber breaks in one fiber affect the occurrence, specifically the location of breaks in adjacent fibers [90]. Even though there are no quantitative data, however, the debond surface area estimates derived from the tests on single fiber specimens likely also apply to multiple fiber specimens. The order of magnitude is of a few tens up to about 100 micrometers. This range is consistent with the matrix crack sizes (diameters) derived from the AE monitoring of quasi-static fracture tests of FRP composites.

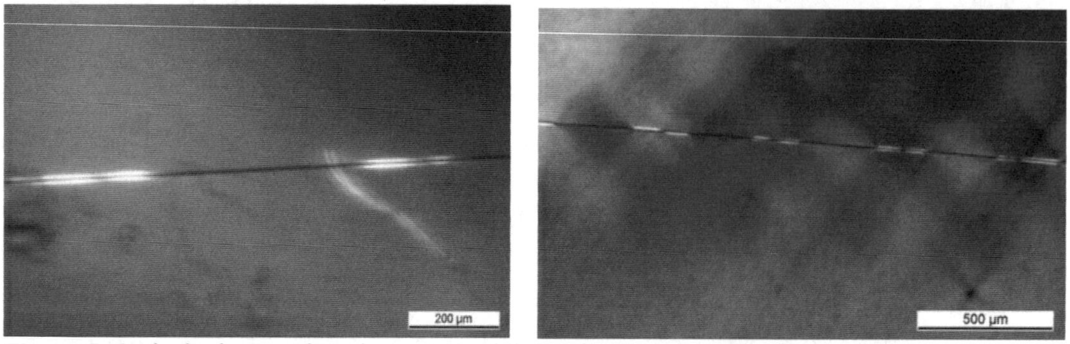

Figure 5. Optical micrographs using polarized light of AS4 carbon fiber breaks after tensile testing single fibers embedded in epoxy matrix, carbon fiber diameter is about 6 µm.

If the applied loading on the embedded fiber is compressive, the so-called single fiber push-in or push-out test yields again debonding between fiber and matrix. The test equipment and set-up is described in detail by Battisti et al. [89]. For carbon fibers embedded in epoxy with a length around 30 micrometers, full debonding was achieved and the fibers effectively pushed-out. With fiber diameters around 6 micrometer and roughly 19 micrometer circumference, the debond area in this case amounts to about 560 µm^2. In analogous experiments with carbon fibers of roughly 300 micrometer length in thicker carbon-fiber epoxy slices, only partial debonding was observed and the fibers failed when increasing the compressive load (unpublished data from the authors' laboratory). Fig. 5 shows examples of this. If the fiber would still debond along the full length of 300 micrometer in the push-in test, the debond area would amount to about 5'600 µm^2 which constitutes an upper limit. Likely, the debond area accumulated up to the point where the fiber fails is less. It has to be noted that Fig. 5 shows a case where the fiber that was pushed in by the indenter is surrounded by a hexagonal close packing of six fibers. The same test was also performed on "isolated" single fibers where the nearest neighbor fiber was at least one fiber diameter (roughly 6 µm) distant from the fiber that was pushed-in. The results for the "isolated" fibers indicate a slightly more compliant behavior and slightly lower failure loads than for the hexagonal closed packed fibers, but no significant difference in failure mode.

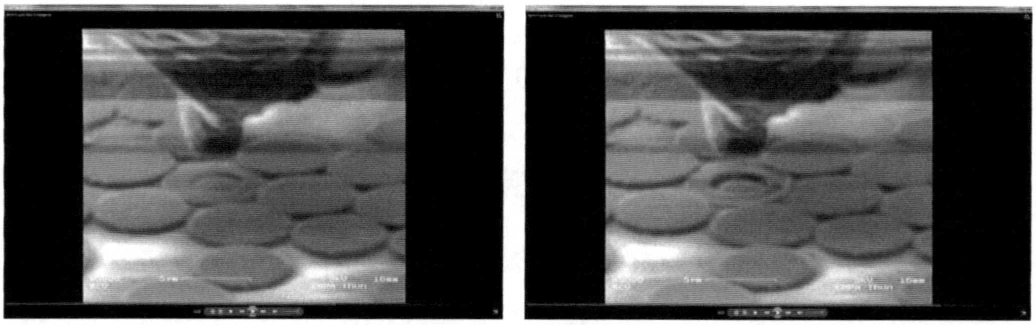

Figure 6. Images from in-situ SEM observation of cyclic push-in loading of a carbon fiber surrounded by a hexagonal group of carbon fibers embedded in a slice of epoxy (about 300 µm thick), first, the indenter tip produces an imprint on the fiber cross-section (left) and finally induces failure (right).

Finally, the question can be asked whether the different types of microscopic damage mechanisms acting in FRP composites and the relevant size or time scales are affected by the micro- or meso-scale structure or morphology of the materials. Considering the discussion of fiber-matrix debonding above, it could be hypothesized that one factor is the fiber diameter (beside the fiber-matrix adhesion and possible residual stresses from processing [91]) or the average fiber-fiber distance in the composite which is again on the order of a few micrometer or up to a few ten of micrometer at most (in composites with high fiber volume fractions exceeding 60%). The lower size bound of matrix cracks in FRP composites is likely given by the sensitivity of the method used for detection, i.e., of the AE or the X-ray computed micro-tomography. It has to be noted that sub-micron size cracks may not be detected by these methods in all cases. The upper bound estimate, on the other hand, possibly correlates with the typical ply level thickness. Therefore, it could be hypothesized that the micro- and/or meso-scale morphology of the FRP composites, i.e., on scales ranging from a few micrometer up to about 100-150 micrometer, may affect the crack or damage size to some extent, possibly contrary to the case of metals or metal alloys [18]. The evidence so far, however, is too scant to draw firm conclusions and it is definitely worthwhile to investigate this further.

7. Summary and open questions

From the data presented and from the discussion above it can be concluded that (1) there is still no full understanding of the effects that require consideration when testing FRP composite materials under cyclic fatigue loads with the aim of using the data for design limits, but that (2) at least for Mode I tensile opening loads certain extrinsic sources of scatter have been identified (e.g., load cell range used in the test set-up, operator experience) and finally that (3) again at least for Mode I, approaches have been proposed for quantifying the fiber bridging effects intrinsic to the test specimens with unidirectional fiber lay-up for estimating safe design limits for FRP composite components and structures. These approaches may still have to be validated, and possibly optimized, in order to not be too conservative. Likely, the data analysis will remain "empirical", even when the Paris-type graphs are replaced by master curves based on modified Hartman-Schijve equations [9,15,41]. Getting sufficiently accurate estimates of material- and processing related intrinsic scatter and reducing any extrinsic scatter to a value below that are essential steps towards the application of the data in fracture mechanics based FRP structural design.

The question whether a future damage tolerant design approach for FRP structures can be based on identification and location of micro- and mesoscopic damage events is still open. It is currently not even clear whether the micro- and meso-scale morphology of FRP composites plays a significant role in activating different damage mechanisms. Therefore, attempts at reinforcing the FRP composites with fillers, e.g., nanoparticles or with combinations of particles covering several scales in order to obtain hierarchical FRP composites, e.g., [92], essentially remain empirical and still pose problems for optimization. Therefore, also these issues require further investigation.

The often cited observation that Mode I quasi-static and cyclic fatigue fracture data constitute a conservative limit may also be questioned. To date (to the best knowledge of the author) there is only one example (Fig. 7, left) showing Mode II in-plane shear fatigue fracture data to be more conservative than the corresponding Mode I data. Two tests performed independently by different groups and at different times on the same type of material (AS4/PEEK) agree very well [37,93]. Another question is the behavior of fatigue fracture data obtained under different modes (Mode I tensile opening, Mode II in-plane shear, and Mixed Mode I/II) for a carbon-fiber epoxy composite (Fig. 7 right). Please note that Fig. 7 shows Paris-type graphs of da/dN versus G_{max} for the respective FRP composites and Modes.

In view of the effects of fiber-bridging discussed above, it could be hypothesized that the Mode I fatigue fracture data for AS4/PEEK (Fig. 7 left) would significantly change position and/or slope in the graph, if the test was run, e.g., using the procedure proposed in [41,75], and the data extrapolated to zero fiber bridging. It can, at this stage, however, not be excluded that Mode II would prove to be more conservative than Mode I for this carbon-fiber thermoplastic matrix composite material even after accounting for fiber bridging. So far (to the best knowledge of the author), there are no published data for other carbon-fiber thermoplastic composites (e.g., PPS, or PES) under cyclic fatigue Mode II for comparison with cyclic fatigue Mode I.

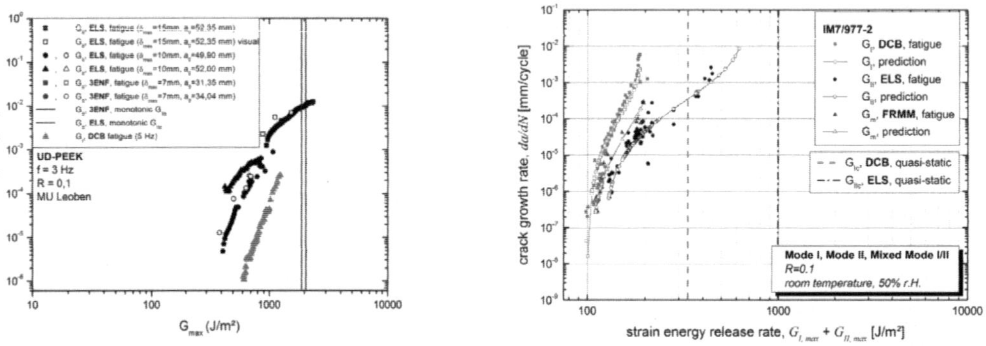

Figure 7: Paris-type graphs of (left) unidirectional AS4/PEEK carbon-fiber thermoplastic composite tested under Mode I and Mode II quasi-static and cyclic fatigue fracture, the quasi-static results are represented by the vertical lines, the cyclic fracture with open and closed symbols according to the insert, (right) the same plot for IM7/977-2 carbon-fiber duromer composite includes also Fixed-Ratio Mixed Mode I/II data (see text for details and discussion).

Fig. 7 right indicates that at low values of G_{max} the data for the different modes (Mode I, Mode II, here with a fixed ratio of 4:3 for Mixed Mode I/II, the blue line is the prediction for this ratio from the fitted Mode I and Mode II data) seem to converge near a value of 100 J/m^2 for low values of da/dN. It is not clear whether this is partly or fully due to limited measurement resolution, i.e. scatter, or representing the material behavior. Independent of this, it is likely that the threshold value for this material is 100 J/m^2 or less. If the Mode I data were determined without effects of fiber bridging, the resulting curves would likely be more conservative, i.e., move to the left in Fig. 7 or to the left and upward on the da/dN axis. This would result in an even lower (average) threshold value, unless the slope of the curve is changed significantly also by this. Therefore, the question whether FRP structural designs can be based on finite threshold values and their associated scatter (e.g., estimated by a given multiple of the standard deviation), or whether damage-tolerant designs (including appropriate non-destructive inspections) will be required for future safe operation of such structures, remains open. Both paths will require further research based on the basics outlined in this paper.

Acknowledgments
The contributions of Ms Đurđija Džodan (Empa, Mechanical Systems Engineering) and Mrs Anja Huch (Empa, Applied Wood Materials) for manufacturing, testing and analyzing the single-fiber tensile specimens, of Dr Gaurav Mohanty (Empa, Mechanics of Materials & Nanostructures) for the in-situ fiber push-in tests, Dr Rolf Kaufmann, Dr. Selina Kolokytha, and Mr Jürgen Hofmann (Empa, Center for X-ray Analytics) for the X-ray Computed Tomography and the 2D Fourier Transform analysis of the X-ray image slices of CFRP specimens after fracture tests as well as data analysis and graphical presentation by Dr. Steffen Stelzer (Montanuniversität Leoben) and extensive discussions of various aspects of fracture and fatigue fracture testing of FRP composites with Prof. René Alderliesten (Technical University Delft), Prof. Leslie Banks Sills and Dr. Ido Simon (Tel Aviv University), Prof. Alfredo de Morais (University of Aveiro), Prof. Rhys Jones (Monash University), Prof. Anthony J. Kinloch (Imperial College), and Prof. Markus G.R. Sause (Augsburg University) are gratefully acknolwedged.

References

[1] Paris PC, Gomez MP, Anderson WE 1961 *The Trend in Engineering* **13** 9-14.
[2] ASTM Standard E647 2015^{e1}, Measurement of Fatigue Crack Growth Rates, ASTM International, West Conshohocken, PA 1-49.
[3] Bathias C, Laksimi A 1985 *ASTM STP* **876** Delamination and Debonding of Materials, Johnson WS Ed. 217-237.
[4] Adams DF, Zimmerman RS, Odom EM 1987 *ASTM STP* **937** Toughened Composites, Johnston NJ Ed. 242-259.
[5] Hojo M, Gustafson CG, Tanaka K, Hayashi R 1987 *Compos. Sci. Technol.* **29** 273-292.
[6] O'Brien TK 1988 *J. Reinf. Plastics Compos.* **7** 341-359.
[7] Mall S, Yun KT, Kochhar NK 1989 *ASTM STP* **1012** Composite Materials: Fatigue and Fracture, Lagace PA Ed. 296-310.
[8] Newaz GM, Lustiger A, Yung JY 1989 *ASTM STP* **1044** Advances in Thermoplastic Matrix Composite Materials, Newaz GM Ed. 264-278.
[9] Hartman A, Schijve J 1970 *Eng. Fract. Mech.* **1** 615-631.
[10] Beden SM, Abdullah S,Ariffin AK, Al-Asady NA 2010 *Materials and Design* **31** 3449–3456.
[11] Sander M, Richard HA 2003 *Int. J. Fat.* **25** 999–1005.
[12] Zhang W, Wang Q, Li XY, He JJ 2016 *Mathematical Problems in Engineering* **2016** Article ID 4298507 1-8
[13] Maierhofer J, Pippan R, Gänser HP 2014 *Int. J. Fat.* **59** 200–207.
[14] Jones R, Pitt S, Brunner AJ, Hui D 2012 *Compos. Structs.* **94** 1343-1351.
[15] Jones R, Stelzer S, Brunner AJ 2014 *Compos. Structs.* **110** 317-324.
[16] Jones R, Kinloch AJ, Michopoulos JG, Brunner AJ, Phan N 2017 *Compos. Structs.* **180** 316-333
[17] Molent L, Forrester C 2017 *Compos. Structs.* **166** 22–26.
[18] Jones R. Raman, RKS, McMillan AJ 2018 *Eng. Fract. Mech.* **187** 190-210.
[19] Alderliesten RC, Brunner AJ, Pascoe JA 2018 submitted to *Eng. Fract. Mech.*
[20] ASTM D6115 1997 (reapproved 2011) Mode I Fatigue Delamination Growth Onset of Unidirectional Fiber-Reinforced Polymer Matrix Composites, ASTM International, West Conshohocken, PA 1-7.
[21] ASTM D5528 2013 Mode I Interlaminar Fracture Toughness of Unidirectional Fiber-Reinforced Polymer Matrix Composites, ASTM International, West Conshohocken, PA 1-13.
[22] Composites Materials Handbook CMH-17-3G 2017 **3** Materials Usage, Design and Analysis, Society of Automotive Engineers SAE.
[23] Composites Materials Handbook MIL-HDBK-17-3F 2002 **3** Polymer Matrix Composites Materials Usage, Design and Analysis.
[24] Davies P, Brunner AJ, Blackman BRK 1998 *Appl. Compos. Mats.* **5** 345-364.
[25] Tay TE 2003 *Appl. Mech. Rev.* **56** 1-31.
[26] Brunner AJ, Blackman BRK, Davies P 2008 *Eng. Fract. Mech.* **75** 2779-2794.
[27] Brunner AJ 2015 Chapt. 8 in *Polymer Composites in the Aerospace Industry*, Irving PE, Soutis C Eds, Woodhead Publishing, Series in Composites Science and Engineering **50** 191-230.
[28] JIS K7086 1993 Testing methods for interlarminar fracture toughness of carbon fibre reinforced plastics, Japanese Standards Association, Tokyo, 1-35.
[29] ISO 15024 2001 Fibre-reinforced plastic composites - Determination of mode I interlaminar fracture toughness, G_{IC}, for unidirectionally reinforced materials, International Organization for Standardization ISO, Geneva 1-24.
[30] ISO 15114 2014 Fibre-reinforced plastic composites - Determination of the mode II fracture resistance for unidirectionally reinforced materials using the calibrated end-loaded split (C-ELS) test and an effective crack length approach, International Organization for Standardization ISO, Geneva 1-18.
[31] ASTM D7905 2014 Determination of the Mode II Interlaminar Fracture Toughness of

Unidirectional Fiber-Reinforced Polymer Matrix Composites, ASTM International, West Conshohocken, PA 1-18.
[32] ASTM D6671/D6671M 2013^{e1} Mixed Mode I-Mode II Interlaminar Fracture Toughness of Unidirectional Fiber Reinforced Polymer Matrix Composites, ASTM International, West Conshohocken, PA 1-15.
[33] Murri G 2013 NASA-TM-2013-217996 Evaluation of Delamination Onset and Growth Characterization Methods under Mode I Fatigue Loading, Langley Research Center, Hampton VA 1-32.
[34] Stelzer S, Jones R, Brunner AJ 2013 Proc. 19th International Conference on Composites ICCM-19 Hoa SV, Hubert P Eds. Canadian Association for Composite Structures and Materials, Association Canadienne pour les Structures et Matériaux Composites, 1689-1697.
[35] Vinciquerra AJ, Davidson BD, Schaff JR, Smith SL 2002 *J. Reinf. Plastics Compos.* **217** 663-677.
[36] Brunner AJ, S. Stelzer S, G. Pinter G, Terrasi GP 2013 *Int. J. Fat.* **50** 57-62.
[37] Stelzer S, Brunner AJ, Pinter G 2015 Proc. 20th International Conference on Composite Materials ICCM-20, Paper-ID 3216-4, 1-7.
[38] Jaeck I, L. Carreras L, Renart J, Turon A, Martin de la Escalera F, Essa Y 2018 *Int. J. Fat.* **110** 63–70.
[39] Stelzer S, Brunner AJ, Argüelles A, Murphy N, Pinter G 2012 *Compos. Sci. Technol.* **72** 1102-1107.
[40] Stelzer S, Brunner AJ, Argüelles A, Murphy N, Cano GM, Pinter G 2014 *Eng. Fract. Mech.* **116**, 92-107.
[41] Yao LJ, Alderliesten RC, Jones R, Kinloch AJ 2018 *Compos. Structs.* **196** 8–20.
[42] Brunner AJ, Mujtaba A, Stelzer S, Jones R 2016 *Proc. Struct. Integ.* **2** 088-095.
[43] Gustafson CG, Hojo M 1987 *J. Reinf. Plastics Compos.* **6** 36-52.
[44] Rans C, Alderliesten R, Benedictus R 2011 *Compos. Sci. Technol.* **71** 230–238.
[45] Alderliesten RC 2013 *Eng. Fail. Anal.* **35** 370-379.
[46] Martin RH 2000 *Proc. Instn. Mech. Engrs. Part L* **214** 91-97.
[47] Murri GB 2006 *Int. J. Fat.* **28** 1124–1135.
[48] Murri GB, Schaff JR 2006 *Compos. Sci. Technol.* **66** 499–508.
[49] Cantwell WJ, Kausch HH 1993 *Mech. Compos. Mats.* **28** 328-333.
[50] Cantwell WJ 1996 *J. Mater.Sci. Lett.* **15** 639-641.
[51] Murri GB 2014 *J. Compos. Mats.* **48** 2413-2424.
[52] Blackman BRK, Brunner AJ, Williams JG 2006 *Eng. Fract. Mech.* **73** 2443–2455.
[53] Paruchuru SP, Kumar VU, Jain A, Dong XN 2017 *J. Test. Eval.* **45** 1139-1149.
[54] Hietschold N, Reinhardt R, Gurtner S 2014 *International Journal of Production Research* **52** 6254-6272.
[55] Davies P 1996 *Appl. Compos. Mats.* **3** 135-140.
[56] Brunner AJ, Tanner S, Davies P, Wittich H 1994 Proc. 2nd ECCM Conference on Composites Testing and Standardization ECCM-CTS 2, Hogg PJ, Schulte K, Wittich K Eds. Woodhead Publishing 523-532.
[57] de Moura MFSF, Morais JJL, Dourado N 2008 *Eng. Fract. Mech.* **75** 3852–3865.Ray BC, Hasan ST, Clegg DW 2007 *J. Reinf. Plast. Comp.* **26** 1187-1192.
[58] Ray BC, Hasan ST, Clegg DW 2007 *J. Reinf. Plast. Comp.* **26** 1187-1192.
[59] -Hsiao KT, Heider D 2017 Chapt. 10 in Manufacturing Techniques for Polymer Matrix Composites (PMCs), S.G. Advani AG, K.-T. Hsiao KT Eds. Woodhead Publishing 310-347.
[60] Yurgartis SW 1987 *Compos. Sci. Technol.* **30** 279-293.
[61] Makeev A, Seon G, Lee E 2010 *J. Compos. Mats.* **44** 95-112.
[62] Airoldi A, Dávila CG 2012 *Compos. Structs.* **94** 3240–3249.
[63] Manshadi BD, Farmand-Ashtiani E, Botsis J, Vassilopoulos AP 2014 *Compos.: Part A* **61** 43–50.

[64] Yao LJ, Sun Y, Guo LC, Jia LY, Zhao MY 2018 *Compos. Part B* **132** 97-106.
[65] Nicholls DJ, Gallagher JP 1983 *J. Reinf. Plast. Comp.* **2** 2-17.
[66] Sebaey TA, Blanco N, Costa J, Lopes, CS 2012 *Compos. Sci. Technol.* **72** 1251–1256.
[67] Chou I, Kimpara I, Kageyama K, Ohsawa I 1995 *ASTM STP* **1230** Composite Materials: Fatigue and Fracture--Fifth Volume, Martin RH Ed, American Society for Testing and Materials, Philadelphia, 132-151.
[68] Pelegri AA, Tekkam A 2003 *J. Compos. Mats.* **37** 579-596.
[69] Polaha JJ, Davidson BD, Hudson RC, Pieracci A. *J. Reinf. Plastics Compos.* **15** 141–173.
[70] Tao JX, Sun CT 1998 *J. Compos. Mats.* **32** 1933-1947.
[71] Shi YB, Hull D, Price JN 1993 *Compos. Sci. Technol.* **47** 173-184.
[72] Brunner AJ, Flüeler P 2005 *Eng. Fract. Mech.* **72** 899-908.
[73] Brunner AJ 2008 Chapt. 9 in *Delamination Behaviour of Composites*, Sridharan S Ed. Woodhead Publishing, Series in Composites Science and Engineering **25** 281-309.
[74] Choi NS, Kinloch AJ, Williams JG 1999 *J. Compos. Mats.* **33** 73-100.
[75] Yao LJ, Sun Y, Guo LC, Zhao MY, Jia LY, Alderliesten RC, Benedictus R 2017 *Compos. Structs.* **176** 556–564.
[76] Yao LJ, Sun Y, Guo LC, Lyu XQ, Zhao MY, Jia LY, Alderliesten RC, Benedictus R 2018 *Eng. Fract. Mech.* **189** 221–231.
[77] Alderliesten RC 2018 Proc. 39th Risø International Symposium on Materials Science 1-14.
[78] Hojo M, Ochiai S, Aoki T, Ito H 1994 Proc. 2nd ECCM Conference on Composites Testing and Standardization ECCM-CTS 2, Hogg PJ, Schulte K, Wittich K Eds. Woodhead Publishing 553-561.
[79] Hojo M, Aoki T 2015 Proc. 20th International Conference on Composite Materials ICCM-20 Paper-ID 2116-1 1-8.
[80] de Morais AB 2015 *J. Compos. Mats.* **49** 1681–1688.
[81] Sause MGR, Horn S 2010 J. Nondestruct. Eval. **29** 123–142.
[82] Bohse J 2000 *Compos. Sci. Technol.* **60** 1213-1226.
[83] Brunner AJ 2016 *J. Acoustic Emission* **33** S41-S46.
[84] Sause MGR, Gribov A, Unwin AR, Horn S 2012 *Pattern Recogn. Lett.* **33** 17–23.
[85] Sause MGR, Müller T, Horoschenkoff TA, Horn S 2012 *Compos. Sci. Technol.* **72** 167–174.
[86] Schubert L, Schulze E, Frankenstein B. 2011 Proc. 8th International Conference on Structural Dynamics, EURODYN 2011, De Roeck G, Degrande G, Lombaert G, Müller G. Eds. 3462-3468.
[87] Brunner AJ 2018 *Constr. Build. Mats.* **173** 629-637.
[88] Nordstrom RA 1995 ETH Zürich Ph.D. Thesis **11402** 1-168.
[89] Battisti A, Esqué-de los Ojos D, Ghisleni R, Brunner AJ 2014 *Compos. Sci. Technol.* **95** 121-127.
[90] Suzuki H, Takemoto M, Ono K 1987 *J. Acoust. Emiss.* **14** 35-50.
[91] Crasto AS, Corey R, J. T. Dickinson JT, Subramanian RV, Eckstein Y 1987 *Compos. Sci. Technol.* **30** 35-58.
[92] Dikshit V, Bhudolia SK, Joshi SC 2017 Fibers **2017 5, 38** 1-27.
[93] Martin RH, Murri GB 1990 *ASTM STP* **1059** Composite Materials: Testing and Design (Ninth Volume), Garbo SP Ed. American Society for Testing and Materials, Philadelphia 251-270.

IOP Conf. Series: Materials Science and Engineering **388** (2018) 012004 doi:10.1088/1757-899X/388/1/012004

Very High Cycle Fatigue - Testing Methods

P Horst, T J Adam, M Lewandrowski, B Begemann and F Nolte

Institute of Aircraft Design and Lightweight Structures (IFL), Technische Universität Braunschweig, Hermann-Blenk-Straße 35, D-38108 Braunschweig, Germany

E-mail: p.horst@tu-braunschweig.de, tj.adam@tu-braunschweig.de, m.lewandrowski@tu-braunschweig.de, b.begemann@tu-braunschweig.de, f.nolte@tu-braunschweig.de

Abstract. This paper presents several approaches for investigating the Very High Cycle Fatigue (VHCF) range of composite materials. Starting from the special challenges associated with high frequency testing of composite specimens, the paper shows some solutions as well as results for such investigations. The test set-up used in the research of the authors is a four point bending rig, which allows for relatively high load frequencies and widely avoids overheating. These bending tests provide fatigue limits for the development of microcracking, while showing no specific fatigue damage phenomena in the very high cycle fatigue range. The approaches found both, in literature and at the actual method utilised here, are discussed. The paper relates to glass fibre reinforced plastics (GFRP), and does not include either simulation approaches or results.

1. Introduction

Very High Cycle Fatigue (VHCF) characterises a fatigue life range in the order of 10^8 loading cycles. It therefore exceeds the usual High Cycle Fatigue (HCF) range by approximately two orders of magnitude, which in turn results in significant challenges regarding appropriate test set-ups or test duration.

Usual HCF tests of composite specimens are conducted at a frequency of 3 to 10 Hz; mainly by making use of servo-hydraulic testing machines. But, even at 10 Hz a test up to 10^8 cycles would last 116 days net duration, and by this would require high energy costs at the same time.

The main reason for restricting the frequency in such a drastic way is mostly the danger of overheating, which is much smaller for metallic specimens, and will to a certain degree depend on the actual lay-up of the specimen.

Other challenges are:

- self fatigue of the equipment
- resistance of the clamping/load introduction areas
- unsuitability of conventional measurement techniques like strain gauges

1.1. *Motivation*

The motivation for investigations on composite materials in the VHCF range is manifold. One - more or less intrinsic - motivation is the fact that specific Very High Cycle Fatigue phenomena have been found in the case of certain metallic materials [12]. These effects result in a drop of the fatigue endurance limit much later than the ordinary assumption of approximately 10^6 cycles.

The phenomena are mainly found, if the preparation of the outer surface is performed so well that damage initiation will not occur at the surface, but instead at extremely small disturbances of the inner structure. This effect is called *internal fatigue limit* [26]. It is one of the questions with respect to composites, whether such effects may be found here, too.

The second, more extrinsic motivation, originates from windpower rotor blades and similar applications (helicopters etc.), where very high numbers of load cycles may occur. The question is, whether there are methods to gather reliable data by tests of limited cost and duration.

1.2. Literature

The number of publications on Very High Cycle Fatigue of composites is relatively low, compared with the number existing on High Cycle Fatigue.

Although there is a strong need, the VHCF of fibre reinforced plastics has not been sufficiently investigated yet [8]. The main reason for this is - as already said - that testing at standard frequencies requires very long testing times. Unfortunately, as polymers show relatively high material damping, an increase of frequency results in undesired specimen heating. A possible solution pursued in literature is the use of very thin specimens of thicknesses about 1 mm. E.g. Mandell et al. [23] to [25] conducted first comprehensive tests of wind turbine blade materials under the auspices of the U.S. Department of Energys (DOE) Wind Energy Program. Most tests were constant-amplitude cyclic stress against cycle to failure (SN) tests of thin test coupons. Additionally to SN-behavior, Hosoi et al. [15] to [18] investigated the effect of load level on transverse cracking and delamination. According to Hosoi et al., tensile load amplitudes below 30% of the average static tensile stress shift failure into the VHCF range. Furthermore, the order of appearance of transverse cracking and delamination changes. At laminate stresses lower than 20% of the static strength, no damage at all is found up to $2.0\,x\,10^8$ cycles.

Other investigations were focused on the fatigue of UD-layers, loaded e.g. in fibre direction only. Gamstedt and Talreja [14] found specific types of damage and damage propagation in carbon fibre reinforced plastics (CFRP) layers, if the level of loading was very low. The paper presented here solely focusses on the fatigue and microcracking in off-axis layers of the GFRP material.

Some further approaches for VHCF composite testing are introduced in Chapter 2

1.3. Outline

The paper starts by discussing the approaches to investigate VHCF of composites found in literature, and tries to show that these are mainly quite different from those used in the case of metallic specimens (see Chapter 2). It then presents the set-up developed by the authors in previous years and discusses their pro's and con's (see Chapter 3). A separate chapter on online monitoring and offline analysis is added (see Chapter 4), although this is mainly not very different from HCF tests, but there are good reasons for VHCF tests to perform most of the monitoring online.

Some typical results of the current test method are presented (see Chapter 5). This has the advantage that some of the features may be discussed in a more direct way, and it also offers the opportunity to present the approach that is currently prepared for the next steps (see Chapter 6). A concise discussion is given in the conclusions.

This paper clearly does not include simulation approaches, although such results may help to understand the results of experiments in a better way. For further information on such subjects, please refer to Adam et al. [4], [2] or Adam [3].

2. Options for VHCF test set-ups

Several test principles for Very High Cycle Fatigue are found in literature. The simplest of course is to carry on with the usual servo-hydraulic testing of standard specimens as in HCF testing, having the disadvantage that testing will last long and cost a lot at least for energy consumption. This approach is not really followed in the next two subsections.

2.1. Set-ups used for metallic materials
The main approach to test metal specimens in the VHCF regime uses tapered rotational probes at very high (ultrasonic) frequencies, i.e. 20 kHz, see e.g. Christ et al. [12]. Kolyshkin [20] shows in his PhD-thesis a set-up similar to the one depicted in Figure 1 (such set-ups are commercially available from BOKU Wien).

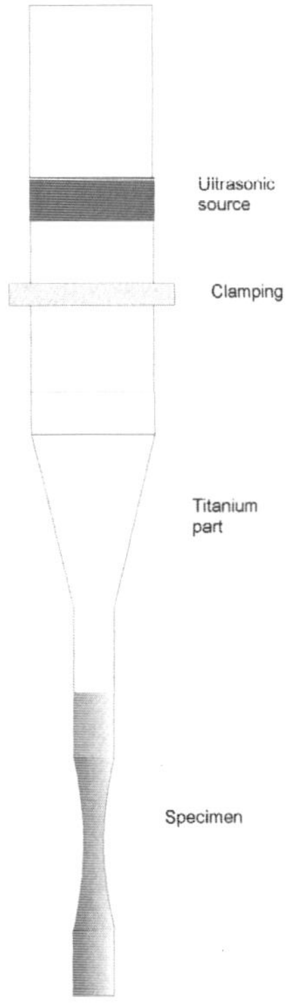

Figure 1. Ultrasonic test set-up (after Kolyshkin [20]).

An elastic wave is sent through the entire set-up which produces high stresses in the minimum cross-section of the specimen. Even though, this concerns a metallic specimen, heating is a problem at 20 kHz. Due to this fact a kind of pulse-pause-tactic is used.

2.2. Set-ups used for composite materials
There are several approaches available to perform VHCF tests on composite materials. One that directly follows the method used for metal materials, as discussed in section 2.1, is proposed by Backe and Balle [7]. They used a piezo-ceramic load actuator, which was able to reach 20 kHz, as in the above mentioned tests. The general principle of the test set-up though is a three-point bending test, where the piezo-electric actuator acts as central load introduction point. Due to the fact that the actuator only allows for very small deflections (about $50\,\mu m$), the bending stiffness of the specimens must be chosen as high as possible. In this case the material was CF-PPS fabric with a thickness of approximately 4 mm and only a very small distance of the two outer load introduction points. As pointed out later in chapter 3, three point bending bears

the disadvantage that the maximum longitudinal strains/stresses are located directly below the central load introduction point.

The test procedure reflects in a way the procedure sketched in section 2.1. Since the surface temperature increased by 4°C after only approximately 100 ms, a type of pulse-pause procedure had to be followed, in order to control the temperature.

The geometry of the specimens has a direct impact on the internal loading of the specimens and in therefore the damage type found in the tests. The length-to-thickness ratio suggests that the main internal loading is due to transverse shear. Damages found during testing support this analysis. Delaminations are found, especially in the region of the load introduction points.

While the Backe and Balle project used bending, where the high frequency was imposed by the actuator alone, Lorsch et al. [22] tried a set-up, where resonance of a piezo-electric actuator, specimen and mass system was used. This is in a way near to the set-up shown in Figure 1, but it needs an additional mass at the lower end to reach reasonable stresses. Both, flat and tube specimens have been used. It turned out that it was quite difficult to find parameters for the masses etc. which resulted in a reasonable frequency range for the entire life of the specimen, from virgin state to a highly damaged state. Additionally, non-symmetric damages could result in off-axis deflections of the specimen, which was not really controllable. In his PhD-thesis [21], Lorsch discusses these issues in a critical way, but he shows that this is a possible method.

Trappe et al. [30] developed a small ±9kN-test-rig, which was fitted into an x-ray-scanner. This meant that an online x-ray-diffraction measurement at flat specimens was applicable without demounting the specimen. The test-rig allowed for a frequency of 50 Hz without significant overheating at $R = 0.1$. Interesting insights were found. One is that in cases, where micro damages, and therefore initiation, were detected only after 10^6 cycles, no final failure of the specimens occurred after 10^8 cycles.

Further methods may be found in literature, namely, bending tests with larger specimens, as e.g. used in Koch et al. [19].

3. The set-up used at IFL for VHCF testing

In the frame of the Priority Programme SPP1466 *Infinite Life for Cyclically Loaded High Performance Materials* the IFL of Technische Universität Braunschweig developed a test-rig for VHCF testing of GFRP specimens in the project *Experimental and numerical investigation of very high cycle fatigue of high performance fibre-reinforced plastics.* A very comprehensive version is given in the PhD-thesis of Adam [3] (partly is a coverage in English by Adam et al. in [5]).

The developed test-rig should meet a set of requirements, namely:

- primarily intralaminar stresses or strains as prime trigger of damages; not, as in the Backe and Balle case, interlaminar shear and transverse shear
- a frequency range between 50 and 100 Hz; which means test durations of about a fortnight for 10^8 cycles
- only marginal temperature increase after long cycle intervals
- relatively cheap specimens, i.e. flat specimens; different from the tube specimens mainly tested at the IFL up to that point (see e.g. [6] or [28]). This also has the sideeffect that cooling of the specimen is quite complex in the tube specimens case.
- high durability of both, the test-rig itself and the load introduction interface between specimen and load introduction/boundary
- a reasonable price of such a device was desirable, offering the chance to multiply it, and use for example 4 rigs in parallel; which in turn reduces the test duration for the completion of a test series considerably

IOP Conf. Series: Materials Science and Engineering **388** (2018) 012004 doi:10.1088/1757-899X/388/1/012004

- low energy consumption
- the opportunity to use as many online monitoring methods as possible, meaning that at least one complete surface should be accessible for optical and thermal *Ganzfeld*-measurements. This has the additional benefit that cooling by air may be used to restrict the temperature increase to an acceptable value.
- ...

The general idea occurred that bending of relatively thin flat specimens, which primarily causes high strains in the outer layers of the laminate could serve as a good solution with respect to the challenge of low temperature increases.

In addition, bending only requires relatively low forces to obtain high stresses in the outer layer of the laminate, resulting in turn in lower cost for both, the device itself and energy consumption.

The general principles offered here for bending are: a cantilever beam, three-point and four-point bending, as depicted in Figure 2.

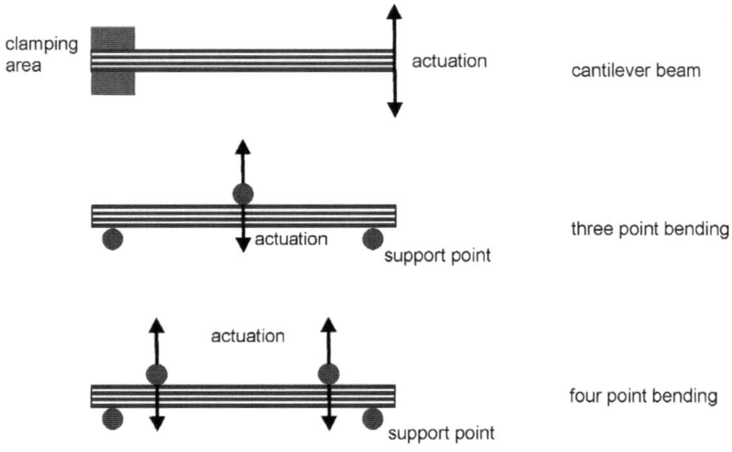

Figure 2. Possible Bending Principles.

From Figure 2 it becomes clear that the cantilever beam configuration will only offer the maximum internal loading very localized in the clamping area itself, where in addition to the bending stresses three-dimensional stresses from clamping occur. For this reason this configuration was discarded. The only positive aspect would be that, due to the clamping the danger of undesired specimen movements is low.

Three-point bending offers a few more key aspects, namely a much lower three-dimensional stress state in the direct contact between actuation and support points, but it still bears the problem that the highest longitudinal stresses occur localized directly under the load introduction point. For this reason this type of general test-rig lay-out was also not chosen.

The four point bending configuration seemed to circumvent the problems arising with the two afore mentioned solutions. Although high stresses still exist below the actuation points, they are only half, and - as main advantage - the complete surface between the two actuation points is loaded by a constant bending moment, as first course technical mechanics reveals. Surely, there is a set of detailed problems, if the four point bending set-up is more closely analysed, as discussed in the next paragraphs. But, from an overall perspective this solution seems to be the most attractive one, and has been chosen for this reason.

One of the difficulties arising with four point bending is the fact that, due to Poisson ratio effects, the bended flat specimen shows a so-called *anti-clastic* bending in the perpendicular direction. This effect leads to the problem that the straight tubular pivots at the support and load actuation points will locally force the specimen to remain flat, while in the free surface area the specimen bends to a small extend, which causes a disturbance in the load introduction area and its vicinity.

In addition, it is a well-known effect that a clamping in longitudinal direction occurs, if the bending deflection gets larger, due to the four points. This results in a certain amount of superimposed longitudinal force. Therefore, it must be the aim to limit the deflection as much as possible, also in this case.

A third effect has to be taken into account for specimens, where microcracking already exists. Since the overall stress under bending is tension on one and compression on the other side of the specimen, and micro-cracks only significantly open under tension, the neutral axis moves accordingly. This of cause will in turn affect local stresses and strains. For detailed analyses this has to be taken into account. Further issues are discussed in chapter 5 in conjunction with experimental results.

Obviously, the type of actuator used in the set-up has a large impact on the design of the test-rig, and may allow for certain configurations, while others have to be excluded from the choice. In addition, they involve the question of achievable frequency, force, costs, duration etc. Adam [3] scrutinized several options both, from literature and industrial offers. Some of the possible means are named, as e.g. Bezazi et al. [10], Belingardi et al. [11], Couillard et al. [13], Bellenger et al. [9], Shenoi et al. [29], Sakin et al. [27], Van Paepegem et al. [31] and Zineb et al. [32]. He looked at servo-hydraulic, electro-linear, electro-mechanic and *voice coil* solutions for the actuation.

The final decision was to use an electro-magnetic actuator. This decision has the consequence that a load ratio R = -1 is imposed, if no special additional measures are taken. Since this is the case, the original four-point-bending configuration depicted in Figure 2 had to be altered in the way shown in Figure 3.

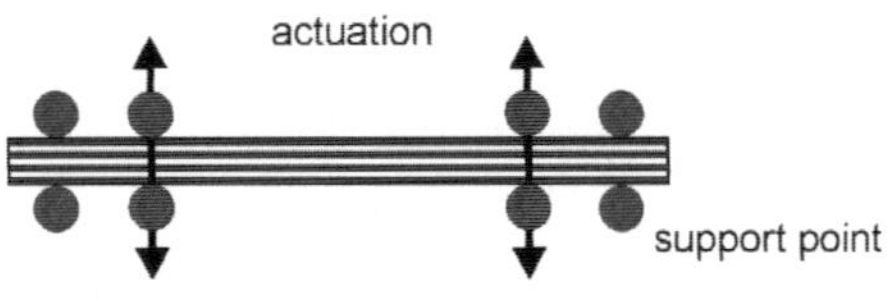

Figure 3. Chosen 4-point-bending configuration.

For sure, this configuration increases the clamping problem mentioned above to a certain extent, but it is very robust, also with respect to movements of the specimen. In order to reduce the clamping effect a very small gap between the tubular support load introduction points and the specimen surface is utilized.

Figure 4 shows a more detailed view of the set-up for a single four-point-bending device. The direct four-point-bending is mounted in a vertical orientation. The four tubular load introduction and support points on the outer surface are visible. Below the aluminium fitting, which holds these tubular parts, the two 1 kN load cells are visible. On the right hand side a laser triangulator is mounted in the way that the central deflection of the flat specimen can continuously be monitored. The specimen istself is only barely visible in this figure. The dimensions of the specimens are $82\,x\,25\,x\,2\,mm^3$, where the distance between the support points is 80 mm.

Two fans are used to cool the surface by blowing air with an ambient temperature of 20° C

Figure 4. Actual set-up of a single test device. Laser triangulator (center right), fans (top and bottom right), four point bending points (center left), below the four-point bending device the two load cells are visible

$\pm 2^o$ C along the surface. In order to keep this ambient temperature the entire rig is located in a housing (see Figure 6) which is cooled down by conventional airconditioning.

The single device is multiplied by a factor of 4, as visible in Figure 5. The figure additionally shows two horizontal rails including a fitting each to mount a camera, which is used to take high resolution pictures of the microcracking state of each device. Furthermore, a thermographic camera mounted on a tripod is visible in the lower left corner of the figure.

The four devices are mounted together on one massive concrete block. It turned out that such a massive basis was needed to assure that no acoustic interference of the four actuators with other parts of the testing hall occurred. This block is located in the housing shown in Figure 6.

The housing in Figure 6 has a further task, i.e. it is isolated in order to reduce noise emission significantly, since under certain conditions the four devices were quite loud. Figure 6 also shows some further features on the lefthand side. In the rear the control cabinet is visible and in the front the computers where Labview is running.

4. Online monitoring and offline damage analysis methods

It should be the intention of all researchers to measure many different parameters during the tests, as there are:

- stiffness changes

Figure 5. Multiple four-point-bending devices

Figure 6. Housing and control cabinet

- internal damages, like microckacking and microdelamination
- damage propagation
- temperature development, both on average as well as localized

Many of these parameters may be monitored in an online procedure. It is advisable to do so, since testing at high frequencies always will introduce at least a small increase in temperature, which in turn means that stopping the test always changes the parameters a bit. Sometimes it is not possible to circumvent such inreruptions, e.g. to measure the initial stiffness change of a specimen. But, the disassembly of the specimen from the test rig in order to perform an external measurement, e.g. by ultrasonics, would not only mean a large impact on the testing period, it also bears the danger that boundary conditions of the test are slightly changing.

Some of the parameters which are monitored in the IFL set-up already has been mentioned in chapter 3. Force and deflection may be monitored by using the load-cells and the individual laser triangulator. Both sensors may be utilized to control loading conditions in a displacement or force controlled manner. In Figure 7 some further points may be seen. Here the constantly stressed part of the specimen is visible and the red dot in the center of the surface marks the laser point.

Figure 7. Actual set-up of two test devices. Laser triangulator, fans and four-point bending device visible, plus Thermocamera

Not visible in the figure is the fact that there exists a background illumination behind each specimen and an optical camera is used to take photographs of the microcracking, where the transparency of the GFRP material is exploited. Stiffness changes in terms of changes in bending stiffness are measured in variable intervals, as mentioned already.

Obviously the amount of data is huge. Especially the characterisation of the microcracking state is cumbersome, if done *manually*. For this reason attempts are made to utilize automated procedures. This is not a simple task, mainly because crack shapes may be complex and micro-delaminations may occur. At IFL a set of researchers jointly developed a method to distinguish between microcracking and micro-delaminations by using laserlight illumination along the edges of the specimen; see Adam et al. [1].

Figure 8 shows three photographs taken from the same specimen at the same number of cycles, namely, about 103000 tested at a stress level of 70 % of the longitudinal strain to inter-fiber fracture of the outer 90 layer. The part depicted shows the evenly bend central part of the four-point bending specimen, described above. Depending on the illumination direction the selective laser illuminations allows to distinguish between microcracking and micro-delaminations within the plies in contrast to usually used backlight illumination. Thus, allowing automated image processing using a Matlab tool to determine crack- and delamination propagation. In order to determine i.e. the crack propagation a set of pictures showing the crack propagation during the experiment is superimposed and aligned using previously applied markings on the specimen. Next the pictures are transformed to negatives where the matrix appears black and any damages appear grey/white. By subdividing these pictures into predefined rectangular sections cracking can be detected by processing them using Hough-Transformations in a loop for crack line detection. Further steps connect the lines of different sections and remove parallel duplicates of neighboring sections which might occur for thick cracks. Using the total crack line length and the initial markings on the specimen the crack density can be automatically evaluated.

5. Some exemplary results

VHCF results of the tests performed at the test-rig described above have been published e.g. in Adam et al. [4] for $[90/0]_s$-specimen (cross-ply) and in Adam et al. [2] for $[\pm 45]_s$-specimen

a.) b.) c.)

Figure 8. Microcracking of a $[90/0]_S$-specimen; (a) optical photograph by transmitted light; (b) line laser lighting from the righthand side to illuminate delaminations (small surface scratches visible, too); (c) line laser lighting from upper edge to illuminate microcracks

(angle-ply). Only a small portion of these results is discussed here. For more detailed information please refer to the two papers mentioned above.

All tests have been performed with specimens made of 1200 tex glass fibre rovings Owens Corning OC111AX and Momentive RIM135i/RIMH137 epoxy resin, where the resulting layer thickness is approximately 0.5 mm. This has the effect that, in case of the cross-ply laminate, the inner 0°-layers mainly serve as some kind of backing-layer, while the 90°-layer is the highly stressed one, where microcracking occurs. In the case of the angle-ply laminate the situation is not as simple as that, due to the fact that very early delamination occurs between the layers. Attempts to test cross-ply laminates with outer 0°-layers have been made, too, but they were not leading to the intended results. Cracking of the 90°-layers occurred only at very high loading; they did not qualify as backing-layers at all. Cracking in the 0°-ayers did not occur. Results of the type found in Gamstedt et al. [14] for pure UD-specimens, as mentioned in section 1.2, were therefore not found.

5.1. $[90/0]_s$ specimens

Tests have been performed at different load or strain levels, respectively. Table 1 lists these levels. By far not all of these load levels lead to real VHCF. Here, w^0_{max} is the maximum deflection at the first load cycle, and $\varepsilon^{90^o}_{22,0}$ is the strain in x-direction of the (outer) 90°-layer.

Figures 9 to 11 show three generally visible effects. All diagrams got 1 to 10^8 load cycles at the abscissa. The first figure depicts the change in (normalized) stiffness over the load cycles, i.e. the normalized bending stiffness in this case. Bending stiffness has been defined by

$$\bar{E}_{xb} = \frac{\hat{F}}{\hat{w}_{max}} \frac{(l_P - l_L)}{bh^3} \left(\frac{1}{4}(l_P - l_L)^2 + \frac{3}{4} l_L (l_P - l_L) + \frac{3}{8} l_L^2\right) \tag{1}$$

where $\hat{F}$ is the maximum force, $\hat{w}_{max}$ is the maximum deflection at the centre of the specimen,

Table 1. Load Levels of $[90/0]_s]$ tests.

Load Level	t_{spec} mm	w^0_{max} mm	$\varepsilon^{90^o}_{22,0}$ %
LL1	2.06	3.05	0.43
LL2	2.07	2.55	0.36
LL3	2.09	1.79	0.25
LL4	2.01	1.61	0.22
LL5	2.01	1.35	0.19
LL6	2.01	0.95	0.13

l_P the distance of the outer support points, l_L the distance of the load introduction points and b and h width and height, respectively. In this sense, the ordinate simply represents the ratio of current and initial displacement.

From this figure it becomes obvious that at load level LL1 and LL2 more or less initial microcracking and as a consequence stiffness reduction sets in. This is also underlined by Figure 10, where the crack density is plotted vs. the number of load cycles.

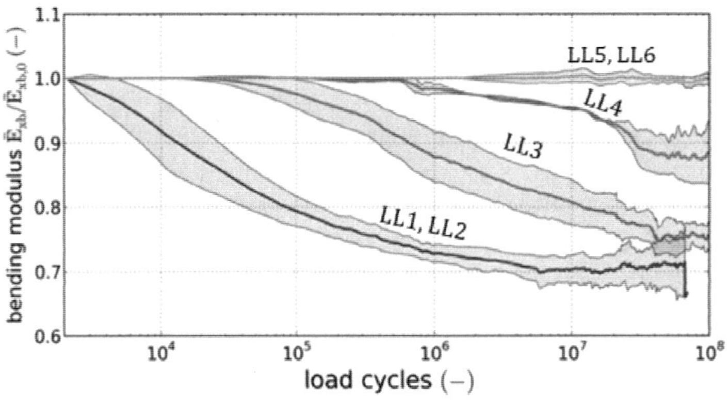

Figure 9. Normalized bending stiffness vs. number of load cycles at different load levels

Load level LL3 and LL4 still show damage accumulation by the decrease of stiffness and increase of microcracking, although this happens after a significant number of load cycles, and after initiation of the microcracks, obviously some kind of crack propagation happens. This is different for LL5 and LL6, where only marginal cracking is found, while the stiffness does not really change. Here, only occational crack initiation occurs; Adam interprets the results in the way that only weak areas, i.e. slightly less durable points/locations, allow for crack initiation. Crack propagation is extremely low or even does not take place.

This result is interesting in the way that the change in character between LL4 and LL5 seems to be at about 10^6 cycles, which is exactly what was found by Trappe et al. [30], as mentioned above, by means of flat specimens and using x-ray-diffraction for detection of small first damages.

Another effect mixes up with microcracking in lowering the bending stiffness. This type of microdelamination occurs at the tips of the microcracks, as already seen in Figure 8. In the

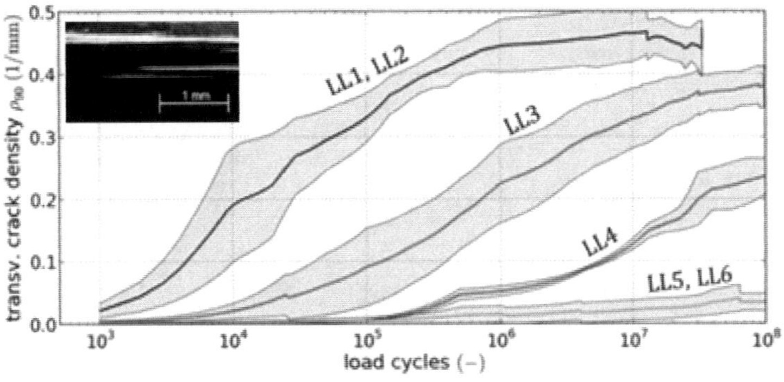

Figure 10. Microcrack density vs. number of load cycles at different load levels

case of load levels LL1, LL2 and LL3 such delaminations occur between the 90°-layer and the backing layer. Several observations on the type of delamination growth are described in Adam et al. [4]. At load levels LL4 to LL6 this does not occur.

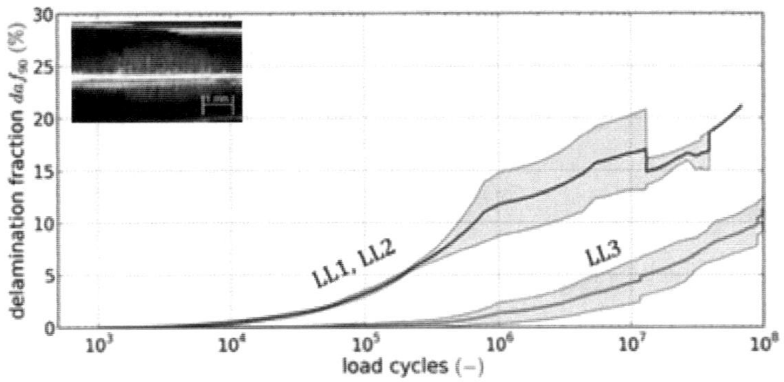

Figure 11. Delamination fraction vs. load cycles

Another effect, which has to be discussed is not visible in the preceding Figures. This is the fact that most of the microcrack initiation happens at the edge, i.e. obviously the stress state is usuall more damaging at the edges also in this case. In Figure 8 c.) this can be observed in a few cases. Different from microcracking in flat specimen under tension etc. the microcrack growth is very slow for specimens at such low load levels, which lead to the VHFC range. As mentioned already, it nearly stops completely in the case of the VHCF load levels LL5 and LL6. In a sense this result shows a big difference between the a.m. observations in metal specimens and those found here in composites. In the metal case, the surface preparation was so well done that no surface initiation occurred, while in all cases with composites the microcracking mostly still occurred at the edges, i.e. the surface.

In addition, all three figures indicate an amount of scatter between the different individual tests. This must be regarded as normal.

5.2. $[+45/-45]_S$ specimens

A quite comprehensive description of the results of $[+45/-45]_S$-specimens are found in Adam et al. [2] and again Adam's PhD-thesis [3]. The following results are not comprehensive at all; they just show some typical problems. Again many tests at several load levels (LL1 to LL7) have been performed (see Table 2). Here ε_x is the strain component in x-direction of the outer layer.

Table 2. Load Levels of $[+45/-45]_s$ tests.

Load Level	t_{spec} mm	w^0_{max} mm	ε_x %
LL1	1.99	3.94	0.54
LL2	1.99	3.51	0.48
LL3	2.01	3.25	0.44
LL4	2.05	2.62	0.37
LL5	1.99	2.49	0.34
LL6	2.03	2.22	0.31
LL7	2.04	1.84	0.25

Figure 12 depicts a typical example for the succession of damage types occurring in a $[+45/-45]_S$-specimen under four-point-bending, and the drop of bending stiffness at the same time. The load level is a medium one in this case.

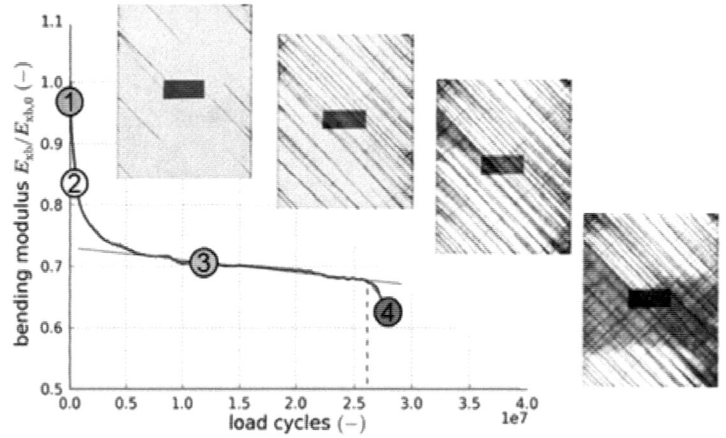

Figure 12. Damage evolution and stiffness decrease in a $[+45/-45]_S$-specimen at a medium load level

The damage types indicated by numbers in Figure 12 are:

① initiation, matrix cracking in outer plies [+45]-plies
② crack initiation in [-45]-plies, first edge delamination
③ steady and predominating delamination growth
④ accelerated delamination growth, final failure

A photograph of a corresponding damage state is also given in the figure. Damages ③ and ④ lead to a very early and massive damage.

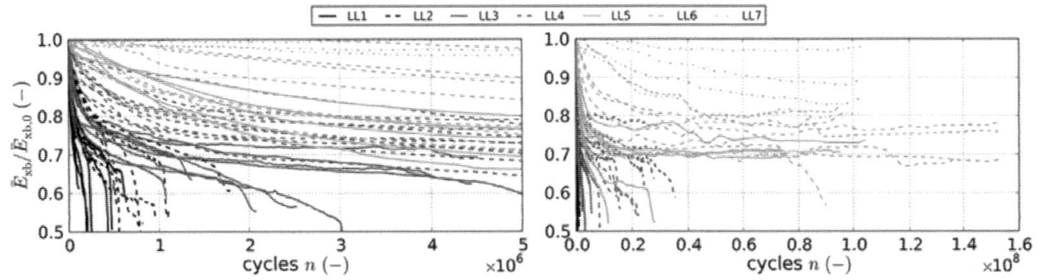

Figure 13. Normalized bending stiffness vs. load cycles for $+45/-45]_S$ specimens.

Figure 13 again shows the decrease of normalized bending stiffness vs. two different scales of load cycles. From this figure the damage behaviour at different load levels may be concluded. Table 3 summarizes the results. For the cases, where delamination does not become the primary damage, i.e. LL5 to LL7, the test procedure obviously is still of interest with respect to VHCF. But, the interpretation is more complex than in the $[90/0]_S$-case.

Table 3. Failure at Load Levels of $[+45/-45]_s$ tests.

LL1, LL2	steep decrease, abrupt final failure between 60 % -70 % of initial stiffness failure at $N < 2x\,10^6$ cycles
LL3, LL4	initial stiffness decrease levels off before accelerating to final failure LL3: $2x\,10^6 < N < 8x\,10^6$ cycles LL4: $8x\,10^6 < N < 3.5x\,10^7$ cycles
LL5, LL6	extremely slow stiffness decrease, degradation comes to rest, mostly no failure up to $1.5x\,10^8$ cycles
LL7	no failure at all up to $1.0x\,10^8$ cycles, residual stiffness between 80 % - 98 %

6. Next steps

Currently, several modifications are being performed and tested on the VHCF-test-rig. To enhance the temperature boundary conditions of the specimen, the strain rate i.e. the testing frequency shall be controlled by monitoring the specimen temperature. Furthermore, efforts are made to modify the test-rig to conduct tests with swelling loads (R = 0). With this testing method it becomes possible to differentiate between tension- and compression-related fatigue damage phenomena. Furthermore, it is planned to make use of the laser light illumination discussed in section 4, to better distiguish between microcracking and microdelamination.

Besides the modifications to the test-rig, a thinner state of the art roving (SE2020) will be used for further test series to maintain the technological relevance. This will result in thinner layers (t = 0.25 ; 600 tex) to evaluate the already found approaches on fatigue behaviour of thicker layers tested before (t = 0.5mm ; 1200 tex), where the inner layers acted as backing layers. Thinner layers may show a different delamination mechanism, due to the different stacking sequence and resulting stress levels. Also, known stackings from literature (Mandell et al. [25] and Hosoi et al. [17]) will be tested; the results compared and the characterization of damage mechanisms added. Investigations on endurance limits, damage initiation, influence of stress ratio.

7. Conclusions

Some conclusions may be drawn from the statements given above, namely

- four-point bending is obviously a principle that allows for VHCF testing up to high numbers of cycles at frequencies between 50 and 100 Hz.
- edges in (bended) flat composite specimens are the primary initiation points in these tests, which is totally different from the type of internal damages in metals in VHCF
- also, no special damage phenomena have been found in the VHCF tests of the flat bending specimens, the phenomena widely reflected the type of microcracking and -delamination, as they are known from HCF.
- at stress levels below a certain limit, no growth of microcracking was found
- ...

All of these data were found for $R = -1$, i.e. alternating constant amplitude testing.

Acknowledgements

The authors thankfully acknowledge the funding of their research in the frame of the Priority Programme SPP1466 *Experimental and numerical investigation of very high cycle fatigue of high performance fibre-reinforced plastics* and the current projects HO2122/24-1, HO2122/28-1 and the Project B3 in the frame of the Collaborative Research Centre SFB880 by the German Research Foundation (DFG).

References

[1] Adam, T J, Nolte F, Begemann B and Horst P 2018 Selective laser illumination method for enhanced damage monitoring of micro cracking and delamination in GFRP laminates *Polymer Testing* **65** 125

[2] Adam, T J and Horst, P 2017 Fatigue Damage and Fatigue Limits of a GFRP Angle-Ply Laminate tested under Very High Cycle Fatigue Loading *International Journal of Fatigue* **99** 202

[3] Adam, T J 2016 *Ermüdungsverhalten faserverstärkter Kunststoffe im Very High Cycle Fatigue-Bereich* (Niedersächsisches Forschungszentrum für Luftfahrt - Forschungsbericht 2016-10)

[4] Adam T J and Horst P 2014 Experimental investigation of the very high cycle fatigue of GFRP [90/0]s cross-ply specimens subjected to high-frequency four-point bending, *Comp. Sci & Techn.* **101** 62

[5] Adam T J, Horst P, Lorsch P and Sinapius M 2012 Experimental Investigation of VHCF of Polymer Composites: Two Alternative Approaches *Materials Testing* **54** 734

[6] Adden S and Horst P 2010 Stiffness degradation under fatigue in multiaxially loaded non-crimped-fabrics *International Journal of Fatigue* **32** 108

[7] Backe D and Balle F 2016 Ultrasonic fatigue and microstructural characterization of carbon fiber fabric reinforced polyphenylene sulfide in the very high cycle fatigue regime *Comp Sci & Techn.* **126**, 115

[8] Bathias C 2006 An engineering point of view about fatigue of polymer matrix composite materials *International Journal of Fatigue* **28** 1094

[9] Bellenger V, Tcharkhtchi A and Castaing P 2006 Thermal and mechanical fatigue of a PA66/glass fibers composite material *International journal of Fatigue* **28** 1348

[10] Bezazi A R, El Mahi A, Berthelot J M and Bezzazi B 2003 Flexural fatigue behavior of cross-ply laminates: An experimental approach *Strength of materials* **35** 149

[11] Belingardi G, Cavatorta M P and Frasca C 2006 Bending fatigue behavior of glasscarbon/epoxy hybrid composites *Composites Science and Technology* **66** 222

[12] Krupp U, Giertler A, Söker M, Fu H, Dönges B, Christ H-J and Ludwig W 2015 The behavior of short fatigue cracks during Very High Cycle (VHCF) Fatigue of duplex stainless steel *Engineering Fracture Mechanics* **145** 197

[13] Couillard R A A and Schwartz P 1997 Bending fatigue of carbon-fiber-reinforced epoxy composite strands *Composites science and technology* **57** 229

[14] Gamstedt E K and Talreja R 1999 Fatigue damage mechanisms in unidirectional carbon-fibre-reinforced plastics, *J Materials Science*, **34** 2525

[15] Hosoi A, Sato N, Kusumoto Y, Fujiwara K and Kawada H 2010 High-cycle fatigue characteristics of quasi-isotropic CFRP laminates over 108 cycles (Initiation and propagation of delamination considering interaction with transverse cracks) *International Journal of Fatigue* **32** 29

[16] Hosoi A, Takamura K, Sato N and Kawada H 2011 Quantitative evaluation of fatigue damage growth in CFRP laminates that changes due to applied stress level *International Journal of Fatigue* **32**, 781

[17] Hosoi A, Arao Y, Karasawa H, and Kawada H 2007 High-cycle fatigue characteristics of quasi-isotropic CFRP laminates *Advanced Composite Materials* **16** 151

[18] Hosoi A, Arao Y and Kawada H 2009 Transverse crack growth behavior considering free-edge effect in quasi-isotropic CFRP laminates under high-cycle fatigue loading *Composites Science and Technology* **69** 1388

[19] Koch I, Just G, Koschichow R, Hanke U and Gude M 2016 Guided bending experiment for the characterisation of CFRP in VHCF-loading *Polymer Testing* **54** 12

[20] Kolyshkin A 2017 *Entwicklung eines Lebensdauervorhersagekonzepts im VHCF-Bereich auf Basis kovariater mikrostruktureller Merkmalsgren* Siegener Werkstoffkundlicher Berichte, Vol. 15/2017.

[21] Lorsch P 2016 Methodik für eine hochfrequente Ermüdungsprüfung an Faserverbundwerkstoffen *NFL-Report No 2016-15*

[22] Lorsch P, Sinapius M and Wierach P 2016 A resonant testing method for high frequency testing of composites in the VHCF regime *Materials Science and Engineering (MSE) 2016*

[23] Mandell JF, Reed RM, Samborsky DD and Pan Q 1993 Fatigue performance of wind turbine blade composite materials *SED Wind Energy ASME* **14**.

[24] Mandell JF, Samborsky DD. DOE/MSU composite material fatigue database: test methods, materials and analysis, Contractor Report, SAND97-3002; 1997.

[25] Mandell JF, Samborsky DD, Wang L. New fatigue data for wind turbine blade materials, Montana State University, AIAA-2003-0692; 2003.

[26] Mughrabi H 2002 On multistage fatigue life diagrams and the relevant life-controlling mechanisms in ultrahigh-cycle fatigue *Fatigue & Fracture of Engineering Materials & Structures* **25** 755.

[27] Sakin R, Ay I and Yaman R 2008 An investigation of bending fatigue behavior for glass-fiber reinforced polyester composite materials *Materials & Design* **29** 212

[28] Schmidt F, Rheinfurth M, Horst P and Busse G 2012 Effects of local fibre waviness on damage mechanisms and fatigue behaviour of biaxially loaded tube specimens *Composites science and technology* **72** 1075

[29] Shenoi R A, Aksu S and Allen H G 1993 Flexural fatigue characteristics of FRP sandwich beams *Fatigue & Fracture of Engineering Materials & Structures* **16** 649

[30] Trappe V, Ortwein H-P and Hickmann S 2013 Infinite life of CFRP evaluated nondestructively with X-Ray-Refraction Topography, *ICCM19*

[31] Van Paepegem W and Degrieck J 2001 Experimental set-up for and numerical modelling of bending fatigue experiments on plain woven glass/epoxy composites *Composite structures* **51** 1

[32] Zineb T B, Sedrakian A and Billoet J L 2003 An original pure bending device with large displacements and rotations for static and fatigue tests of composite structures *Composites Part B: Engineering* **34** 447

IOP Conf. Series: Materials Science and Engineering **388** (2018) 012005 doi:10.1088/1757-899X/388/1/012005

Multiaxial and Multiscale Implications of Dissipative Behavior of Composites

J.G. Michopoulos[a1], **A.P. Iliopoulos**[a], **J.C. Steuben**[a], **J.C. Hermanson**[b]

[a] U.S. Naval Research Laboratory, Computational Multiphysics System Lab., Code 6394, Washington DC, 20375, USA

[b] U.S. Forest Products Lab., Madison WI, 53726, USA

Abstract. A brief description of the progress on mechatronic multi-degree of freedom material testers developed at the US Naval Research Laboratory (US-NRL) is given to describe the enabling technology for acquiring data that encapsulate both the recoverable and the dissipative behaviors of various composite material systems under quasi-static multiaxial loading conditions. The notion of material equivalence is then explored from the perspective of the data-driven dissipative behavior of multiaxially tested composite laminates. Emphasis is given by considering the same fiber within the same resins and vice versa. The effect of different curing conditions is also examined. As a follow-up, an overview of a multiscale computational framework is also outlined for the purpose of introducing multiscale implications of how macro-scale behavior in terms of experimental data can yield information about the behavior of materials at lower length scales.

1. Introduction

Contemporary cradle-to-grave engineering requirements placed on system products include safety, economy, maximum functionality, manufacturability, and maintainability. Combined with the need for qualification, certification, and utilization of such systems, these requirements have significantly raised the demand for validated, efficient, and rapid simulation of the behavior of complex whole systems. In the particularly complex category of material and structural systems known as Polymer Matrix Composites (PMCs), which exhibit a time-varying degradation behavior over large time scales where aging and maintenance are critical, life extension for usability purposes has become a focal area of interest. On the other hand, simulations inherit all of their utility and economic properties from those of the analytical and computational models they are based upon. Data-driven determination of analytical and computational models is an activity that contributes to the objectives mentioned previously by promoting the embedding of validated realism originating from the measured behavioral characteristics of PMCs encoded in the systematically collected data.

In an effort to be consistent with this view, our team at the U.S. Naval Research Laboratory (US-NRL) has developed a system identification approach utilizing custom-made robotic testing machines. These machines possess the unique capability to expose material specimens to loading paths spanning the representative loading space of the specimen structure in order to achieve

[1] corresponding author

data-driven characterization of the constitutive response of PMC systems. A short history of the multi-decade evolution of multi-degree of freedom robotic testing machines is given elsewhere [1, 2, 3].

The motivating goals of the work presented herein are two-fold. First, to demonstrate that if structural response of composites is recorded in terms of their exposure to multi-axial loading in a manner that accounts for their dissipative behavior due to strain-induced damage, then comparing materials and designing composite structures may require a more enriched approach than the one followed currently in many industrial applications by the relevant stakeholders. It should be noted here that other investigators have focused their attention towards the dissipative aspects of PMCs such as [4, 5, 6, 7, 8, 9], but none approaches the topic from the perspective of mutliaxiality of loading and strain induced damage as it will be described here.

The second motivating goal is to demonstrate some of the opportunities and difficulties in inferring the micro-scale behavior of constituent materials from both the conservative and non-conservative perspectives, when experimental data are available at the macro-scale of actual composite parts.

2. Overview of NRL's Multiaxial Loading Systems

To apply multiaxial loading conditions on PMC coupons with the purpose of identifying the bulk lamina constitutive characteristics both in terms of their elastic and inelastic responses, our group has embarked in an effort of prototyping and using custom made multi-degree of freedom (m-DoF) mechatronic testing systems. The most recent families of these systems involves various evolutionary incarnations of the 3-DoF NRL "In-Plane Loader System" (NRL-IPLS) and the 6-DoF NRL66.x systems. The most recent versions of these mechatronic systems as shown in Fig. 1.

In addition to these systems, high performance full-field displacement and strain measurement technologies were also developed to alleviate the inadequate sensitivity, accuracy and efficiency of other full field methods such as Digital Image Correlation. These technologies are the Meshless Random Grid (MRG) [10] and Direct Strain Imaging (DSI) [11, 12] methods [13]. These were integrated with the automated multiaxial testing frameworks and the appropriate inverse problem solution methodology to collect the necessary displacement and strain field data [14, 15, 16, 17, 3].

3. Characterization via the 3-DoF In-Plane Loader System

3.1. Specimen Considerations

PMCs associated with various applications range through a wide variety of constituent materials. Every distinct combination of matrix, fiber, fiber coating (for matrix-fiber interphase), layup angle, stacking sequence, etc. corresponds to a different material. Approximately 100 material systems with fibers ranging from Kevlar to IM7 graphite and several thermoset resins and thermoplastic organic polymers have been tested and characterized with the approach discussed here. A partial list of all the materials tested up to now can be found in [18]. The specimen geometry is described in detail [18, 19] elsewhere and was designed to satisfy the following requirements:

- The characteristic dimensions should be large enough relative to the fiber diameter and the lamina thickness to ensure that the material could be analyzed as either a single mechanically equivalent homogeneous anisotropic monolithic material, or a collection of layers of varying orientations of such materials.
- The overall specimen size should be small enough to keep material costs at a manageable level.

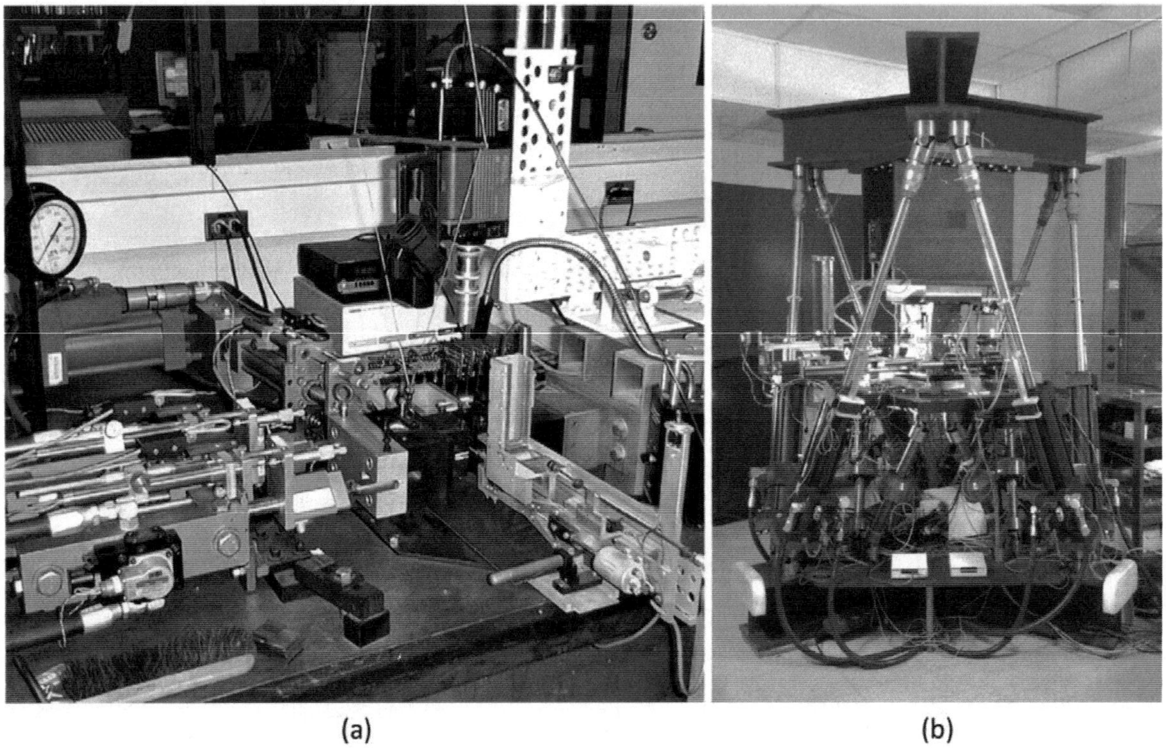

Figure 1. Long view images of the most recent versions of the 3-DoF NRL-IPLS (a) and 6-DoF NRL66.3 (b) multiaxial robotic testing systems.

- Strain riser(s) (such as notches, holes etc.) should be present to guarantee that high strain regions occur away from all gripping boundaries of the specimen.

3.2. Experimental Procedure

The main design objective of the IPLS is to control the rigid body motion of the one boundary of the specimen that is held by the movable grip while another grip holds the other boundary fixed. At the same time sensing subsystems measure the boundary displacements and tractions. Because the actuators of the IPLS are constrained to move in a plane parallel to the specimen, the resulting motion involves only three degrees of freedom relative to any frame of reference on that plane. As it can be seen in Fig. 2 the grip motion can be resolved into three basic components: sliding (shearing) u_0, elongation (opening/closing) u_1, and in-plane rotation u_2.

Specified combinations of actuator displacements, therefore, map into particular combinations of these three basic motions. In order to visualize the loading space it is advantageous to think in terms of a three dimensional displacement space with coordinates (u_0, u_1, u_2). The issue then is how to select a representative family of paths that cover the space and how to sample along each path. It was decided to cover the boundary displacement space with a set of 15 uniformly distributed radial loading paths as indicated in Fig. 3. Note that because of geometry

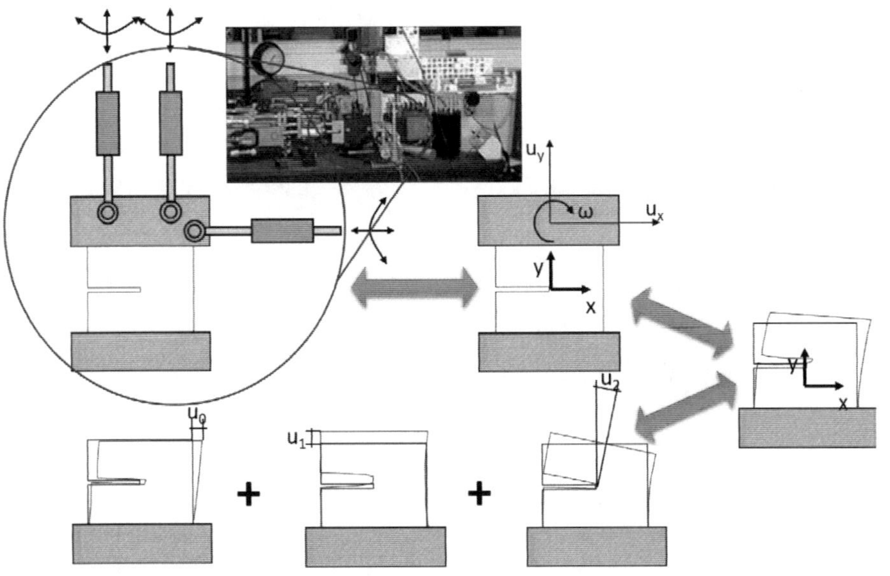

Figure 2. Loading Modes of the In Plane Loader Systems.

and material symmetry about the axis along the notch(es), only the half space corresponding to positive sliding displacement ($u_0 > 0$) need be considered. The required set of observation points is generated by sampling boundary displacement and force data along each path. A particular test in which the actuator motions are continuously varied corresponds to a specific path in this space. Assuming using one specimen per loading path implies that only 15 specimens are required, and 50 observations per loading path are obtained from a single specimen along the respective loading path. The locus of the end points of all loading paths for the same increment is a sphere as shown in Fig. 3(a), where three arbitrary loading paths are presented in a Cartesian frame spanned by with the polar coordinates in Fig. 3(b).

The process of computing the total dissipated energy (DE) due to strain induced damage is based on the boundary displacements and tractions that are measured at each increment imposed by the IPLS along each loading path. More details of the experimental procedure are presented elsewhere [18, 19, 20, 21]. One specimen per loading path is used initially and the procedure is then repeated for a total of two specimens per loading path.

3.3. Analytical and Computational Approach

The analytical approach followed is based on the conservation of energy and the additive decomposition of internal energy density of the material system under test. For a specimen of volume V in the absence of rigid body motion, for the case of quasistatic loads at the boundary that permit the elimination of inertial terms and in the absence of body forces, the conservation of energy can be reduced in the equivalence between the change of the internal energy being equivalent to the external work done as expressed by the form

$$\Delta U^{internal} = W^{external} \Longleftrightarrow \int_{\partial V} w dV = \int_0^{u_i} t_i dq_i, \tag{1}$$

where U is the total internal strain energy and w is the internal strain energy density (energy per unit volume) and t_i, u_i are the tractions and associated displacements at the boundary, that are considered to be reduced to point forces for example measured by load cells. Allowing an

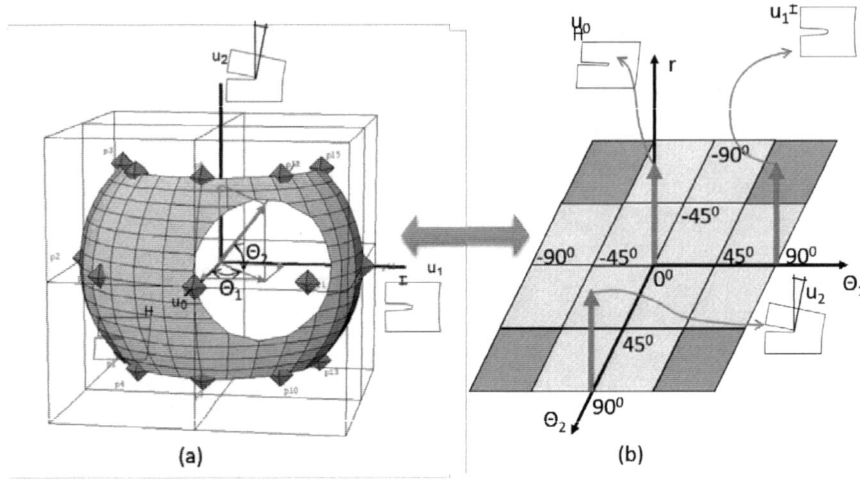

Figure 3. Loading Paths defined in the Cartesian space spanned by the basis modes (a) and corresponding unfolding of the truncated path surface in the spherical coordinates into a corresponding Cartesian system (b).

additive decomposition of the strain energy density to a recoverable w^R part that is assumed linearly elastic and an irrecoverable w^D (due to strain induced damage) or dissipated part ψ yields

$$w = w^R + w^D \Longleftrightarrow w = \frac{1}{2}\sigma_{ij}\varepsilon_{ij} + \psi(\boldsymbol{d};\varepsilon_{ij}), \tag{2}$$

where σ_{ij} and ε_{ij} represent the components of the Cauchy stress and infinitesimal strain second order tensors respectively and $\boldsymbol{d}$ represents a vector of material specific constants related to how the strains contribute to damage. Introducing Eq. 2 to Eq. 1 yields,

$$\int_{\partial V} \frac{1}{2}\sigma_{ij}\varepsilon_{ij} dV + \int_{\partial V} \psi(\boldsymbol{d};\varepsilon_{ij}) dV = \int_0^{u_i} t_i dq_i. \tag{3}$$

Assuming that the equivalency between the internal energy and external work of the conservative part still holds (i.e. $\int_{\partial V} \frac{1}{2}\sigma_{ij}\varepsilon_{ij} dV = \frac{1}{2}t_k u_l$), the expression in Eq. 3 can be reduced to

$$\int_{\partial V} \psi(\boldsymbol{d};\varepsilon_{ij}) dV = \int_0^{u_i} t_k dq_l - \frac{1}{2}t_k u_l = W^D, \tag{4}$$

where W^D represents the external work spent for dissipative processes due to damage evolution in the interior of the material and is measurable experimentally via force and displacement quantities that are measurable via the IPLS sensors. The expression in Eq. 4 enables the formation of an inverse problem for the determination of the material parameters in $\boldsymbol{d}$ that require the minimization of the objective function

$$J = \sum_{p=1}^{n_{Lam}} \sum_{l=1}^{n_{LP}} \left[W_{lp}^D - \left(\sum_{k=1}^{m} \psi(\boldsymbol{d};\varepsilon_{ij})\Delta V_k \right)_{lp} \right]^2, \tag{5}$$

where after a proper Finite Element Analysis (FEA) discretization of the domain in elements of volume ΔV_k with k indexing the elements, l indexing the n_{LP} loading paths and p indexing the

n_{Lam} distinct laminates made off the constitutive lamina under consideration. As mentioned earlier for the case of the IPLS based characterization, $n_{LP} = 2 \times 15 = 30$ and $n_{Lam} = 4$.

There are multiple ways to construct an analytical expression for the dissipated energy density (DED) function utilizing basis functions of the strain tensor components. The simplest is to consider it as a linear combination of strain components weighted by the n_B components d_r of the material parameter in $\boldsymbol{d}$ in the form

$$\psi(\boldsymbol{d}; \varepsilon_{ij}) = \sum_{r=1}^{n_B} d_r \chi_r(\varepsilon_{ij}). \tag{6}$$

Introduction of this expression in the objective function of Eq. 5 reduces the minimization problem to a typical L_2 norm optimization problem with the extra constrain that $\psi(\boldsymbol{d}; \varepsilon_{ij}) \geq 0$ imposed by the second law of thermodynamics. Computational implementation of this optimization enables the determination of all d_r that essentially determine the dissipative character of the material. Consequently, utilizing Eq. 6 the distribution of $\psi(\boldsymbol{d}; \varepsilon_{ij})$ can be plotted over the volume of the specimen. Typical distributions for two arbitrary loading paths and two basis load paths defined in the in plane load space is shown in Fig. 4. Two of them are

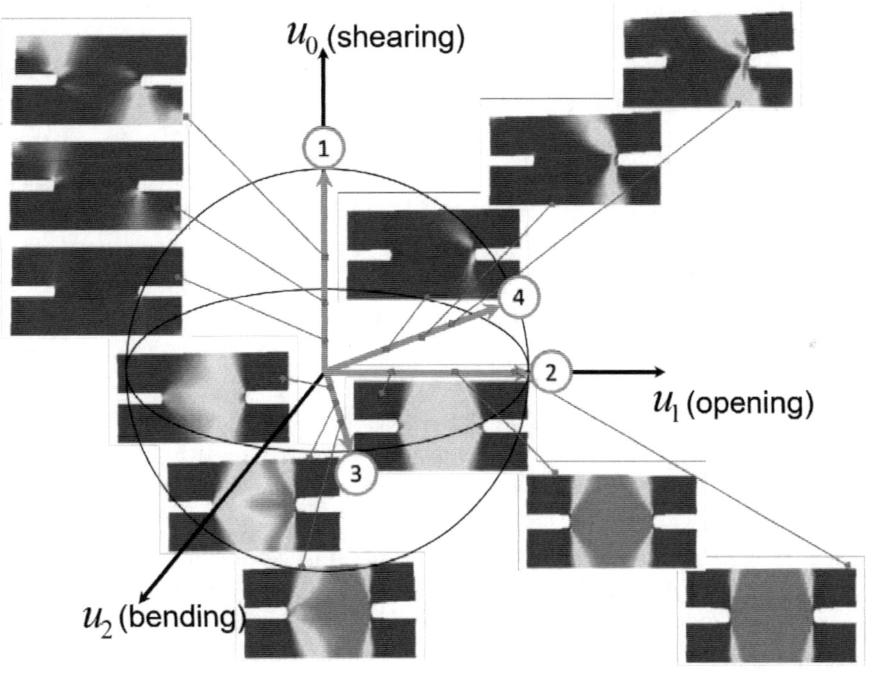

Figure 4. In plane loading space with post computed distributions of dissipated energy density of specimens at designated points of selected loading paths. Path (1) involves only shearing mode, path (2) involves only opening elongation, path (3) involves a linear combination of bending and elongation modes while path (4) involves all three modes.

the basis case for pure shearing (u_0) and pure opening elongation (u_1), one of them lies on the ($u_1 - u_2$) plane and one involves all three modes.

It is now trivial to compute the constitutive law as

$$\sigma_{ij} = \frac{\partial w^R}{\partial \varepsilon_{ij}} + \frac{\partial w^D}{\partial \varepsilon_{ij}} = C_{ijkl}\varepsilon_{kl} + \sum_{r=1}^{n_B} d_r \frac{\partial \chi_r(\varepsilon_{ij})}{\partial \varepsilon_{ij}}, \tag{7}$$

where C_{ijkl} represents the fourth order Hooke's tensor containing all the elastic constants of the material. Implementation of this law in a FEA code enables computing all quantities of interest involved in the structural mechanics problem associated with any composite that use laminae with the properties encapsulated by the material parameters C_{ijkl} and d_r. For a typical transversely isotropic material in a plane state of strain and for bilinear basis functions for ψ, the set of material parameters involves 4 elastic and 125 inelastic (i.e. dissipative) constants for a total of 129 material property constants.

4. Characterization via the 6-DoF NRL66.3 Loader

While the NRL-IPLS enables the dissipative characterization of composites dominated by in-plane states of strain and stress it does not provide any information about the out-of-plane behavior of thick composites exposed to simultaneous in-plane and out-of-plane states of strain induced by generalized 6-DoF loading conditions. Addressing this issue motivated the design of 6-DoF loader frames the most recent (third) version of them being the NRL66.3 as depicted in Fig 1(b).

The main differences in the analysis relative to that followed by the in plane approach exploiting the NRL-IPLS data, are the following:

- Instead of assuming that the elastic constants are known, the solution scheme was expanded to consider them as unknowns to be determined by the minimization of the objective function that in this case is formed in terms of the total strain energy density.
- To reduce the number of the unknown material parameters associated with the dissipative portion of the material behavior (i.e. the 125 coefficients mentioned earlier), we allowed certain material symmetries to hold beyond the elastic limit. This was achieved by employing a damage tensor that contains terms that correspond to those of Hooke's elastic tensor but that are weakening as a function of strain component magnitude.
- To enable the possibility of non-linear elastic behavior, in addition to the infinitesimal strain formalism, we also introduced a strain energy density hyperelastic formalism with damage involving large strains (Green strains) by integrating the two families of fibers model in [22] into the inverse characterization methodology.
- The space for defining the objective function required for the optimization for the parameter determination was expanded to include objective functions expressed in terms of strain tensor component distributions. This was enabled due to the fact that we developed high performance full field measurement technologies such as the MRG and DSI methods [13].
- The loading path space was chosen to be a 4-dimensional subspace of the full 6-dimensional (3 translations and 3 rotations) space that can be realized by the NRL66.3. This subspace is spanned by the 3 rotations and the opening-closing elongation. The hypersphere defined by the locus of the end points of the definable loading paths, is sampled homogeneously by 72 loading paths by proper programming of the NRL66.3.

These differences are in addition to the architectural differences between the NRL-IPLS and the NRL66.3. For more details on the constitutive characterization of PMCs by the use of the NRL66.3 system and associated solution of the relevant inverse problems, the reader is encouraged to explore [14, 15, 16, 17, 3].

The approach was validated against experimental data generated by testing of coupons that were used and data that were not used for the inversion process, and data from experiments performed on composite specimens and parts of completely different geometries and stacking sequences than those of the characterization coupons [16, 17]. A typical example demonstrating the quantitative and qualitative accuracy of developed approach is shown in Fig. 5. The top row of images shows the strain tensor component field distributions from the front side of the specimen evaluated experimentally via the MRG method. The specimen has been loaded by the

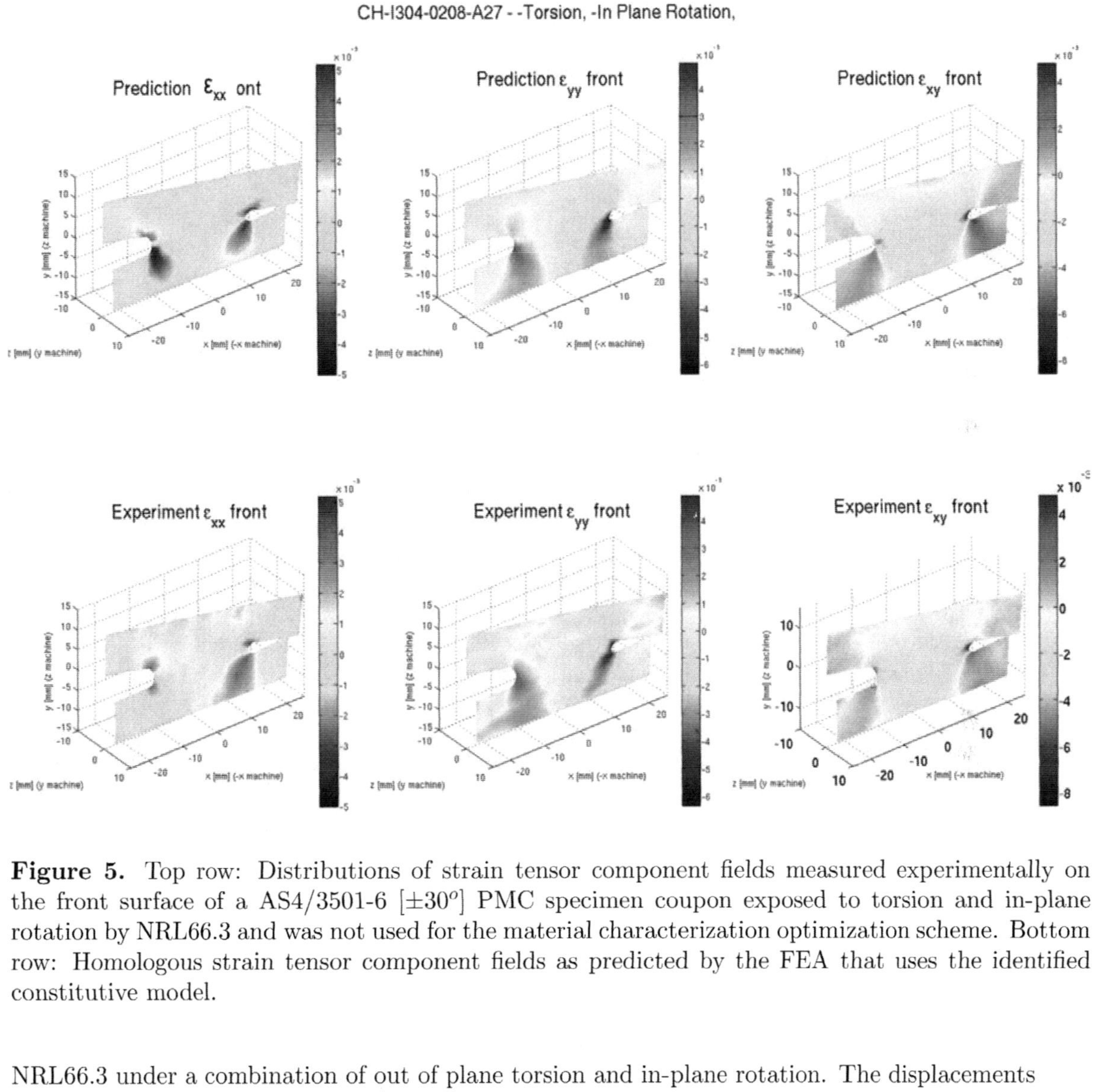

Figure 5. Top row: Distributions of strain tensor component fields measured experimentally on the front surface of a AS4/3501-6 [$\pm30^o$] PMC specimen coupon exposed to torsion and in-plane rotation by NRL66.3 and was not used for the material characterization optimization scheme. Bottom row: Homologous strain tensor component fields as predicted by the FEA that uses the identified constitutive model.

NRL66.3 under a combination of out of plane torsion and in-plane rotation. The displacements have been exaggerated by a scaling factor to reveal the deformation of the specimen that is a [$\pm30^o$] laminate of a AS4/3501-6 PMC system. The bottom row shows the corresponding distributions of the homologous strain component fields as predicted by FEA that uses the identified constitutive behavior. The detected deviation between experimental and predicted strain fields does not exceed 2% at any point of the domain an is on the order of the noise level of the experimental method.

5. Material Comparison based on Data-driven Dissipative Behavior

There are many processing parameters that can have significant effects on the structural properties of an advanced PMC, and on its suitability for various applications. Subtle variations between materials from different vendors can result in supposedly identical material systems exhibiting markedly different performance capabilities. Because of the limited ability to characterize and predict performance of composite materials under the combined loads experienced in structures, the US Navy has developed and follows an extensive series of characterization and certification tests as part of the development of a PMC material structure. This test series is valid only for the material system (and vendor) actually tested for a relevant application. Use of another vendor's material system requires repeating at least part of the qualification and certification testing. Because it is not clear which of these tests will show differences significant to the performance of the structure, often much of the test series must be repeated at great cost.

The US Navy has recognized the need to provide cost effective alternative sources for composite material systems. Alternative sources are essential to ensure competition in procurement and lower costs. It is also necessary to provide replacement materials in support of operational systems still in service, but made of materials no longer commercially available. The Office of Naval Research (ONR) specifically addressed this issue through the "New Materials, New Processes And Alternative Second Source Materials Data Base Generation And Qualification Protocol Development" program deployed in year 2000. This program was performed by a joint industry team lead by carbon fiber supplier BP-AMOCO. The team included prime airframe manufacturer Boeing, material suppliers Hexcel and Cytec-Fiberite, composite resin transfer molding part manufacturer GKN Westland, test laboratory Delsen, and data analysis company Materials Sciences Corporation in addition to US-NRL.

The BP team developed guidance in the form of a protocol defining the test matrix to characterize a new material for use on an existing structure. The guidance reflected judgment as to what constitutes a rational, cost effective set of tests sufficient to qualify the material for the application. The protocol considered the significance of differences between the substitution candidate and the qualified material. The greater the differences, the higher the apparent risk associated with the substitution and the more extensive the test matrix required by the protocol. A checklist of potential material application issues is provided, as well as guidance on test methods, number of replicates, and property values the replacement material must demonstrate for a successful qualification. The guidance considers not just mechanical properties, but also issues such as chemical compatibility and cost. In developing the protocol, the team performed selected material characterization tests to provide data and examples.

US-NRL was asked to apply its DE methodology to test materials manufactured in conjunction with the BP-AMOCO protocol development. The DE approach, through the automated multi-axial test machines, can rapidly generate a material characterization database. This was considered a good opportunity for US-NRL to demonstrate the convenience and utility of the methodology to the problem of second source materials.

5.1. Establishing Material Equivalence in Terms of Dissipative Behavior

The problem of determining whether two material systems are considered mechanically equivalent is different from (and much simpler than) the problem of deciding if a similar but perceptibly different material can safely be substituted in a specific structural application. Material equivalence has traditionally been approached by comparing characterization data. Identical characterization data has been taken to imply identical performance in any structural application. The term "identical" is used here in the sense of no statistically significant differences, given the repeatability of the material and test methods. If the characterization data is not identical, the new material may still be suitable for use in the structure. The question then becomes how much further analysis and testing is required to demonstrate the material can safely be used in the structure and to re-certify the structure with the new material.

The traditional approach rests on an implicit assumption that the characterization methodology adequately captures all material responses critical to success or failure in the application. This assumption is rarely challenged; after all, we have been building successful composite structures for over five decades. However, a brief review of the evolution of materials characterization current practices, and the assumptions underlying them, does not necessarily support this assumption. The final report [23] of the previously outlined project has exposed insight as to why material characterization and second source certification have been proved such difficult and costly issues for composites.

Here we are focusing on material comparison aspects that are directly related to their dissipative behavior, in a way that does not rely on comparison of load-displacement curves along all the loading paths tested that would be analogous to the traditional approaches. We are rather focusing on a comparison based on an expressive data-reduction methodology that accounts for the multiaxiality mode interactions first at the entire structure level and then in the material level from the perspective of their dissipative behavior. This is achieved first by studying the distribution of DE at the structural level in terms of the loading mode combinations and secondly, at the material level in terms of the DED as a function of the strain state.

5.2. Dissipative Behavior Comparison at the Structural Level

For the purpose of identifying materials performance differences at the structural level a data reduction scheme is needed first. This data reduction and analysis approach is based on the multi-axial tests via the NRL-IPLS, we selected the data from testing specimens cut from Cytek's lot 372633, plate P2 made from T300 (RH) fiber and 977-3 resin laminae at a $[\pm 15^o]_8$ layup for the second source protocol project mentioned earlier.

As a first step the eight load paths that do not involve in-plane bending and therefore represent the cases of combinations of elongation (tension-compression) and in-plane shear that all lie on the 2D $(u_0 - u_1)$ plane are considered. Then the load displacement curves are constructed on the resultant planes $(||f|| - ||u||)$, where $(||f|| = \sqrt{(f_0)^2 + (f_1)^2})$ and $(||u|| = \sqrt{(u_0)^2 + (u_1)^2})$ are the norms of the component vectors for the forces and displacements respectively. Then we distribute these curves azimuthally on the $(u_0 - u_1)$ plane such as they all share the resultant force axis being normal to this plane. This operation defines a 3D space spanned by $(u_0, u_1, ||f||)$ as shown in the upper left image of Fig. 6. The black lines represent the individual $(||f||, ||u||)$ plotted on the azimuthally distributed planes corresponding to the respective paths defined on the $(u_0 - u_1)$ plane. An interpolation surface is the constructed that passes from the $(||f||, ||u||)$ distributions. This surface represents the continuous distribution of $||f||$ as a function of (u_0, u_1). The upper right image in Fig. 6 shows the same surface color coded as a function of the force magnitude. Contours of equal force levels are also superimposed on this surface.

As a second step the tangential stiffness distribution surface is constructed by taking the derivative of the force interpolation surface with respect to $||u||$. This distribution along with the equal stiffness contours are depicted at the left bottom image of Fig. 6. Finally, as a third step the process is repeated to construct the surface of evolution of the DED W^D as a function of (u_0, u_1) and the associated contours as shown in the lower right image of Fig. 6.

A material comparison can now be performed by comparing the contours at a specific level of the corresponding quantities by using any of the three distributions depicted in Fig. 6. However, for the sake of brevity and to underline the difference in terms of the dissipative behavior we restrict the comparative analysis by using the DED distribution and specifically comparing the contours for a specific energy level corresponding to the materials of interest. Figure 7 shows the 0.1 lb.in contours of DE for two families of PMCs sharing the same 8552 resin but with different fibers (IM7 G and T650 G2). It is clear that not only there are very few areas that

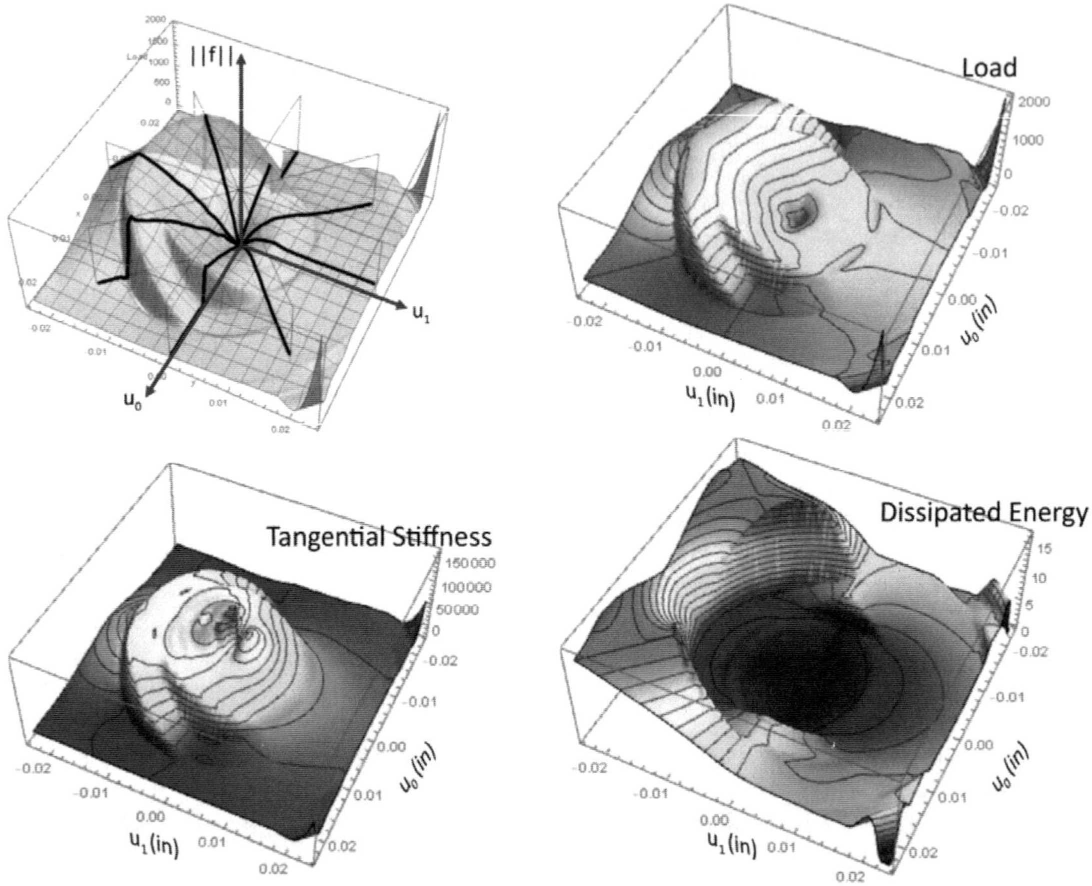

Figure 6. Top Left: Force resultants distributions along with corresponding interpolation surface. Top Right: Color coded and contoured load surface distribution. Bottom Left: Color coded and contoured tangential stiffness surface distribution. Bottom Left: Color coded and contoured dissipated energy density surface distribution.

the two PMCs are similar but the mixed mode behavior for the $[\pm 15^o]$ and $[\pm 30^o]$ layups is morphologically very different from each other, while the ones for the $[\pm 60^o]$ and $[\pm 75^o]$ layups present mostly scaling behavior for the mixed mode area. It is apparent that the difference of the elastic properties of the fibers are not sufficient for explaining the significant differences in these DE behaviors.

Figure 8 shows the 0.1 lb.in contours of DE for two families of PMCs sharing the same T650 G2 fiber but with different resins (3501-6, 8552, 977-3). This comparison indicates that the fiber properties can be associated with the close values in pure tension for most layups (for some cases for pure compression), but the resin differences have a much larger impact on the mixed mode cases, resulting to both morphological and scaling effects.

Figure 9 shows the 0.1 lb.in contours of DE for two families of PMCs sharing the same T300 fiber and the same 977-3 resin, but originating from three different production batches (Cy372633, Cy372636, Cy372637). Given that material is the same in all regards on would anticipate very small deviation of the respective DE contours. However, this is not the case and in particular, the most significant discrepancies appear in the region of pure closing compression)

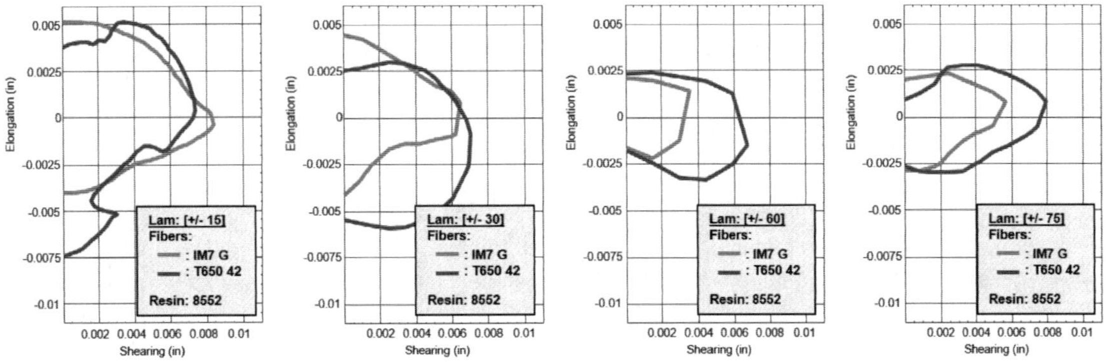

Figure 7. Contours of constant dissipated energy at 0.1 lb.in showing the effect of fiber variations within the same resin for four distinct layups.

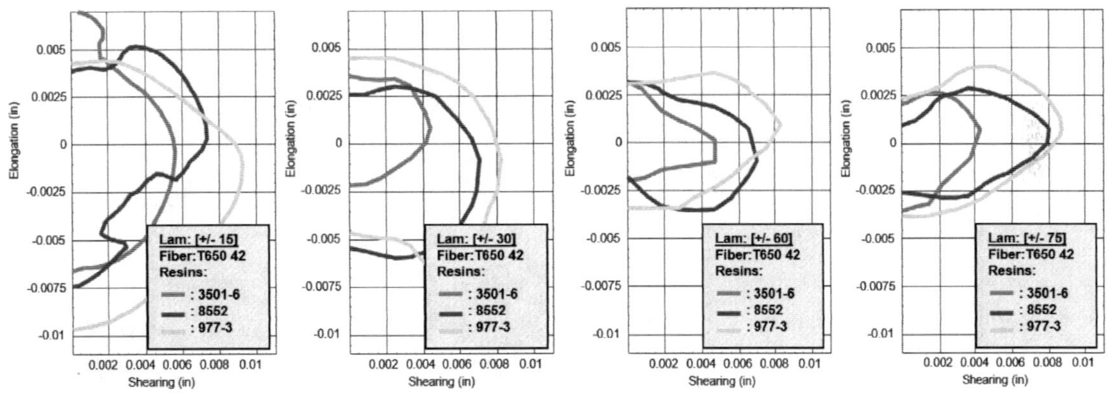

Figure 8. Contours of constant dissipated energy at 0.1 lb.in showing the effect of resin variations for the same fiber system for four distinct layups.

and the regions of closing-shearing combinations for the $[\pm 15^o]$ and $[\pm 30^o]$ layups.

These findings imply that if the level of DE is calibrated to correspond to the failure of the material in tension, batches Cy372633, Cy372636 could produce easily A- and B-basis allowables for the cases of $[\pm 15^o, 60^o, 75^o]$ since the contours at pure tension point coincide. However, that cannot be said for other areas where combined loading modes are present. This introduces the important consideration that even if one derives A- and B-basis allowables from uniaxial tests, it does not mean that the same specimens under biaxial load will also produce equivalent A- and B-basis allowables.

5.3. Dissipative Behavior Comparison at Material Level

For comparing materials without the effect of the structural geometry, it is necessary to turn attention on a quantity that describes the dissipative nature of the material in terms of quantities that do not depend on the shape of the structure but rather depend on the state of strain. Clearly, the DED function ψ satisfies this description. Since ψ is constructed to be a function of the local strains and captures the collective non-linear material behavior, it can be plotted in various strain planes to generate DED surfaces, in fashion similar to the one we followed for the DE in the structure. A typical example of the distribution of DED for three different laminates

IOP Conf. Series: Materials Science and Engineering **388** (2018) 012005 doi:10.1088/1757-899X/388/1/012005

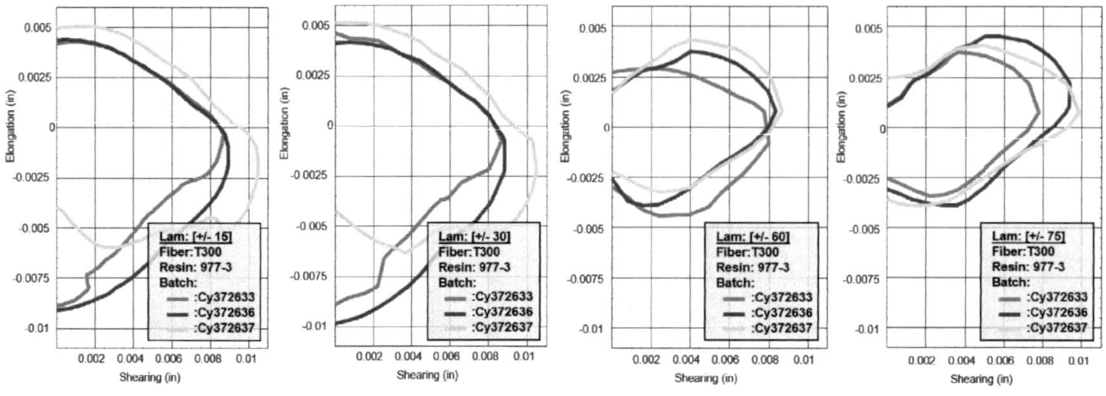

Figure 9. Contours of constant dissipated energy at 0.1 lb.in showing the effect of batch to batch variability for 3 different batches sharing the same resin and fiber systems for four distinct layups.

with plies from AS4/3501-6 system as shown in the top row of Fig. 10.

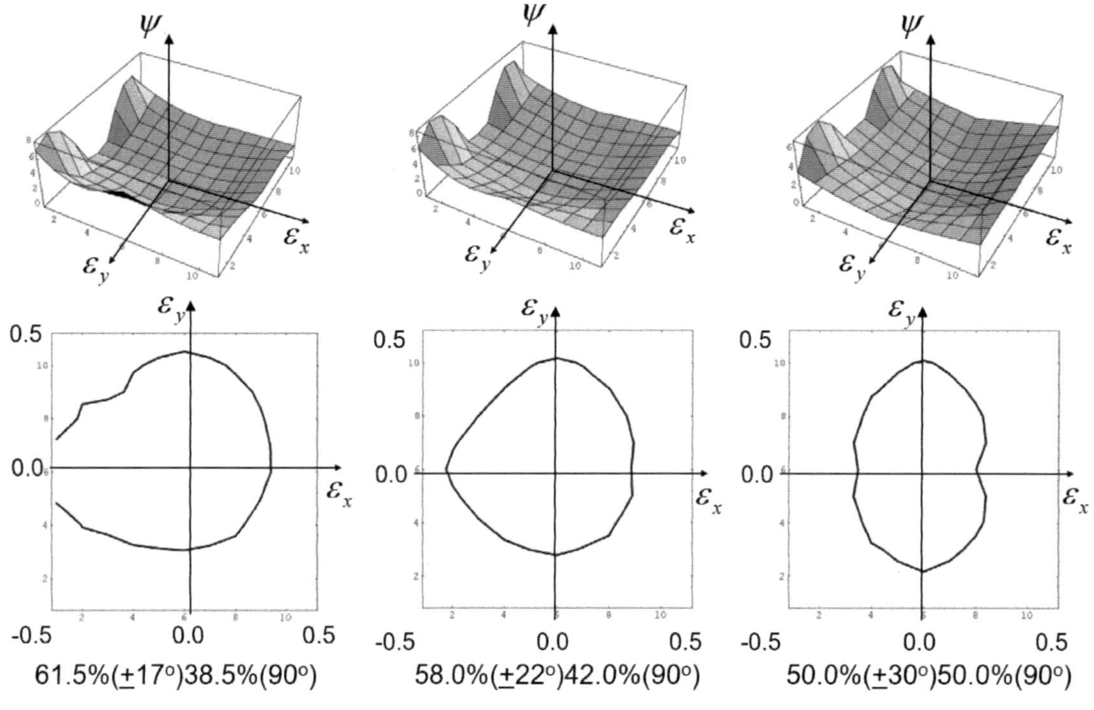

Figure 10. Dissipated energy density distribution as a function of he strains $\varepsilon_x, \varepsilon_y$ for the case of $\varepsilon_{xy} = 0$ (top row). Respective contours of constant dissipated energy density at 1 $lb.in/in^3$ for the same layups.

Contours of constant DED can be extracted from the DED surfaces to obtain envelopes of criticality or failure (if one wishes to associate failure with a critical amount of DED) for each material system, as also shown in Fig. 10. Comparisons of these surfaces and/or contours (or failure envelopes) can be used to determine the adequacy of the candidate materials to replace the qualified material in a particular application with respect to a particular metric for characterizing failure, namely a critical level of DED. It is evident that if the strain envelopes of the structure under consideration are available, then the suitability of the candidate material to safely satisfy the required strain performance can be easily determined without necessarily having to match the qualified materials performance.

If the suitability of the candidate material is still unclear, i.e. it is different from the qualified material but may be good enough to meet the design needs of the structure, then a detailed analysis of the structure of interest must be performed. At this point, the divergence between the two materials is such that the risk associated with the substitution becomes great. A database containing these types of DED representations would be applied to the structure through a structural simulation of the structure/material/load system. The structural performance of the system would be evaluated over all critical load conditions and the ability of the candidate material to satisfy the structural requirements determined. A successful simulation of the candidate materials ability to meet the requirement of the structure would greatly reduce the risk associated with the substitution. Further considerations would then need to be made to determine what appropriate structural verification testing would be required.

6. Towards Multiscale Inverse Problem

In an effort to assess how much information can be derived for material behavior at the micro-scale based on experimental data associated with the macro-scale tests, an inverse method for a cascaded model chain is presented. In order to achieve the inverse characterization of the composite material across multiple length scales, a representative volume element (RVE) of the composite structure at the micro scale was modeled, aiming to estimate the average properties of the heterogeneous media in the meso-scale (lamina). Consequently, the forward problem must first be formulated to enable the calculation of the homogenized properties of the given composite material. This material is assumed to have a structure consisting of parallel, cylindrical fibers, embedded in a polymer matrix and arranged in a periodic distribution, and packed in either a square or hexagonal array with a density reflecting realistic volume fraction of the fibers.

Subsequently, the properties estimated by the aforementioned homogenization process are considered as the properties of the lamina at the meso-scale, which in turn forms a balanced, angle-plied laminate at the macro scale. Synthetic experiments performed on the laminate include mechanical loading under 6-DoF conditions. These synthetic (simulated) experiments provide predictions of the associated strain fields. Solution of the inverse structural problem to estimate the micro-scale properties of the composite material may be accomplished using appropriate optimization algorithms, in order to enable the multiscale characterization of the composite material. This requires consideration of the homogenized response of the lamina at the mesoscale, as well as the laminate at the macroscale, utilizing experimental data collected at the macro scale.

6.1. Forward problem and mathematical homogenization

The three dimensional modeling of the composite microstructure outlined previously utilizes a micromechanics model that allows for the estimation of the full set of the elastic properties using a single model. This is in contrast to the conventional approach utilizing a collection of forward models, accompanied by scale-dependent assumptions, in order to estimate the full set of the properties needed. The problem of determining the material properties of heterogeneous materials such as composites, has long been studied [24, 25, 26] and many analytical homogenization techniques have been built based on the equivalent eigenstrain method [24]. These methods provide approximate estimates of the average elastic properties of the macroscopically isotropic and sometimes anisotropic heterogeneous material. However, in order

IOP Conf. Series: Materials Science and Engineering **388** (2018) 012005 doi:10.1088/1757-899X/388/1/012005

to comply with the transversely isotropic behavior that most composites exhibit due to the random distribution of fibers in the cross-section, an appropriate averaging procedure is used to provide a stiffness tensor with five independent constants corresponding to the transverse isotropy of the individual laminae.

The current implementation involves only numerical homogenization [27] through the use of finite element analysis, in order to work towards the inverse problem of composite characterization at the micro-scale of constituents (fiber and matrix) in a manner that follows the established methodologies as described in [28, 29]. The composite is assumed to be comprised of parallel cylindrical transversely isotropic fibers embedded in an polymer isotropic matrix. The microstructure is considered to be periodic and in either a square or a hexagonal arrangement, as shown in Fig. 11. In order to estimate the effective elastic properties at the meso-scale, the

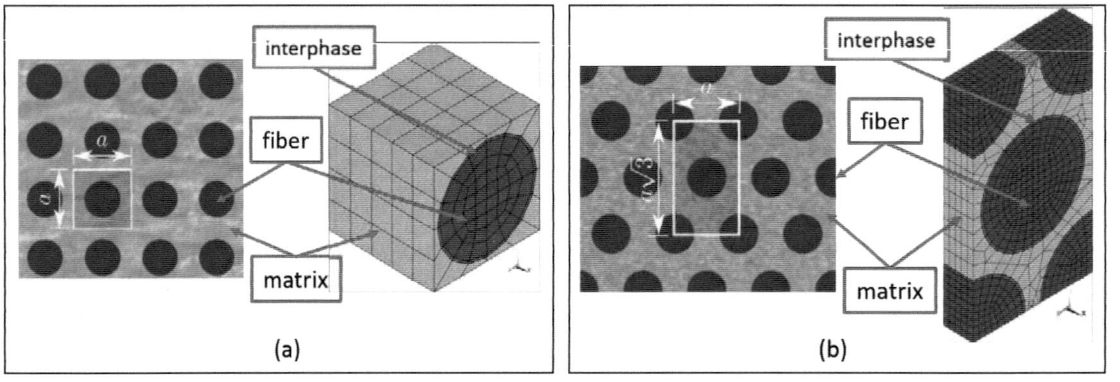

Figure 11. Mesoscale representation of composite lamina with tetragonal (a) and hexagonal (b) periodic fiber distributions along with equivalent FEA discretizations of respective RVEs required for homogenization.

composite is represented at the micro-scale by a 3D representative volume element (RVE), that is consistent with the arrangement of the constituents and is subjected to periodic boundary conditions [27].

Each RVE is subjected to an average strain $\bar{\varepsilon}$ [30] and each of the six components of strain are applied by enforcing appropriate displacement boundary conditions through constraint equations at the faces, edges and the vertices of the RVE [28]. The volume average of the strain ($\bar{\varepsilon}_{ij}$) is equal to the applied strain (ε_{ij}^{app}):

$$\bar{\varepsilon}_{ij} = \frac{1}{V}\int_V \varepsilon_{ij} dV = \varepsilon_{ij}^{app}. \tag{8}$$

For the homogeneous composite, the average stress is calculated (in Voigt notation) by:

$$\bar{\sigma}_\alpha = C_{\alpha\beta}\bar{\varepsilon}_\beta. \tag{9}$$

Consequently, the components of the Hooke tensor $C_{\alpha\beta}$ of the homogenized material can be determined by:

$$C_{\alpha\beta} = \bar{\sigma}_\alpha = \frac{1}{V}\int_V \sigma_\alpha(x,y,z)\, dV \text{ , with } \varepsilon_\beta^{app} = 1, \tag{10}$$

where x, y, z are the spatial coordinates. A detailed description of the process alongside the necessary constraint equations that satisfy periodicity for each loading case can be found in [28]. The components of C_{ij} for the case of transverse isotropy are given by [31, 27, 32].

6.2. Multiscale Inverse Problem

As indicated earlier, in the single scale characterization framework [16, 33, 34] the inverse problem takes the form of an optimization problem. In that framework the forward evaluation is applied through a composed form of the models across various scales such that the usage of experimental measurements at the macro scale is naturally applicable. In the multiscale framework the constituents of the composite at the micro-scale are considered to be the matrix, the fiber and the interphase between the matrix and the fiber. The material properties for the matrix and the interphase have been assumed to be isotropic and transversely isotropic for the fiber. The properties in consideration are then defined as: E_m, ν_m are the Young's modulus and Poisson's ratio of the matrix, E_i, ν_i are the Young's modulus and Poisson's ratio of the interphase and E_{xf}, E_{yf}, G_{xyf}, ν_{xyf}, ν_{yzf} are the five independent engineering properties of the fiber. The set of properties is denoted as:

$$P = \{E_m, \nu_m, E_i, \nu_i, E_{xf}, E_{yf}, G_{xyf}, \nu_{xyf}, \nu_{yzf}\} \tag{11}$$

Since the aim of this work was to investigate the feasibility of identifying any subset of P: $U \subset P$ using an appropriate inversion scheme we can now define the set difference $K = P \setminus U$ as the set of known properties prior to characterization. When needed, the values of the properties in the set U can be identified by simpler experiments, for example a tension experiment to identify the matrix E_m and ν_m, or the fiber axial Young's modulus E_{xyf}.

It is anticipated that there will always exist elements of P that cannot be known a-priori. In the specific case of an RVE with an interphase layer between the fiber and the matrix these are the properties of the interphase (E_i, ν_i) that manifest only after the composite layers have been manufactured. In addition, sometimes determination of simple elastic material constants like the ν_{yzf} cannot be determined experimentally with high confidence.

In the form described herein, and with the aid of the definition of the forward problem, a block diagram of the multiscale inverse characterization process is presented in Fig. 12. The entire process is inside an optimization loop that takes the homogenized parameters and performs yet another forward evaluation of the meso- and macro-scale models. This process is performed for each load case that full field experimental data exist for. The strain results of the multi-scale forward analysis are then incorporated in an additive operation of objective functions. These objective functions are selected to be of the form,

$$f_k = \frac{1}{3N} \sum_{l=1}^{N} \sum_{i=x,y} \sum_{j=i,y} \left| \left[\varepsilon_{ij}^{exp_k} \right]_l - \left[\varepsilon_{ij}^{fea_k} \right]_l \right|, \tag{12}$$

where $|\cdot|$ denotes the absolute value, $\varepsilon_{ij}^{exp_k}$ are the surface strains measured at predetermined locations via a full-field measurement method [35, 36, 37, 38, 39, 40, 41, 42, 12] for the load case k, and $\varepsilon_{ij}^{fea_k}$ are the surface strains calculated from the forward finite element analysis of the load case k. l is the id number of the location the strain measurement takes place and N is the total number of strain measurement locations. An aggregate objective function $f = 1/K \sum_{k=1}^{K} f_k$ is used to collect the values of each of the objective functions evaluated for the specific load cases. It is important to note here that this approach still holds for the case of damage induced dissipative behavior of the structure that leads to reduction of the elastic properties.

6.3. Computational framework and synthetic experiments

The inverse methodology described in the previous section was implemented using MATLAB®[2] as the control language and ANSYS®Mechanical™ [3] as the forward multi-scale evaluator.

[2] MATLAB® is a registered trademark of The MathWorks, Inc.
[3] ANSYS, is a registered trademark of ANSYS, Inc.

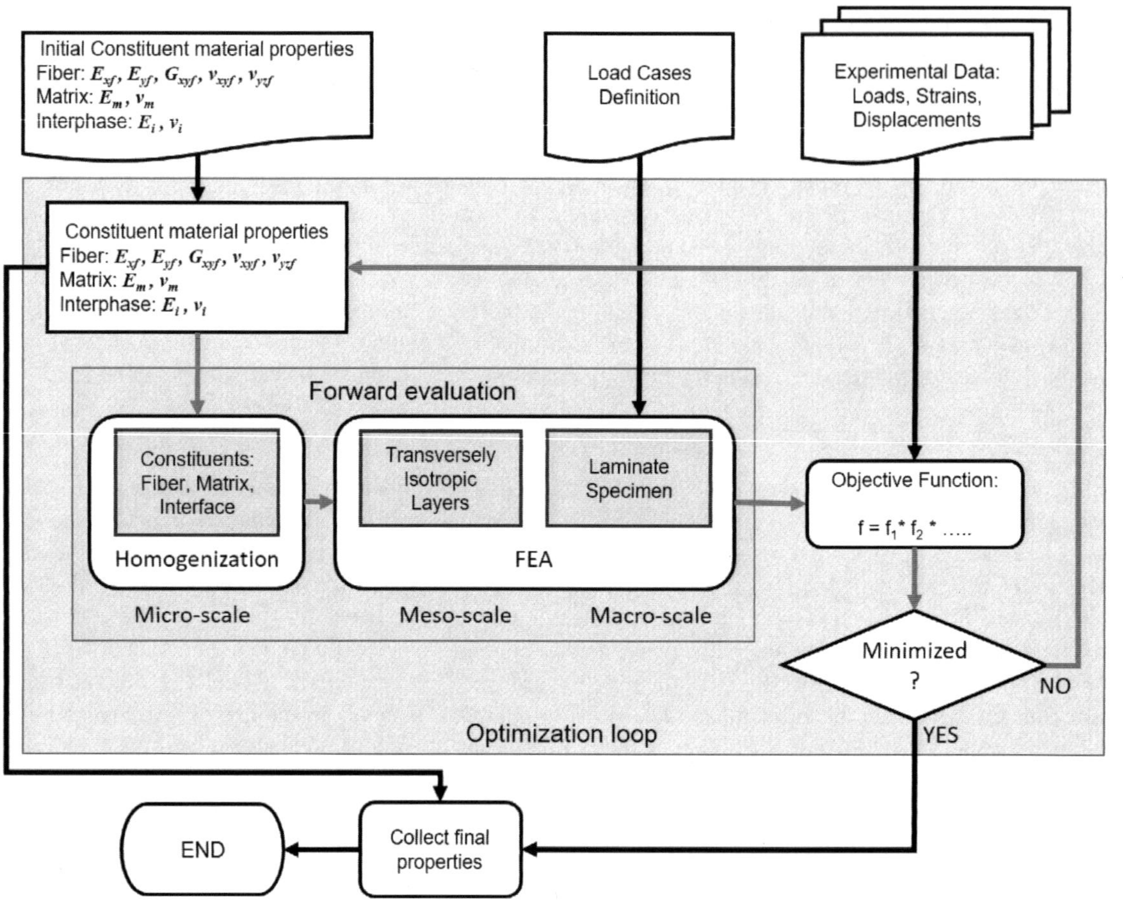

Figure 12. Block diagram of the multiscale inverse characterization process.

The connection between the two was achieved through an appropriate serial approach using file-system based exchange of information. Many optimization algorithms were explored as described in [32]. A non-linear derivative free global optimization algorithm with proper modifications to constrain the search space of the optimization parameters [32] was selected. Figure 11 shows the mesoscale geometry of both the square and the hex cells utilized with an interphase layer present between the resin and the fiber. The macro scale geometry is presented in Fig. 13(a). It has been discretized with hexahedral elements (SOLID185). A closeup at the area of the hole is depicted in 13(b) indicating that the composite material layers are modeled with a single element through their thickness.

6.4. Synthetic Experiments

In order to investigate the capabilities of the presented multiscale inverse method for characterizing a composite material at the constituents level by using data and the laminate level, we evaluated how accurately the elastic properties of the isotropic matrix and these of the transversely isotropic fiber can be estimated. This estimation was based on comparisons made relative to synthetic data that were generated by running the forward problem with known constituent properties.

All 128 possible combinations of properties appearing in Eqn. 11 were examined assuming

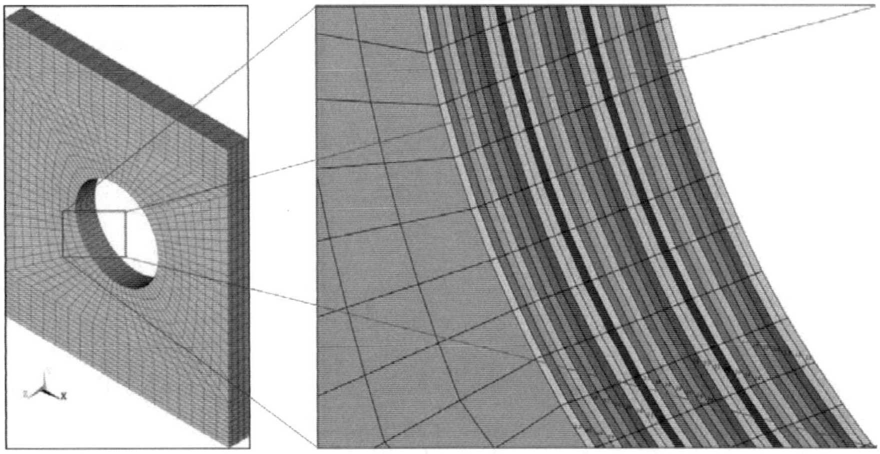

(a) (b)

Figure 13. General view of the meshed macro-scale specimen (a) and close-up showing the meso-scale layer configuration of the meshed PMC specimen geometry with an open hole utilized for the synthetic experiments.

Table 1. Constituents material properties and optimization bounds

	E_m	ν_m	E_{xf}	E_{yf}	
Value	$4.3GPa$	0.35	$235GPa$	$15GPa$	
min	$0.5GPa$	0.0	$10GPa$	$1GPa$	
max	$15.0GPa$	0.45	$600GPa$	$60GPa$	
	ν_{xyf}	ν_{yzf}	G_{xyf}	E_i	ν_i
Value	0.2	0.07	$27GPa$	$2.3GPa$	0.25
min	0.0	0.0	$1GPa$	$0.1GPa$	0.0
max	0.45	0.45	$75GPa$	$5GPa$	0.45

that the interphase properties (E_i, ν_i) are always unknown. With regard to the macro-scale model, the laminate specimen dimensions where chosen to be: height 45 mm, width: 40 mm, thickness: 3.5 mm and hole diameter: 15 mm. For all the synthetic experiments, a single loading case was selected, defined in the full loading space of a 6 degrees of freedom loading vector acting upon the laminate at macro scale ([43]). The values of the components of this vector for the three force components and the three components of moment were: $F_x = 2241\ N$, $F_y = 4310\ N$, $F_z = 517\ N$, $M_x = 3.45\ Nm$, $M_y = 6.9\ Nm$, $M_z = 69\ Nm$. The magnitude of these loads was chosen to ensure that the developed corresponding levels of strain respect the linear elastic limits of the material model.

The composite material used to generate the synthetic results is the AS4/3501-6 carbon epoxy, with values of elastic properties of the fiber, the matrix and the lamina as they appear in [31] and also presented in Table 1, together with assumed properties for the interphase layer. The errors for each material property reported in the relevant tables were calculated using the formula:

$$e = \left| \frac{P_{inverse} - P_{nominal}}{P_{nominal}} \right| \times 100\,[\%] \tag{13}$$

where $P_{inverse}$ is the value of the property identified by the multiscale inverse characterization

Table 2. Multiscale inverse characterization for selected cases of the square volume element with interphase layer synthetic experiments. Case IDs represent different choices of unknown parameters

ID	E_m	ν_m	E_{xf}	E_{yf}	ν_{xyf}	ν_{yzf}	G_{xyf}	E_i	ν_i	Evaluations	f $(pStrain)$
sia	-	-	-	-	-	-	-	< 0.1%	< 0.1%	124	2.3
sib	-	-	< 0.1%	-	-	< 0.1%	< 0.1%	< 0.1%	< 0.1%	1346	2.1
sic	-	0.13%	-	< 0.1%	< 0.1%	-	< 0.1%	< 0.1%	0.8%	3231	1.7
sid	0.3%	4.2%	-	-	-	1.2%	4.6%	0.65%	35.0%	2605	52.8

Table 3. Multiscale inverse characterization for selected cases of the hex volume element with interphase layer synthetic experiments

ID	E_m	ν_m	E_{xf}	E_{yf}	ν_{xyf}	ν_{yzf}	G_{xyf}	E_i	ν_i	Evaluations	f $(pStrain)$
hia	-	-	-	-	-	-	-	< 0.1%	< 0.1%	139	2.0
hib	-	-	< 1.3%	-	-	19%	10.4%	4.2%	52%	485	509910
hic	-	1.2%	-	< 0.1%	1.6%	-	0.6%	0.1%	6.3%	1430	2.1
hid	2.2%	7.1%	-	-	-	8.1%	4.6%	2.4%	35.4%	1485	347679

and $P_{nominal}$ is the value used to generate the synthetic data.

Starting by deploying the methodology described above and utilizing the associated computational framework for only one parameter and working our way up to the total number of unknown parameters and their possible combinations, we were able to obtain a maximum number of unknown parameters that can be concurrently defined with high accuracy. We have assumed an isotropic interphase zone, under the condition that the matrix and fiber properties are known. The parameter identification was successful, even for values of the interphase properties as low as 0.5 % of the matrix's elastic properties. In absence of noise in the synthetically generated data, the estimation of the interphase properties can be achieved with very low error.

In table 2 some interesting inversion results are presented for the case of the square RVE. In this table we observe that the interphase properties can be found with high accuracy if the properties of the other constituents are known, which may indeed reflect a realistic situation when constituent properties can be determined experimentally via separate testing. The same is true for other cases like the *sib* case. As the number of unknown increases the results are getting worse. The analysis of all 128 cases revealed that no case with over 6 unknowns was inverted efficiently.

The results of the hex RVE are presented in Table 3. It is observed that in this case the interphase properties are identified when they are the only unknowns. For other cases though the properties cannot be identified with satisfactory accuracy, besides the *hic* case. It should be noted that this is only a subset of interesting results and other cases have been solved with high accuracy. The value of this exercise lies on the fact that if strain induced damage causes certain reduction of the material properties at the micro-scale, then the approach proposed here has the potential to identify the values of the effective properties even for the case of these damaged materials.

7. Conclusions and Plans

The present work has outlined a data-driven methodology for executing multi-degree of freedom testing on composites, and subsequently identifying the constitutive behavior of such composites at the bulk lamina level by solving a properly formed inverse problem. As a natural extension of this process, the acquired data can enable the description of the dissipative behavior of composites both at the coupon level (via dissipated energy) and at the material level (via dissipated energy density).

As an example of the utility of this view on the non-linear behavior of composites, it was demonstrated that material comparison has a considerably more complex phenomenology than can be expressed using only design allowables established by uniaxial experiments. Furthermore, it was shown that multiaxially loaded composites can exhibit drastically different behaviors in terms of dissipated energy despite appearing identical under uniaxial loading conditions.

In an effort to potentially identify damage effects at the micro-scale manifested as reduced mechanical properties in terms of experimental data acquired at the macro-scale, the present work also described a methodological framework along with its computational implementation. The feasibility of identifying mechanical properties of constituents of composite materials at the micro scale from experiments conducted at the macro scale was demonstrated by proper application of the proposed framework. The inversion methodology was applied to identify the properties of the interphase layer that develops between the matrix and the fibers of a PMC system. Both the Young's modulus and the Poisson's ratio were identified for the synthetic experiments conducted. Furthermore, it was also shown that it is possible to identify properties of other constituents as long as some of them are known via independently conducted characterization.

Plans for the future include a sensitivity analysis with respect to the constituent material properties, as well as the investigation of additional geometrical parameters as measures of damage evolution. It is also sought to incorporate the uncertainty of realistic (random) fiber distributions to eliminate the assumptions of periodic fiber arrays. The proposed methodology will be further enhanced by utilizing various global optimization methodologies in an attempt to establish a more detailed and efficient implementation of the inversion problem. The effect of adding additional loading paths in the optimization scheme (and the formation of the relevant objective function) will also be examined as potential means for both reducing the error of the characterized properties, as well as increasing the total number of the potentially identifiable properties. Finally, the methodology must be expanded to allow spatial variation of the degraded properties in a manner that frees the analysis from the implicit assumption that the identified properties are the same across the entire domain of the structure.

Acknowledgments

The authors acknowledge support for this work by the Office of Naval Research (ONR) through the Naval Research Laboratory's core funding as well as direct support from ONR under grant N0001417WX01510. The authors also would like to express their gratitude to Ms. C. Komnineli of George Mason University for her leadership and computational support for the multiscale inversion and homogenization activities reported in the present paper.

References

[1] Michopoulos J 2003 *Computational and Mechatronic Automation of Multiphysics Research for Structural and Material Systems* (Kluwer Academic Press) pp 9–21

[2] Michopoulos J 2004 *Proceedings of the 6th World Congress on Computational Mechanics (WCCM-VI)* (Beijing China: Tsinghua University Press and Springer) pp 486–491

[3] Michopoulos J, Hermanson J C and Iliopoulos A 2014 *Advances in Computers and Information in Engineering Research* vol 1 ed Michopoulos J G, Rosen D W, Paredis C J J and Vance J M (New York: ASME) pp 73–95

[4] Reifsnider K L and Williams R S 1974 *Experimental Mechanics* **14** 479–485 ISSN 1741-2765 URL `https://doi.org/10.1007/BF02323148`

[5] Ermakov; Y N and Zinoviev P A 1994 *Energy dissipation in composite materials* (Lancaster, PA: Technomic Publishing Company, Inc.)

[6] Tarnopol'sku Y and Beyle A 1995 *Mechanics of Composite Materials and Structures* **2** 177–178 (*Preprint* `https://doi.org/10.1080/10759419508945837`) URL `https://doi.org/10.1080/10759419508945837`

[7] Einav I, Houlsby G and Nguyen G 2007 *International Journal of Solids and Structures* **44** 2487 – 2508 ISSN 0020-7683 URL `http://www.sciencedirect.com/science/article/pii/S0020768306002939`
[8] Jalalvand M, Czl G, Fuller J D, Wisnom M R, Canal L P, Gonzlez C D and LLorca J 2016 *Composites Science and Technology* **134** 115 – 124 ISSN 0266-3538 URL `http://www.sciencedirect.com/science/article/pii/S0266353816308648`
[9] Chatzigeorgiou G, Charalambakis N, Chemisky Y and Meraghni F 2017 *Thermomechanical Behavior of Dissipative Composite Materials* (ISTE Press & Elsevier Ltd.)
[10] Iliopoulos A P, Michopoulos J G and Andrianopoulos N P 2010 *Strain* In Print ISSN 1475-1305
[11] Iliopoulos A and Michopoulos J 2012 *ASME 2012 International Design Engineering Technical Conferences & Computers and Information in Engineering Conference IDETC/CIE, August 13-15, 2012, Chicago, IL, USA* vol 2 pp 1021–1031
[12] Iliopoulos A and Michopoulos J G 2013 *International Journal for Numerical Methods in Engineering* **95** 313–330 ISSN 1097-0207 URL `http://dx.doi.org/10.1002/nme.4509`
[13] Iliopoulos A and Michopoulos J 2014 *Advances in Computers and Information in Engineering Research* vol 1 ed Michopoulos J G, Rosen D W, Paredis C J J and Vance J M (New York: ASME) pp 97–124
[14] Michopoulos J, Hermanson J and Furukawa T 2008 *Composite Structures* **86** 154–164 ISSN 02638223
[15] Michopoulos J, Hermanson J, Iliopoulos A, Lambrakos S and Furukawa T 2011 *Journal of Computing and Information Science in Engineering* **11**
[16] Michopoulos J, Hermanson J C and Iliopoulos A 2012 *2012 ASME International Design Engineering Technical Conferences & Computers and Information in Engineering Conference,Chicago IL, USA*
[17] Michopoulos J, Iliopoulos A and Hermanson J C 2012 *Proceedings of the ASME 2012 International Design Engineering Technical Conferences & Computers and Information in Engineering Conference IDETC/CIE, August 12-15, 2012, Chicago, IL, USA* submitted
[18] Mast P W, Nash G E, Michopoulos J G, Thomas R W, Badaliance R and Wolock I 1992 Experimental determination of dissipated energy density as a measure of strain-induced damage in composites Tech. Rep. Tec. Rpt. NRL/FR/6383–92-9369 Naval Research Laboratory Washington, DC, USA
[19] Mast P, Nash G, Michopoulos J, Thomas R, Wolock I and Badaliance R 1995 *Theoretical and Applied Fracture Mechanics* **22** 71–96 ISSN 01678442
[20] Mast P, Nash G, Michopoulos J, Thomas R, Wolock I and Badaliance R 1995 *Theoretical and Applied Fracture Mechanics* **22** 97–114 ISSN 01678442
[21] Mast P, Nash G, Michopoulos J, Thomas R, Wolock I and Badaliance R 1995 *Theoretical and Applied Fracture Mechanics* **22** 115–125 ISSN 01678442
[22] Kaliske M 2000 *Comput. Meth. Appl. Mech. Eng.* **185** 225–243
[23] Badaliance R, Chwastyk T, Gause L, Mast P and Michopoulos J 2000 New materials, new processes and alternative second source materials data base generation and qualification protocol development - application of dissipated energy techreport Naval Research Lab and Boeing enclosure 15
[24] Eshelby J D 1957 *Proceedings of the Royal Society of London A: Mathematical, Physical and Engineering Sciences* **241** 376–396 ISSN 0080-4630
[25] Hill R 1965 *Journal of the Mechanics and Physics of Solids* **13** 213 – 222 ISSN 0022-5096
[26] Budiansky B 1965 *J. Mech. Phys. Solids* **13** 223–227
[27] Barbero E J 2008 *Finite Element Analysis of Composite Materials* (CRC Press, Taylor & Francis Group)
[28] Barbero E J 2013 *Finite Element Analysis of Composite Materials Using ANSYS®* (CRC Press, Taylor & Francis Group)
[29] Yuan Z and Fish J 2008 *International Journal for Numerical Methods in Engineering* **73** 361–380 ISSN 1097-0207 URL `http://dx.doi.org/10.1002/nme.2074`
[30] Luciano R and Sacco E 1998 *European Journal of Mechanics - A/Solids* **17** 599 – 617 ISSN 0997-7538 URL `http://www.sciencedirect.com/science/article/pii/S0997753899800242`
[31] Daniel I M and Ishai O 2005 *Engineering Mechanics of Composite Materials* 2nd ed (Oxford University Press, USA) ISBN 019515097X
[32] Komninelli Fand Iliopoulos A and Michopoulos J 2015 *Proccedings of the ASME 2015 International Design Engineering Technical Conferences & Computers and Information in Engineering Conference* (Boston, MA)
[33] Michopoulos J, C H J and Iliopoulos A 2012 *Materials Challenges in Alternative & Renewable Energy, February 26-March 1, 2012, Clearwater, FL, USA*
[34] Steuben J, Michopoulos J, Iliopoulos A and Turner C 2015 *Composite Structures* **132** 694–708
[35] Sevenhuijsen J 1981 *Strain* **17** 20–24 ISSN 00392103
[36] Bruck H, McNeill S, Sutton M and Peters W 1989 *Experimental Mechanics* **29** 261–267 ISSN 0014-4851 10.1007/BF02321405
[37] Sirkis J and Lim T 1991 *Experimental Mechanics* **31** 382–388 ISSN 0014-4851 10.1007/BF02325997

[38] Andrianopoulos N P and Iliopoulos A P 2005 *Strain measurements by a hybrid experimental-numerical method using a mesh-free field function, Honorary Volume for Professor P. S. Theocaris* (Armenian Academy of Sciences) pp 31–41

[39] Andrianopoulos N P and Iliopoulos A P 2006 *Proc. 16th European Conference of Fracture, Alexandroupolis, Greece, July 3–7*

[40] Michopoulos J G and Iliopoulos A P 2009 *ASME Conference Proceedings* **2009** 55–63 URL `http://link.aip.org/link/abstract/ASMECP/v2009/i48999/p55/s1`

[41] Michopoulos J G and Iliopoulos A 2011 *ASME Conference Proceedings* **2011** 489–498

[42] Iliopoulos A, Michopoulos J G, Orifici A C and Thomson R S 2011 *ASME Conference Proceedings* **2011** 513–520

[43] Iliopoulos A, Michopoulos J G and C H J 2010 *Proceedings of the ASME 2010 International Design Engineering Technical Conferences & Computers and Information in Engineering Conference IDETC/CIE 2010* DETC2010/CIE-28741 ed ASME

IOP Conf. Series: Materials Science and Engineering **388** (2018) 012006 doi:10.1088/1757-899X/388/1/012006

Observations of microscale tensile fatigue damage mechanisms of composite materials for wind turbine blades

L P Mikkelsen

Section of Composite Mechanics and Structures, DTU Wind Energy, Technical University of Denmark, 4000 Roskilde, Denmark

E-mail: lapm@dtu.dk

Abstract. In the presented study, it has been studied how tensile/tensile fatigue damage will develop in quasi-unidirectional non-crimp fabric based glass fibre epoxy matrix composites. A material system conventionally used as the load carrying laminates in wind turbine blades. It will be demonstrated by stopping a tensile/tensile fatigue test before the final material failure, how 3D X-ray computer tomography (x-ray CT) can be used for determining fibre failure inside the composite material. Due to the rather coarse bundle-structure of the non-crimp fabrics investigated, a relative large cross section area of the test-sample is necessary in order for testing a representative material volume. Therefore, the x-ray CT technique is used in a scout and zoom test-setup where a scout scan of the 15 mm large cross-section is used for identifying the damage region, while the subsequently zoom scan of a 2-3 mm field of view region is used for determine the fibre failures. This scan-setup is non-destructive, and it is therefore possible to obtain a number of scans during the fatigue damage development. Thereby, it is found that the fibre failure are mainly occurring close to points where the load carrying uni-directional fibre bundles are in close contact with the crossing points of secondary oriented fibre bundles, the so-called backing bundles.

1. Introduction

The load carrying laminates in modern large wind turbine blades are often based on non-crimp fabric reinforced composites. Contrary to a woven fabric, a non-crimp fabric is buildup of straight bundles stitched to a thin backing layer. Thereby, waviness of the stiffness giving fibers are avoided and larger material stiffness are achieved. In a typical quasi-unidirectional non-crimp fabric used in the wind turbine industry, around 90% of the fibers are oriented in the load direction. The remaining 10% is secondary oriented bundles with a much lower tex-values (thinner bundles). The fact that not all fibers are oriented in one direction is the reason for the term "quasi-unidirectional", which we will use in the following. The load carrying laminates in a wind turbine is due to the requirement of large stiffness and fatigue resistance [1] to a large degree buildup of such quasi-unidirectional non-crimp fabrics resulting in wind turbine blades where more than 60% of the fibers are in the axial direction of the blade [2].

Through the last couple of years, comprehensive studies of the fatigue damage mechanism in quasi-unidirectional non-crimp fabrics based composites has been performed non-destructive using 3D x-ray computer tomography [3]–[5]. Studies which has been enhanced the fatigue damage mechanisms obtained in earlier destructive fatigue damage characterizations of similar non-crimp fabric based glass fiber composites [6]. From the studies it has been observed that transverse tunnel cracks in the backing bundles will develops in the initial phase of the fatigue loading. Those cracks will subsequently grow into the load carrying fiber bundles as local fiber breakages. Finally, those fiber breakages will localize into the final failure of the test sample. Each of those three damage mechanisms can also be identified on the measured stiffness degradation where first a small rapid stiffness drop will occur at the initiation phase, a stable stiffness degradation region fiber breakage phase and a fast large stiffness drop for the localization phase. A stiffness degradation curve rather similar to the three main stages for fatigue damage of fiber composites reported by Reifsnider and Jamison [7] for pre-preg composites. Nevertheless, the failure mechanism in the non-crimp fabric differ from that observed for the pre-preg based composites due to the difference in the fiber architecture.

In the present study, a non-crimp fabric based high-module (E_f = 80GPa) glass-fiber epoxy matrix material system used in the wind turbine blade industry is investigated. It is demonstrated how thermovision can be used for identification of the initial damage region in a tensile/tensile fatigue test coupon. In addition, the observed material stiffness degradation can be used for identify how far we are into the fatigue damage region. The fatigue damage is characterized using a scout and zoom 3D x-ray scan sequence. Thereby, it is possible to identify a small damage region inside a field of view of 2-3 mm. Only by using such small a field of view, it is possible to identify the single fiber failure of the 17 μm thick glass fibers.

2. Material system

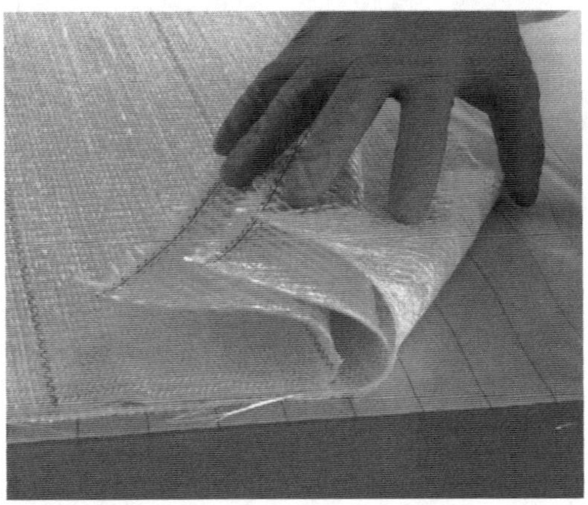

(**a**) Three of the 6 non-crimp fabrics used in the symmetric layup

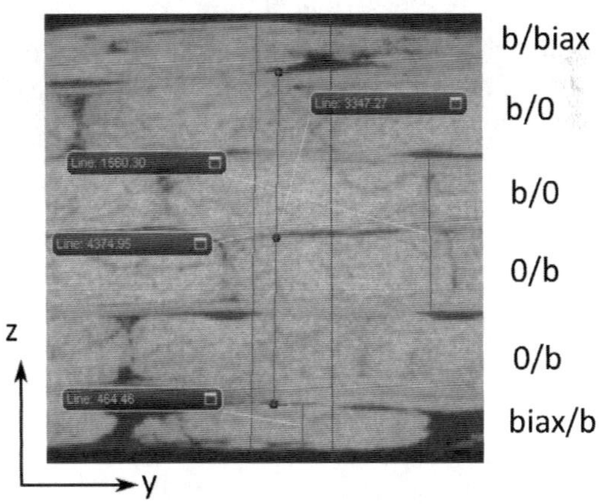

(**b**) A 2D slice from the X-ray scan of the full cross section of the composite

Figure 1. The [b/biax,b/0,b/0]$_s$ glass fibre non-crimp fabric layup with the backing pointing outward from the centre is used in the study

A 6 layer glass fiber composite is studied where the layup is [b/biax,b/0,b/0]$_s$ as seen in Figure 1a. Only the fatigue damage located inside the quasi-unidirectional non-crimp layers in the center of the sample is investigated.

The 4 quasi-unidirectional fabrics denoted 'b/0' is made by 1152g/m^2 unidirectional bundles each with a tex value (weight per km bundle) on 2400 g/km with an average fiber diameter on D_0=17 μm. With a glass fiber density on $\rho_{glass} = 2.6g/cm^3$ this will correspond to approximately 4000 fibers in each UD bundles. The UD bundles are stitched using a polyester yarn to a backing layer consist of ±45 and 90 oriented bundles with a tex value of 200 g/km and an average fiber diameter on D_b=16 μm. The backing bundles are does therefore consist of approximately 380 fibers. The area weight of the ±45° bundles is 100g/m^2 and the 90° bundles is 19g/m^2.

The two biax fabrics denoted 'b/biax' is made of 2x296g/m^2 ±45° oriented fiber bundles each with a tex value on 600g/km with an average fiber diameter on $D_{\pm}$45=16 μm corresponding to 1150 fibers in each bundles. In addition, the biax fabrics does also have 19 g/m^2 90° bundles each with a tex value on

200 g/km. Those bundles are stitched to a Chopped strand Mat (CSM) with an area weight on 100 g/m^2. Except from the CSM mat, where the fibers is standard E-glass with a stiffness on $E_f = 72GPa$ all the other fibers are high-modulus fibers with on $E_f = 80GPa$.

After stacking the non-crimp fabrics, the composite is manufactured using vacuum assisted transfer molding using a Huntsman epoxy denoted: Aralide LY 1564SP/XB3486. The laminates was cured using a two-stage cure profile first with a pre-cure at 50C for 7 hours followed by a post-cure sequence on 90C in 3 hours. The thickness of the resulting composite was approximately 2.3mm which corresponds to a fiber volume fraction on 57%.

3. Tension/tension fatigue testing

(a) Test-setup with back-to-back mounted extensometers for stiffness degradation measurement

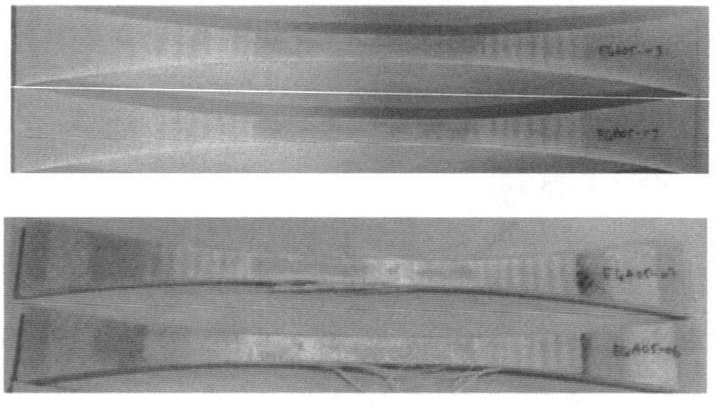

(b) Test sample before and after the test

Figure 2. Tension/tension fatigue testing

The test coupon is fatigue loaded cyclic in tension/tension under prescribed loads with a fixed R-ratio given by $R = \frac{\sigma_{min}}{\sigma_{max}} = 0.1$. Figure 2a show the experimental setup for the samples tested to final failure. In here, two extensometers are mounted back-to-back in order to measure the stiffness degradation of the material. A butterfly shaped specimen [8] is used in the test, as it can be seen in figure 2b. In order to have a representative volume of material in the central part of the test coupons, a constant cross-section on 15×4.5mm^2 over a gauge length of 60mm is used. The total length of the test-sample is 410mm with a maximum sample width in the gripping region of 50mm.

Figure 3a show the SN-curve obtained during the fatigue test of 8 test-samples at different load levels. The load level of the test material is inhere presented as a normalization of the prescribed maximum stress σ_{max} with respect to the initial stiffness E_0 of the test-sample. The normalization maximum stress is denoted by $\varepsilon_{max} = \sigma_{max}/E_0$. A Basquin curve $N_{fail} = K\varepsilon_{max}^m$ is fitted to the measured fatigue failure SN-points. Together with the fatigue test tested until final failure, figure 3 also include the data for a test sample stopped 10.000 cycles before the expected material failure (marked with the red point on the SN-curve). The material damage inside this sample is subsequently studied using 3D x-ray CT scan presented in the next section. Figure 3b show the corresponding stiffness degradation of the test-material. The composite material stiffness degradation is measured over the full load cycle using the back-to-back mounted 50mm/±2 mm extensometers inside the gauge section. The three main stages for fatigue damage evolution as reported by e.g. Reifsnider and Jamison [7] with (stage I) an initial fast stiffness drop during the initial cyclic loadings, (stage II) a rather constant stiffness degradation over the major second stage fatigue life and (stage III) a fast stiffness degradation in the third stage just before the final fatigue damage localization in the material. The fatigue damage evolution has been investigated in details in [5] and [4] where [5] mainly investigated the initial fatigue damage govern by the transverse cracking of the backing bundles related mainly to the initial stage I of the stiffness degradation, while [4] investigate the progression of the fiber breakages in the load carrying fibers related to stage II. The red stiffness degradation curve shown in figure 3b is the stiffness degradation of the x-ray studied sample where the end-point of the red curve correspond to the red point in figure 3a. It should be noted that the stiffness degradation curve for the x-ray studied sample has been obtained without extensometers so it was possible to measure the surface temperature of the test sample as shown in figure 4. The red stiffness degradation curve is obtained from the overall test sample stiffness degradation which also can include effect of tab delamination and damage in the gripping region. Nevertheless, the curve is found to follow the other curves quite well.

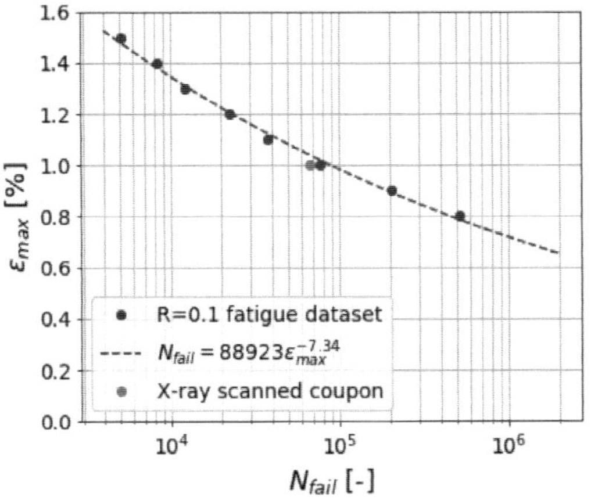

(a) SN-curve for the normalized stress versus number of cycles to failure

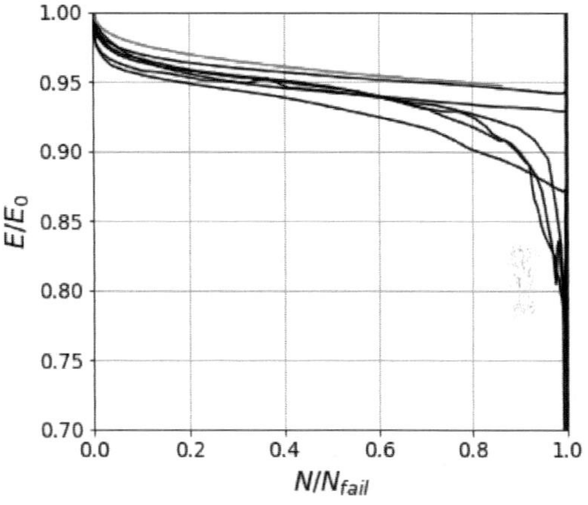

(b) Stiffness degradation caused by the fatigue damage development

Figure 3. Results from the R=0.1 load controlled tension/tension fatigue testing

In figure 4, the surface temperature of the test-sample and the corresponding stiffness degradation curve is shown for a test specimen denoted EQA05-14 which is the test-sample stopped after 67000 load cycles and therefore 10.000 cycles before the expected lifetime of the specimen. In the figure there is a QR-code which are linking to a video showing the development of this temperature profile. A URL of the same link can also be found in the figure caption. From the temperature profile, it can be seen that there are some self-heating going on near the end of the tabs which are properly related to tab-delamination similar to what has been found by [9]. In addition, close to the stopping point (see the video), there was observed a hot spot in the center of the test-sample. It is this point which in the next section will turn out to be an active fatigue damage region with a lot of fiber breakages.

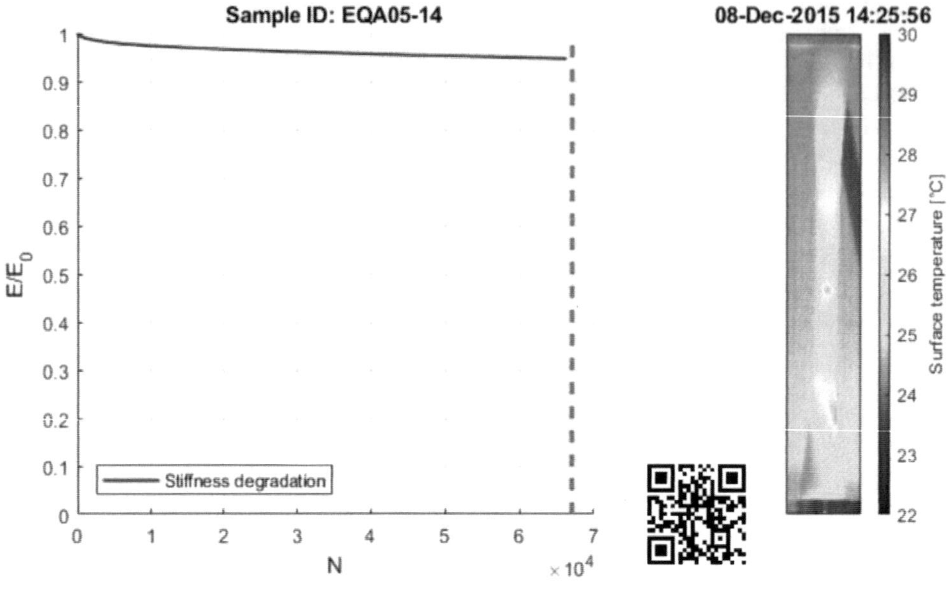

Figure 4. Test-sample stiffness degradation and surface temperature of the x-ray scanned specimen. A link to the video use the QR-code on a mobile phone or use the link (https://youtu.be/UkPjghocsTg).

4. Scout and zoom scan

After 67000 cycles, the central part of the test-sample, see figure 5a was investigated using a so-called scout and zoom scan sequence using a Zeiss Xradia Versa 520 x-ray scanning machine. The scout scan, figure 5b, is used for identifying the region of interest, a region which subsequently is scanned at a much higher resolution, figure 5c. Table 1 show the scan-setting for the two scans. The data and tiff-stack from the two scans can be downloaded from [10].

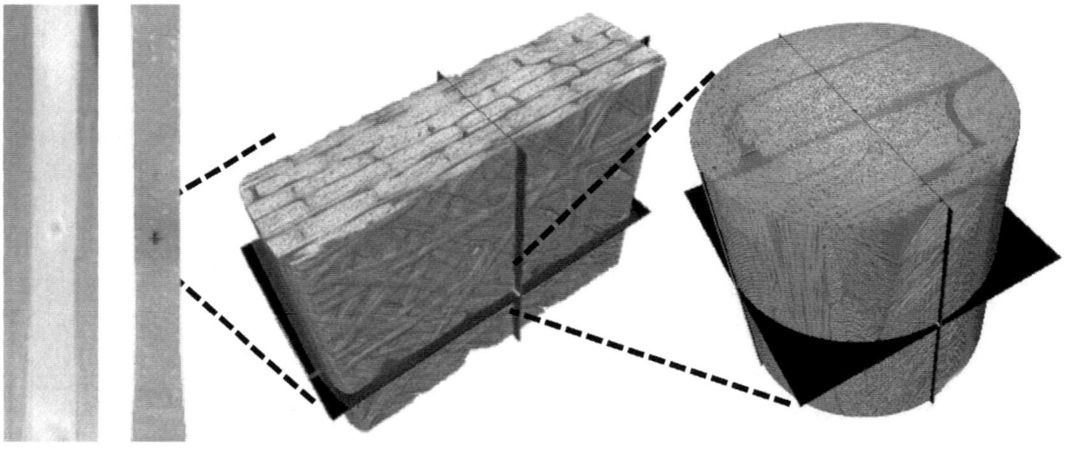

(a) Central part of test sample

(b) Large field of view: $15x15x4.5mm^3$

(c) Small field of view: $2.5x2.5x2.5\ mm^3$

Figure 5. Scout and zoom scan

IOP Conf. Series: Materials Science and Engineering **388** (2018) 012006 doi:10.1088/1757-899X/388/1/012006

Table 1. Scan-setting

	Source to sample [mm]	Detector to sample [mm]	Exposer time [s]	No. of project. [-]	Accel. voltage [keV]	Bin-ning	Pixel size [μm]	Optical magnifi-cation	Scan time [hours]
Scout (LFoV)	28	91	2	801	50	2	16.1114	0.4	0.5
Zoom (SFoV)	28	47	15	3201	70	2	2.5043	4	13.3

Figure 6, present selected slices from the scout scan. Despite the large voxel size on 16 μm, it is possible to see indication on damage in a region of the scan. A region which on the slice in figure 6 is marked by a white box. A reconstructed 3D x-ray scan is a 3-D image which in figure 6 is shown as a set of selected 2D slices. Trying identifying damage region in composite materials it is often must better slicing through the 3D reconstructed image instead of looking on single 2D slices. Therefore, all the images in figure 6 includes a link with a QR-code and a URL-link to videos of slicing through the 3D reconstructed image. From the videos, the damage region show up rather clear and from that, the coordinates for the zoomed scan can be defined. The zoomed scan shown in figure 7 is a scan of the region defined by the white box shown in figure 6.

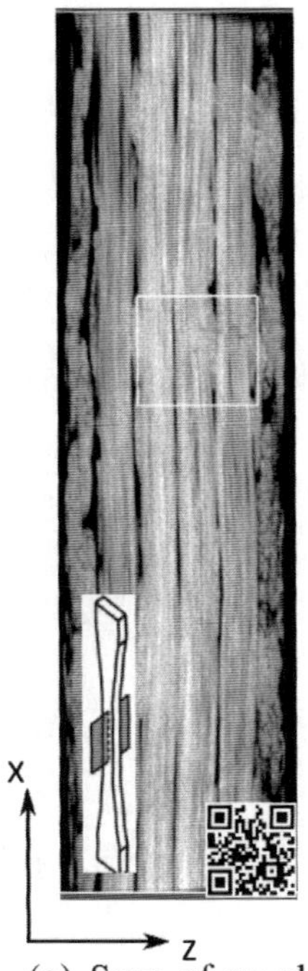

(a) Scan of xz-plane (https://youtu.be/AwW6idKOpTE)

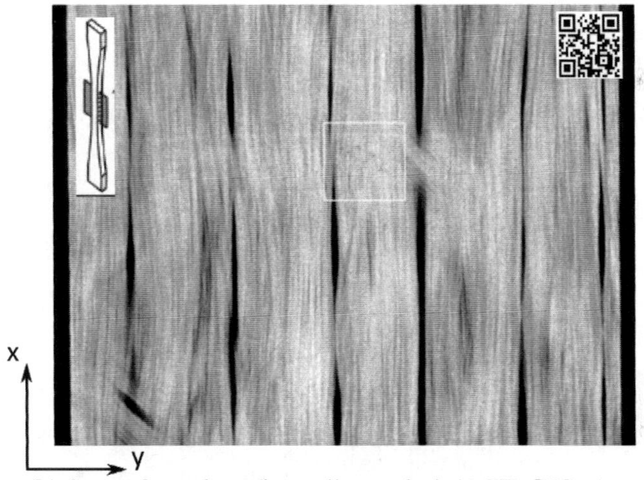

(b) Scan of xy-plane (https://youtu.be/xAMJFcfUCnc)

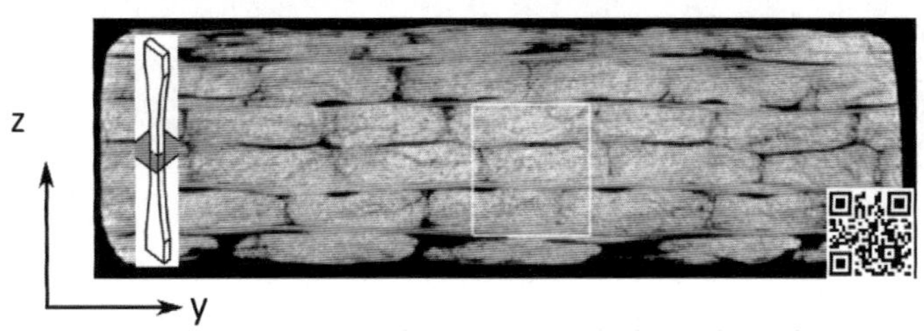

(c) Scan of yz-plane (https://youtu.be/uvPulo7Vfbs)

Figure 6. Scout scan of large field of view (LFoV: 15mm × 15mm × 4.5mm)

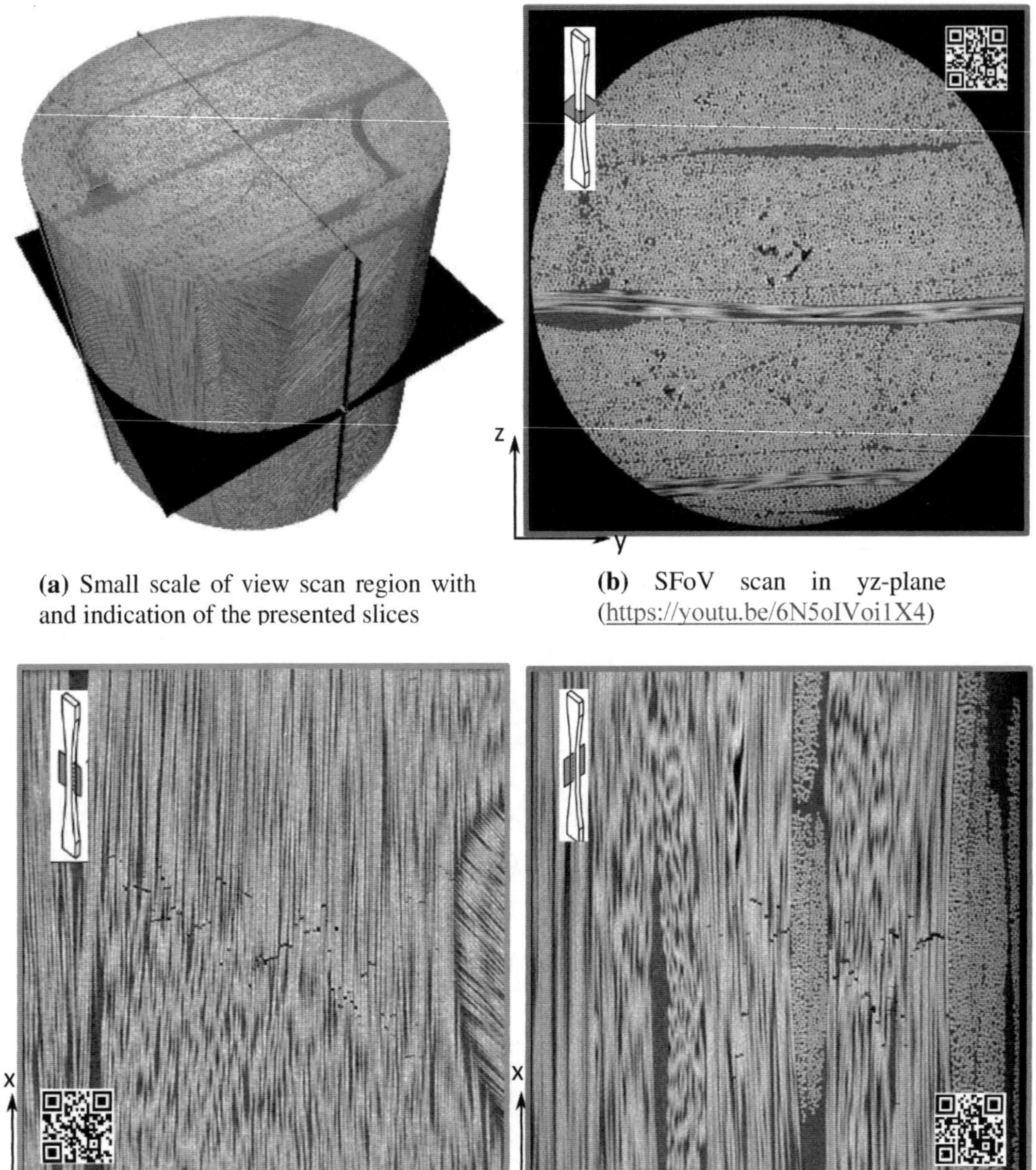

(a) Small scale of view scan region with and indication of the presented slices

(b) SFoV scan in yz-plane (https://youtu.be/6N5oIVoi1X4)

(c) SFoV scan in xy-plane (https://youtu.be/48VZmUs1C-8)

(d) SFoV scan in xz-plane (https://youtu.be/ICjs1Zeg6pU)

Figure 7. Small field of view scan (SFoV: 2.5mm × 2.5mm × 2.5mm)

In figure 7, the results from the zoomed scan is shown. Also here, the selected slice are linked to a QR-code and a URL-link showing a video slicing through the 3D tiff-stack. In figure 7, the voxel-size is 2.5 μm and with an average fibre diameter of D_0=17 μm, it is possible to see the individual fibres. Inside the field of view, a damage region with a lot of fibre failure can be seen. Similar to the scout scan, also for the zoomed scan, the fibre failure is easier to locate in the video than in on selected slice. It can be observed that the fibre failure region, is following the orientation of the backing bundle just beneath the slice shown figure 7c, something indicating that the fibre failure in the load carrying fibre bundles are controlled or strongly influenced by the present of the secondary oriented backing bundles. From the corresponding video of the region around the slice shown in figure 7c, it can be seen that there in addition to the 45° bundles also is a 90° bundle present at the location close to all the fibre failures indicating that a cross over point not only between the backing and the UD bundle but with multiple cross over points is critical fatigue damage initiation regions.

In the investigated case, the fatigue test has been stopped rather close to the final failure of the sample and the fatigue damage region is rather well developed. Therefore, a further evolution of the fatigue damage is difficult to obtain without running into the final failure of the material. In [4], [11], the same material system has been studied where the fatigue test has been interrupted much earlier in the fatigue life. Thereby, it has been possible using the non-destructive nature of the x-ray techniques obtaining four subsequently 3D x-ray based reconstructed dataset of the same region in the material. From that data-set, it was possible to see how the fibre failure was evolving from a region close to the crossing point of the secondary oriented backing bundles into the load-carrying bundles later during the fatigue loading. In addition, subtracting two aligned 3D reconstructed images obtained with 20.000 fatigue load cycles in between, it was possible to get a 3D reconstruction of the fatigue damage region by simply looking at the difference between the two 3D reconstructions. The 3D reconstructed fatigue damage region can be seen in figure 8.

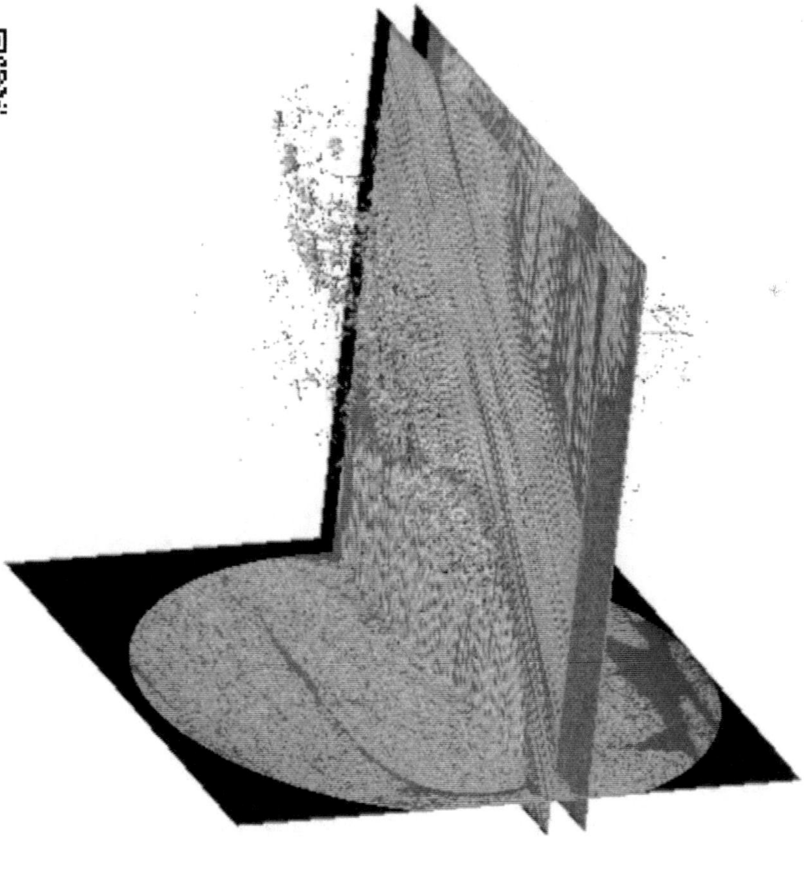

Figure 8. The 3D reconstructed fatigue damage region (https://youtu.be/8tg4aa7Gvks)

5. Conclusion

For the investigated material, it has been found that the tension/tension fatigue damage region is located close to a crossing region of the backing bundles. This region has been located first by investigating a 15mm full cross section area of the test sample. From that scout scan, a fatigue damage region was located. From the small field of view scan, details about the fatigue damage region is obtained such as the orientation of the fiber failure are following the orientation of the backing bundles. This observation for the specific material system has been validated in a number of other similar material system as shown in [3]–[5].

Acknowledgment

Financial support from CINEMA: "the allianCe for ImagiNg of Energy Materials", DSF-grant no. 1305-00032B under "The Danish Council for Strategic Research" is gratefully acknowledged. This research was conducted using x-ray tomograhpy and mechanical testing equipment from DTU Center for Advanced Structural and Material Testing (CASMaT), Grant No. VKR023193 from Villum Fonden.

References

[1] L. P. Mikkelsen, "A simplified model predicting the weight of the load carrying beam in a wind turbine blade," *IOP Conf. Ser. Mater. Sci. Eng.*, vol. 139, pp. 1–8, 2016.

[2] R. P. L. Nijssen and P. Brøndsted, "Fatigue as a design driver for composite wind turbine materials," in *Advances in Wind Turbine Blade Design and Materials*, 2013, pp. 175–209.

[3] K. M. Jespersen, J. Zangenberg, T. Lowe, P. J. Withers, and L. P. Mikkelsen, "Fatigue damage assessment of uni-directional non-crimp fabric reinforced polyester composite using X-ray computed tomography," *Compos. Sci. Technol.*, vol. 136, pp. 94–103, 2016.

[4] K. M. Jespersen and L. P. Mikkelsen, "Three dimensional fatigue damage evolution in non-crimp glass fibre fabric based composites used for wind turbine blades," *Compos. Sci. Technol.*, vol. 153, pp. 261–272, 2017.

[5] K. M. Jespersen, J. A. Glud, J. Zangenberg, A. Hosoi, H. Kawada, and L. P. Mikkelsen, "Uncovering the fatigue damage initiation and progression in uni-directional non-crimp fabric reinforced polyester composite," *Compos. Part A Appl. Sci. Manuf.*, vol. 109, pp. 481–497, 2018.

[6] J. Zangenberg, P. Brøndsted, and J. W. Gillespie, "Fatigue damage propagation in unidirectional glass fibre reinforced composites made of a non-crimp fabric," *J. Compos. Mater.*, vol. 48, pp. 2711–2727, 2013.

[7] K. L. Reifsnider and R. Jamison, "Fracture of fatigue-loaded composite laminates," *Int. J. Fatigue*, vol. 4, no. 4, pp. 187–197, Oct. 1982.

[8] S. Korkiakoski, P. Brøndsted, E. Sarlin, and O. Saarela, "Influence of specimen type and reinforcement on measured tension–tension fatigue life of unidirectional GFRP laminates," *Int. J. Fatigue*, vol. 85, pp. 114–129, 2016.

[9] J. Z. Hansen and R. Østergaard, *The effects of fibre architecture on fatigue life-time of composite materials*. DTU Wind Energy, 2013.

[10] L. P. Mikkelsen, "Observations of microscale tensile fatigue damage mechanisms of composite materials for wind turbine blades [Data-set]." https://doi.org/10.5281/zenodo.1300017, 2018.

[11] K. M. Jespersen and L. P. Mikkelsen, "Ex-situ X-ray computed tomography data for a non-crimp fabric based glass fibre composite under fatigue loading," *Data Br.*, vol. 15, pp. 1003–1005, 2017.

IOP Conf. Series: Materials Science and Engineering **388** (2018) 012007 doi:10.1088/1757-899X/388/1/012007

Predicting damage accumulation and fatigue life of UD composites under longitudinal tension

Soraia Pimenta, Alex Mersch and Marco Alves

Department of Mechanical Engineering, South Kensington Campus, Imperial College London, SW7 2AZ, United Kingdom

E-mail: `soraia.pimenta@imperial.ac.uk`

Abstract. Unidirectional composites under cyclic longitudinal tension develop damage through the accumulation and clustering of fibre–breaks, and through fibre–matrix interface debonding growth; these processes lead to a reduction of the material's load–carrying ability with increasing loading cycles, which raises a challenge to predict the fatigue response of composite structures. This paper proposes the first model in the literature to predict the kinetics of fibre–breakage and their effect on the macroscopic response of unidirectional composites under cyclic longitudinal tension. The model couples (i) a statistical hierarchical scaling law to predict fibre failure with (ii) a Paris law to predict interfacial fatigue damage propagating from broken fibres; due to its analytical formulation, the model predicts the response of composite bundles up to virtually any size and for their entire fatigue life in less than one minute. Model predictions for the accumulation and clustering of fibre–breaks show a good correlation with experiments from the literature; the model also predicts that, although the critical cluster size does not vary significantly between static, low–cycle/high–stress fatigue, and high–cycle/low–stress fatigue, the material can withstand the highest amount of softening under high–cycle/low–stress fatigue.

1. Introduction

Carbon–fibre composite materials are sensitive to cyclic degradation of stiffness and strength, which creates a challenge for the design of composite structures [1]. This challenge is currently addressed mostly by introducing large safety factors and/or extensive experimental campaigns, partially due to the lack of effective and efficient models to predict the accumulation of micromechanical damage with progressive cyclic loading, as well as its effect on macroscopic properties [2].

Ultimate failure of composite structures is often dominated by failure of the load–aligned plies; this makes it particularly important to develop predictive models for the cyclic response of unidirectional composites under longitudinal tension. Experimental work has shown that the load–carrying ability of carbon–fibre composites under longitudinal tension is reduced with the accumulation of loading cycles [1]. It is generally agreed that this is mostly due to degradation of matrix / interface properties with cyclic loading, since carbon–fibres are usually considered fatigue–insensitive [3].

At the micromechanical level, is is widely accepted [4] that unidirectional composites under longitudinal failure undergo accumulation and clustering of fibre–breaks, due to fibre–strength variability; this generates stress concentrations in the neighbouring fibres — at a magnitude and within a distance controlled by the properties of the matrix/interface — which might lead to further accumulation and clustering of fibre–breaks. This process can become unstable once a cluster of fibre–breaks reaches a critical size, as this triggers failure of the entire specimen.

High–resolution *in situ* micro–Computed Tomography (micro–CT) has recently demonstrated that progressive accumulation and clustering of fibre–breaks occurs under both static [5] and cyclic [6] longitudinal tension. It was also observed that cyclic loading favours the formation of small broken–clusters, and the propagation of debonds / matrix–cracks from fibre–breaks [6].

At the moment, there are no models in the literature able to predict how fibre–breaks accumulate and cluster under longitudinal tensile fatigue, nor predict the influence of this damage accumulation on the longitudinal Young's modulus. This lack of models can be justified by two factors: firstly, these are very challenging phenomena to model even under static loading [4]; secondly, most existing models for longitudinal tensile failure are too computationally demanding [4] to be applicable to fatigue loading.

We have recently developed and proposed a new approach to model longitudinal tensile failure of composites [7, 8, 9], based on a *hierarchical scaling law* defined by a statistical analysis of failure in *bundles of bundles*. This approach has allowed us to predict strength distributions and size effects for unidirectional composites under static [7] and cyclic [9] longitudinal loading, and to predict the accumulation and clustering of fibre–breaks under static loading [8]; model predictions have been extensively validated against experiments, with blind predictions achieving an excellent correlation with independently–obtained experiments [4]. However, this approach is not yet able to predict progressive damage accumulation during fatigue loading.

Therefore, the goal of this paper is to develop this hierarchical scaling law further, to enable the prediction of the accumulation and clustering of fibre–breaks under *cyclic* longitudinal tension, and their effect on the apparent longitudinal Young's modulus of the material. An overview of the hierarchical scaling law previously developed to predict the fatigue life of unidirectional composites [9] is provided in Sections 2.1–2.3; this is then used in Sections 2.4–2.6 to predict damage accumulation under longitudinal tensile fatigue, by extending our previously proposed static formulation [8]. Model results are presented and compared against experiments in Section 3, and discussed in Section 4. Finally, Section 5 draws the main conclusions.

2. Model development

2.1. Idealisation of constituents and their constitutive laws

The model considers hierarchical fibre bundles, which are generated by paring two individual fibres (level–[0] bundles) into a level–[1] bundle, and then recursively pairing two level–$[i]$ bundles into one level–$[i+1]$ bundle (Figure 1) [7, 8, 9]. The number of fibres $n^{[i]}$ in a level-$[i]$ bundle is given by

$$n^{[i]} \overset{\text{def}}{=} 2^i \qquad \Leftrightarrow \qquad i = \log_2 n^{[i]}. \tag{1}$$

The fibres (with cross–section A^{f}) resist longitudinal stresses only [10] (with a longitudinal Young's modulus E^{f}) and are fatigue insensitive [3]. The strength of a single fibre under a uniform stress σ^{∞} follows a Weibull distribution with shape parameter m and scale parameter σ_0^{f}, measured at the reference length l_0^{f} [11]. Therefore, the survival probability of a single fibre ($S_{\text{U}}^{[0]}$) under a uniform stress σ^{∞} can be scaled to any length l

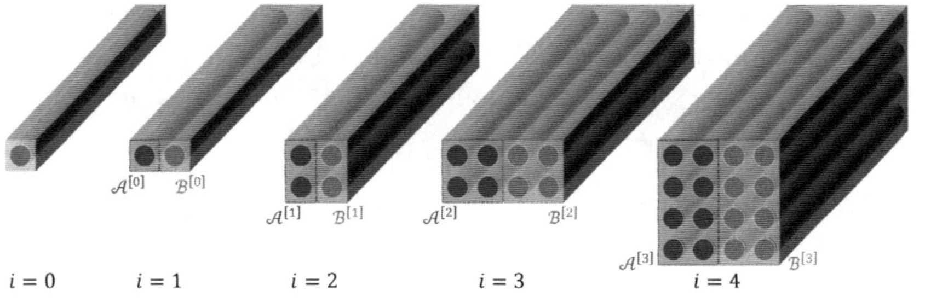

Figure 1: Hierarchical generation of composite fibre–bundles [7, 8, 9].

according to the Weakest Link Theory (WLT):

$$S_{\mathrm{U}}^{[0]}(\sigma^{\infty};l) \stackrel{\mathrm{def}}{=} \exp\left[-\frac{l}{l_0^{\mathrm{f}}}\cdot\left(\frac{\sigma^{\infty}}{\sigma_0^{\mathrm{f}}}\right)^{m}\right]. \quad (2)$$

The fibres are embedded in a soft matrix that redistributes stresses in the neighbourhood of fibre– and sub–bundle–breaks, according to a shear-lag process [10, 12]. This shear–lag process homogenises the contributions of the matrix and fibre–matrix interface, which will hereafter be simply referred to as the *interface* between fibres and/or sub-bundles [9]. The static shear–lag response will be governed by a linear cohesive law, defined by a shear strength (τ^{int}) and mode–II fracture toughness ($\mathcal{G}_{\mathrm{IIc}}^{\mathrm{int}}$); the growth rate of mode–II interface cracks along a level–[i] sub-bundle ($\mathrm{d}A^{[i]}/\mathrm{d}N$) due to cyclic longitudinal tension of the bundle (with peak stress σ^{∞} and R–ratio R^{∞}) will be governed by a Paris law [13]:

$$\frac{\mathrm{d}A^{[i]}}{\mathrm{d}N}(\sigma^{\infty},R^{\infty}) \stackrel{\mathrm{def}}{=} C_{\mathrm{II}}^{\mathrm{int}}\cdot\left(\frac{\Delta\mathcal{G}_{\mathrm{II}}^{[i]}(\sigma^{\infty},R^{\infty})}{\mathcal{G}_{\mathrm{IIc}}^{\mathrm{int}}}\right)^{m_{\mathrm{II}}^{\mathrm{int}}}, \text{ for } \mathcal{G}_{\mathrm{IIth}}^{\mathrm{int}} < \mathcal{G}_{\mathrm{II}}^{[i]}(\sigma^{\infty}) < \mathcal{G}_{\mathrm{IIc}}^{\mathrm{int}}, \quad (3)$$

where $C_{\mathrm{II}}^{\mathrm{int}}$ and $m_{\mathrm{II}}^{\mathrm{int}}$ are the Paris law constants, $\mathcal{G}_{\mathrm{IIth}}^{\mathrm{int}}$ is the threshold model–II energy release rate (below which no fatigue occurs), $\Delta\mathcal{G}_{\mathrm{II}}^{[i]}(\sigma^{\infty},R^{\infty}) \stackrel{\mathrm{def}}{=} \mathcal{G}_{\mathrm{II}}^{[i]}(\sigma^{\infty})-\mathcal{G}_{\mathrm{II}}^{[i]}(R^{\infty}\cdot\sigma^{\infty})$, and $\mathcal{G}_{\mathrm{II}}^{[i]}(\sigma)$ is the energy release rate at the level–[i] sub-bundle interface crack under a remote bundle stress σ (which will be calculated in Equation 4).

2.2. Micromechanical analysis of a level–[1] bundle under cyclic longitudinal tension

2.2.1. Static loading

Consider a level–[1] composite fibre bundle, composed of two level–[0] individual fibres bound together by a level–[0] interface, under a static uniform tensile stress field with magnitude σ^{∞}. The following scenarios may occur [7, 8, 9]:

No level–[0] breaks scenario. The bundle survives the remote peak stress σ^{∞} with both fibres undamaged, in which case there is no fatigue degradation.

Broken level–[1] bundle scenario. The bundle fails under the remote peak stress σ^{∞} (i.e. both fibres break, and an interface split forms between them), in which case no further loading is possible.

One level–[0] break scenario. The bundle survives the remote peak stress σ^{∞} with one broken fibre, while the other fibre survives the results stress concentration field created through shear of the interface (as shown in Figure 2a, with $k = 2$ from equilibrium). Depending of the mode–II energy release rate associated with crack growth along the level–[0] sub-bundle interface

$$\mathcal{G}_{\mathrm{II}}^{[0]}(\sigma^{\infty}) \overset{[9]}{=} \frac{n^{[0]} \cdot A^{\mathrm{f}} \cdot (\sigma^{\infty})^2}{P^{[0]} \cdot E^{\mathrm{f}}} , \tag{4}$$

where $P^{[0]}$ is the perimeter of the level–[0] interface, two sub-scenarios may occur:

Fully–debonded scenario. If $\mathcal{G}_{\mathrm{II}}^{[0]}(\sigma^{\infty}) \geq \mathcal{G}_{\mathrm{IIc}}^{\mathrm{int}}$, an interface crack tip will form and propagate unstably during the static loading cycle, leaving the specimen *fully-debonded* (see Figure 2c);

Fatigue–sensitive scenario. If $\mathcal{G}_{\mathrm{II}}^{[i]}(\sigma^{\infty}) < \mathcal{G}_{\mathrm{IIc}}^{\mathrm{int}}$, the interface will be partially damaged near the fibre–break, within the *effective recovery length* (see Figure 2a):

$$l_{\mathrm{e}}^{[0]}(\sigma^{\infty}, N=0) \overset{[9]}{=} 2 \cdot \frac{1}{\lambda^{[0]}} \cdot \arcsin\left(\sqrt{\frac{\mathcal{G}_{\mathrm{II}}^{[0]}(\sigma^{\infty})}{\mathcal{G}_{\mathrm{IIc}}^{\mathrm{int}}}}\right) ,$$

$$\text{where} \quad \lambda^{[0]} \overset{\text{def}}{=} \frac{\tau^{\mathrm{int}}}{\sqrt{n^{[0]} \cdot A^{\mathrm{f}} / P^{[0]} \cdot E^{\mathrm{f}} \cdot \mathcal{G}_{\mathrm{IIc}}^{\mathrm{int}}}} . \tag{5}$$

2.2.2. Fatigue life

Only the **fatigue–sensitive scenario** defined above is susceptible to fatigue degradation (and only if $\mathcal{G}_{\mathrm{II}}^{[0]}(\sigma^{\infty}) > \mathcal{G}_{\mathrm{IIth}}^{\mathrm{int}}$). In this scenario, a subsequent number of load cycles (N) under a loading ratio R^{∞} will lead to a bi-lateral increase of the effective recovery length governed by the mode–II Paris law of the interface [9]:

$$\frac{\mathrm{d}l_{\mathrm{e}}^{[0]}}{\mathrm{d}N}(\sigma^{\infty}, R^{\infty}, N) \overset{\text{Eq. 3}}{=} \frac{2}{P^{[0]}} \cdot \frac{\mathrm{d}A^{[0]}}{\mathrm{d}N}(\sigma^{\infty}, R^{\infty}) . \tag{6}$$

This growth of the effective recovery length will lead to an evolution of the stress fields within the fibres and interface of the level–[1] bundle, leading to three district fatigue phases, as will be described next.

Crack initiation phase, for $0 < N \leq N_{\mathrm{init}}^{[0]}$. Cyclic loading progressively reduces the ability of the interface to transfer shear–stresses near the fibre–break, up to the moment when a crack–tip is formed. It has been previous shown [9] that the

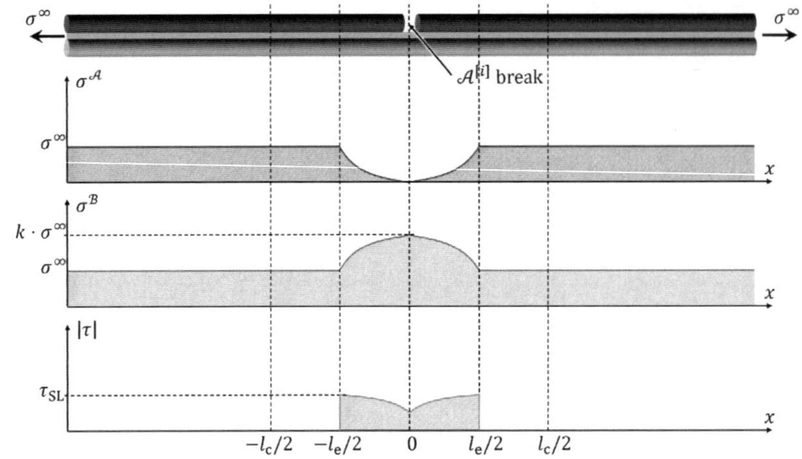

(a) During the crack–initiation phase.

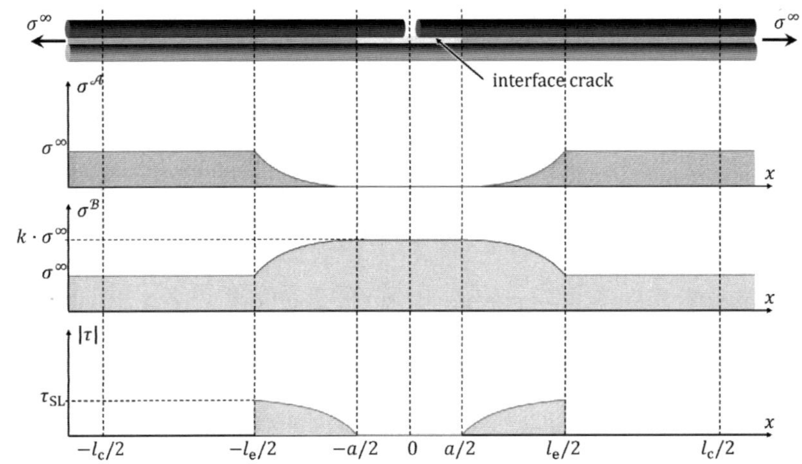

(b) During the crack–propagation phase.

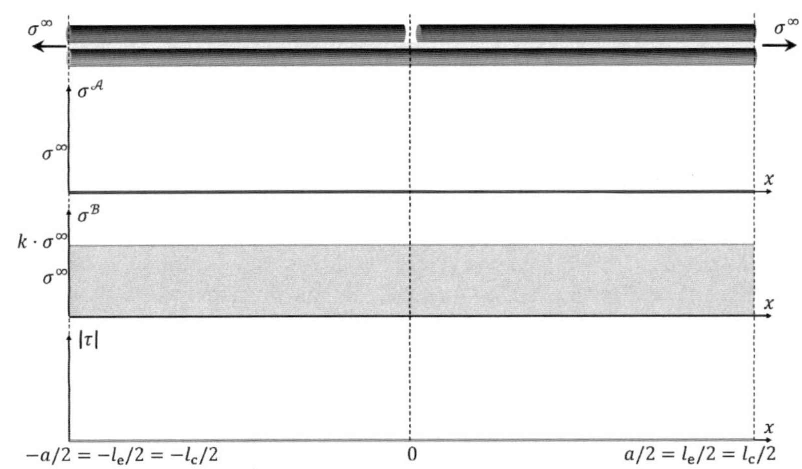

(c) After full debonding of the bundle.

Figure 2: Stress fields in the fibres and interface of a level–$[i + 1 = 1]$ bundle, after failure of the weakest fibre (fibre $\mathcal{A}$) at $x = 0$ [9].

number of cycles required to form this crack tip is

$$N_{\text{init}}^{[0]} \overset{[9]}{=} \frac{l_{\text{K}}^{[0]} - l_{\text{e}}^{[0]}(\sigma^{\infty}, N=0)}{\frac{2}{P^{[0]}} \cdot \frac{\mathrm{d}A^{[0]}}{\mathrm{d}N}(\sigma^{\infty}, R^{\infty})}, \qquad \text{where } l_{\text{K}}^{[0]} \overset{\text{def}}{=} \frac{\pi}{\lambda^{[0]}}. \tag{7}$$

Because the formation of a crack tip marks the end of the initiation phase, the crack length is defined as $a^{[0]} = 0$ for $N \leq N_{\text{init}}^{[0]}$.

Crack propagation phase, for $N_{\text{init}}^{[0]} < N \leq N_{\text{init}}^{[0]} + \Delta N_{\text{prop}}^{[0]}$. Once an interface crack tip is formed at $N = N_{\text{init}}^{[0]}$, further cyclic loading will lead to self–similar crack propagation (see Figure 2b), up the moment when the effective recovery length reaches the end of the specimen. It has been previously shown [9] that the number of cycles required for the crack propagation phase is

$$\Delta N_{\text{prop}}^{[0]} \overset{[9]}{=} \frac{l_{\text{s}} - l_{\text{K}}^{[0]}}{\frac{2}{P^{[0]}} \cdot \frac{\mathrm{d}A^{[0]}}{\mathrm{d}N}(\sigma^{\infty}, R^{\infty})}. \tag{8}$$

The crack length during the propagation phase increases linearly with the number of cycles, according to the Paris law:

$$a^{[0]}(\sigma^{\infty}, R^{\infty}, N) \overset{\text{def}}{=} l_{\text{e}}^{[0]}(\sigma^{\infty}, N) - l_{\text{K}}^{[0]} \overset{\text{Eq. 6}}{=} (N - N_{\text{init}}^{[0]}) \cdot \frac{2}{P^{[0]}} \cdot \frac{\mathrm{d}A^{[0]}}{\mathrm{d}N}(\sigma^{\infty}, R^{\infty}). \tag{9}$$

Fully–debonded phase, for $N \geq N_{\text{init}}^{[0]} + \Delta N_{\text{prop}}^{[0]}$. Once the interface is damage along the entire length of the specimen, further crack degradation is not possible, and the stress fields in the bundle remain equal to the ones shown in Figure 2c.

2.3. Hierarchical scaling law

Considering the scenarios described in the previous section, we have previously derived [9] the following hierarchical scaling law to predict the survival probability of a level–[1] bundle (for clarity, the dependence of the survival probabilities on R^{∞} will be omitted):

$$S_{\text{U,c}}^{[1]}(\sigma^{\infty}, N) \overset{[9]}{=} S_{\mathcal{U}}^{[1]}(\sigma^{\infty}, N) + S_{\mathcal{K}}^{[1]}(\sigma^{\infty}, N). \tag{10a}$$

The first term on the right–hand side of Equation 10a corresponds to the **no level–**[i] **breaks scenario** (under static or fatigue loading), and can be defined as

$$S_{\mathcal{U}}^{[1]}(\sigma^{\infty}, N) \stackrel{[9]}{=} \left[S_{\mathrm{U,e}}^{[0]}(\sigma^{\infty}, N)^{\frac{l_{\mathrm{c}}^{[1]}}{l_{\mathrm{e}}^{[0]}}} \right]^2 . \tag{10b}$$

The second term on the right–hand side of Equation 10a corresponds to the **one level–**[i] **break scenario** (under static or fatigue loading), and can be defined as

$$S_{\mathcal{K}}^{[1]}(\sigma^{\infty}, N) \stackrel{[9]}{=} 2 \cdot \left[1 - S_{\mathrm{U,e}}^{[0]}(\sigma^{\infty}, N)^{\frac{l_{\mathrm{c}}^{[1]}(\sigma^{\infty}, N)}{l_{\mathrm{e}}^{[0]}(\sigma^{\infty}, N)}} \right] \cdot S_{\mathrm{U,e}}^{[0]}(k \cdot \sigma^{\infty}, N)^{\frac{a^{[0]}(\sigma^{\infty}, N)}{l_{\mathrm{e}}^{[0]}(\sigma^{\infty}, N)}}$$
$$\cdot S_{\mathrm{K,e}}^{[0]}(\sigma^{\infty}, N)^{\frac{l_{\mathrm{e}}^{[0]}(\sigma^{\infty}, N) - a^{[0]}(\sigma^{\infty}, N)}{l_{\mathrm{e}}^{[0]}(\sigma^{\infty}, N)}} \cdot S_{\mathrm{U,e}}^{[0]}(\sigma^{\infty}, N)^{\frac{l_{\mathrm{c}}^{[1]}(\sigma^{\infty}, N) - l_{\mathrm{e}}^{[0]}(\sigma^{\infty}, N)}{l_{\mathrm{e}}^{[0]}(\sigma^{\infty}, N)}} . \tag{10c}$$

The level–[0] survival probabilities in Equation 10 are calculated at the level–[0] recovery length $l_{\mathrm{e}}^{[0]}(\sigma^{\infty}, N)$ (defined by Equations 5 and 6); the terms corresponding to survival probabilities under uniform stresses ($S_{\mathrm{U,e}}^{[0]}(\sigma, N)$) are directly defined by the single fibre strength distribution (Equation 2), and the term corresponding to the survival probability under non-uniform stresses ($S_{\mathrm{K,e}}^{[0]}(\sigma, N)$) is calculated based on a generalised weakest link theory [7, 9]. The resulting level–[1] survival probability is calculated at the *level–*[1] *control length* (which represents the shortest statistically independent partition of the bundle) [9]:

$$l_{\mathrm{c}}^{[1]}(\sigma^{\infty}, N) \stackrel{[9]}{=} \min\left\{2 \cdot l_{\mathrm{e}}^{[0]}(\sigma^{\infty}, N), l_{\mathrm{s}}\right\} . \tag{11}$$

The analysis done so far focuses on a level–[1] bundle, composed of two level–[0] sub-bundles; assuming that failure propagates in a hierarchical and self–similar way in larger bundles [7, 8, 9], the hierarchical scaling law in Equation 10 can be extrapolated to any bundle of a generic level [$i+1$] composed of two level–[i] sub-bundles (with $i \geq 0$). This extrapolation requires the following modifications to the derivations shown so far:

- Replacing 'single–fibre/level–[0] sub-bundles' (and their superscript identifiers) with 'level–[i] sub-bundles', and replacing 'level–[1] bundle' (and their superscript identifiers) with 'level–[$i+1$] bundle' in Equations 4–11;

- For $i \geq 0$, replacing the standard effective recovery length (calculated in Equations 5 and 6) with the *averaged* effective recovery length, defined as [9]

$$\bar{l}_{\mathrm{e}}^{[i]}(\sigma^{\infty}, N) \overset{[9]}{=} \frac{\sum_{\eta=1}^{N} \left[l_{\mathrm{e}}^{[i]}(\sigma^{\infty}, N-\eta+1) \cdot \Delta F_{\mathrm{U,e}}^{[i]}(\sigma^{\infty}, \eta) \right]}{F_{\mathrm{U,e}}^{[i]}(\sigma^{\infty}, N)},$$
$$\text{where} \quad \Delta F_{\mathrm{U,e}}^{[i]}(\sigma^{\infty}, \eta) \overset{\text{def}}{=} \begin{cases} 1 - S_{\mathrm{U,e}}^{[i]}(\sigma^{\infty}, 0) & \text{for } \eta = 0, \\ S_{\mathrm{U,e}}^{[i]}(\sigma^{\infty}, \eta-1) - S_{\mathrm{U,e}}^{[i]}(\sigma^{\infty}, \eta) & \text{for } \eta \geq 1, \end{cases} \tag{12}$$

in Equations 9 to 11. This averaging process accounts for the probability of broken clusters of level $i \geq 1$ to be formed at any fatigue cycle $N \geq 0$ (as detailed in a previous publication [9]), and will be used to calculate all effective recovery ($\bar{l}_{\mathrm{e}}$) and control ($\bar{l}_{\mathrm{c}}$) lengths hereafter.

2.4. Formation of clusters of fibre–breaks

According to the analysis done in Section 2.3, and as further detailed in a previous publication [8], the probability of having a level–$[i]$ broken–cluster in a level–$[i+1]$ bundle of length $\bar{l}_{\mathrm{c}}^{[i+1]}$ is exactly $S_{\mathcal{K}}^{[i+1]}(\sigma^{\infty}, N)$. The expected density ($\hat{\rho}_{\mathrm{clust}}^{[i]}$) of level–$[i]$ broken–clusters in a level–$[i+1]$ bundle is therefore

$$\hat{\rho}_{\mathrm{clust}}^{[i]}(\sigma^{\infty}, N) \overset{[8]}{=} \frac{S_{\mathcal{K}}^{[i+1]}(\sigma^{\infty}, N)}{\bar{V}_{\mathrm{c}}^{[i+1]}(\sigma^{\infty}, N)}, \quad \text{where } \bar{V}_{\mathrm{c}}^{[i+1]}(\sigma^{\infty}, N) \overset{\text{def}}{=} A^{[i+1]} \cdot \bar{l}_{\mathrm{c}}^{[i+1]}(\sigma^{\infty}, N). \tag{13}$$

The total number of level–$[i]$ broken–clusters in a specimen of volume V_{s} (where the hierarchical level of the specimen is $i_{\mathrm{s}} > i+1$) will therefore correspond to performing a number $V_{\mathrm{s}}/\bar{V}_{\mathrm{c}}^{[i+1]}$ of Boolean trials with a probability of success per trial equal to $S_{\mathcal{K}}^{[i+1]}(\sigma^{\infty}, N)$. Therefore, the number of level–$[i]$ broken–clusters in the specimen is expected to follow a binomial distribution with the following expected value ($\mathbb{E}$) and variance ($\mathbb{V}$):

$$\begin{aligned} \mathbb{E}\left[N_{\mathrm{clust}}^{[i]}(\sigma^{\infty}, N)\right] &\overset{\text{def}}{=} \frac{V_{\mathrm{s}}}{\bar{V}_{\mathrm{c}}^{[i+1]}(\sigma^{\infty}, N)} \cdot S_{\mathcal{K}}^{[i+1]}(\sigma^{\infty}, N) \overset{\text{Eq. 13}}{=} \hat{\rho}_{\mathrm{clust}}^{[i]}(\sigma^{\infty}, N) \cdot V_{\mathrm{s}} \\ \mathbb{V}\left[(N_{\mathrm{clust}}^{[i]})(\sigma^{\infty}, N)\right] &\overset{\text{def}}{=} \frac{V_{\mathrm{s}}}{\bar{V}_{\mathrm{c}}^{[i+1]}(\sigma^{\infty}, N)} \cdot S_{\mathcal{K}}^{[i+1]}(\sigma^{\infty}, N) \cdot \left[1 - S_{\mathcal{K}}^{[i+1]}(\sigma^{\infty}, N)\right] \end{aligned} \tag{14a}$$

Assuming that $V_{\mathrm{s}} \gg \bar{V}_{\mathrm{c}}^{[i+1]}$ (which is usually valid for all broken–clusters with a non-negligible probability of occurrence in a macroscopic specimen), this binomial distribution of the number of broken–clusters can be approximated by a Normal distribution with the same expected value and variance:

$$N_{\text{clust}}^{[i]}(\sigma^{\infty}, N) \sim \mathcal{N}\left(\mathbb{E}\left[N_{\text{clust}}^{[i]}(\sigma^{\infty}, N)\right], \mathbb{V}\left[N_{\text{clust}}^{[i]}(\sigma^{\infty}, N)\right]\right) \tag{14b}$$

2.5. Total fibre breakage

The total number of fibre–breaks in a specimen of hierarchical level i_{s} may contain contributions from broken–clusters between levels $i = 0$ (i.e. individual fibre–breaks) and $i = i_{\text{s}} - 1$ (i.e. a level–$[i_{\text{s}} - 1]$ broken–cluster) [8]. Therefore, the expected density of fibre–breaks in the specimen can be calculated as

$$\hat{\rho}_{\text{breaks}}(\sigma^{\infty}, N) \overset{[8]}{=} \sum_{i=0}^{i_{\text{s}}-1} n^{[i]} \cdot \hat{\rho}_{\text{clust}}^{[i]}(\sigma^{\infty}, N)\,. \tag{15}$$

Similarly, the total number of fibre–breaks in the level–$[i_{\text{s}}]$ specimen can be calculated by adding the contribution of all clusters:

$$N_{\text{breaks}}(\sigma^{\infty}, N) \overset{[8]}{=} \sum_{i=0}^{i_{\text{s}}-1} n^{[i]} \cdot N_{\text{clust}}^{[i]}(\sigma^{\infty}, N) \tag{16}$$

This shows that the total number of broken fibres in a specimen is a linear combination of the number of broken–clusters of all hierarchical levels $i \leq i_{\text{s}} - 1$; since the numbers of broken–clusters are normally distributed random variables (as approximated in Equation 14b), then the total number of broken fibres in a level–$[i_{\text{s}}]$ specimen follows a Normal distribution as well:

$$N_{\text{breaks}}(\sigma^{\infty}, N) \sim \mathcal{N}\left(\sum_{i=0}^{i_{\text{s}}-1} n^{[i]} \cdot \mathbb{E}\left[N_{\text{clust}}^{[i]}(\sigma^{\infty}, N)\right], \sum_{i=0}^{i_{\text{s}}-1} {n^{[i]}}^{2} \cdot \mathbb{V}\left[N_{\text{clust}}^{[i]}(\sigma^{\infty}, N)\right]\right) \tag{17}$$

2.6. Strain softening under cyclic loading

Consider the different scenarios analysed in Section 2.2, but now focusing on a level–$[i+1]$ bundle, composed of two level–$[i]$ sub-bundles. One can impose compatibility between the expected strains in the level–$[i+1]$ bundle ($\hat{\varepsilon}^{[i+1]}$) and those in the level–$[i]$ sub-bundles ($\hat{\varepsilon}^{[i]}$), as we have previously applied to the static loading (only) [8]; extending that approach to cyclic loading (i.e. considering the dependence of stress fields on N as well as σ^{∞}) results in the following hierarchical scaling law:

$$\hat{\varepsilon}^{[i+1]}(\sigma^{\infty}, N) \overset{[8]}{=} \left[\frac{S_{\mathcal{U}}^{[i+1]}(\sigma^{\infty}, N)}{S_{\mathrm{U,c}}^{[i+1]}(\sigma^{\infty}, N)} \cdot \bar{\varepsilon}_{\mathcal{U}}^{[i+1]}(\sigma^{\infty}, N) + \frac{S_{\mathcal{K}}^{[i+1]}(\sigma^{\infty}, N)}{S_{\mathrm{U,c}}^{[i+1]}(\sigma^{\infty}, N)} \cdot \bar{\varepsilon}_{\mathcal{K}}^{[i+1]}(\sigma^{\infty}, N) \right] . \tag{18a}$$

The first term on the right–hand side of Equation 18a corresponds to the **no levell–[i] breaks scenario** analysed in Section 2.2, which leads exclusively to linear–elastic strains. In this case, the level–[$i+1$] bundle is expected to have an average strain ($\bar{\varepsilon}_{\mathcal{U}}^{[i+1]}$) exactly equal to that of its two level–[i] sub-bundles ($\bar{\varepsilon}_{\mathcal{U}}^{\mathcal{A}^{[i]}}$ and $\bar{\varepsilon}_{\mathcal{U}}^{\mathcal{B}^{[i]}}$), imposed as follows:

$$\bar{\varepsilon}_{\mathcal{U}}^{[i+1]}(\sigma^{\infty}, N) \overset{[8]}{\equiv} \bar{\varepsilon}_{\mathcal{U}}^{\mathcal{A}^{[i]}}(\sigma^{\infty}, N) = \bar{\varepsilon}_{\mathcal{U}}^{\mathcal{B}^{[i]}}(\sigma^{\infty}, N) = \hat{\varepsilon}^{[i]}(\sigma^{\infty}, N) . \tag{18b}$$

The second term on the right–hand side of Equation 18a corresponds to the **one levell–[i] break scenario** (under static or fatigue loading) analysed in Section 2.2, and which leads to non-linear–elastic strains and to a reduction of the apparent Young's modulus. In this case, the level–[$i+1$] bundle is expected to have an average strain ($\bar{\varepsilon}_{\mathcal{K}}^{[i+1]}$) exactly equal to that of its surviving level–[i] sub-bundle $\mathcal{B}$ ($\bar{\varepsilon}_{\mathcal{K}}^{\mathcal{B}^{[i]}}$), imposed as follows:

$$\bar{\varepsilon}_{\mathcal{K}}^{[i+1]}(\sigma^{\infty}, N) \overset{[8]}{\equiv} \bar{\varepsilon}_{\mathcal{K}}^{\mathcal{B}^{[i]}}(\sigma^{\infty}, N) = \frac{1}{\bar{l}_{\mathrm{c}}^{[i+1]}(\sigma^{\infty}, N)} \cdot \int_{x=-\bar{l}_{\mathrm{c}}^{[i+1]}(\sigma^{\infty}, N)/2}^{\bar{l}_{\mathrm{c}}^{[i+1]}(\sigma^{\infty}, N)/2} \hat{\varepsilon}^{[i]}(\sigma^{\mathcal{B}^{[i]}}(x), N)\, \mathrm{d}x \tag{18c}$$

Equation 18 provides a full mathematical description of a hierarchical law which scales the linear–elastic strain of single–fibres (with $\hat{\varepsilon}^{[0]}(\sigma^{\infty}, N) = \sigma^{\infty}/E^{\mathrm{f}}$) to non-linear–elastic strains of a level–[$i+1$] bundle. The only missing element is defining and solving the integral of the non-linear–elastic term in Equation 18c; this is particularly challenging due to the following reasons:

- The solution of the integral in Equation 18c depends on the exact strain fields considered for the **one level–[i] break scenario** (shown in Figure 2). Therefore, the solution previously developed for *static* loading [8] is not applicable to *cyclic* loading, because the strain fields are different in the two cases;
- The stress field $\sigma^{\mathcal{B}}(x; \sigma^{\infty}, N)$ in the surviving fibre/sub-bundle in the **one level–[i] break scenario** is continuously evolving as the number of loading cycles N increases

(as shown in Figure 2). This can be addressed by isolating the different regions of the stress field in the integral of Equation 18c:

$$\bar{\varepsilon}_{\mathcal{K}}^{[i+1]}(\sigma^\infty, N) \overset{\text{Eq. 18c}}{=} \frac{2}{\bar{l}_{\mathrm{c}}^{[i+1]}(\sigma^\infty, N)} \cdot \left[\int_{x=0}^{a^{[i]}(\sigma^\infty,N)/2} \hat{\varepsilon}^{[i]}(k \cdot \sigma^\infty, N)\,\mathrm{d}x \right.$$
$$+ \int_{x=a^{[i]}(\sigma^\infty,N)/2}^{\bar{l}_{\mathrm{e}}^{[i]}(\sigma^\infty,N)/2} \hat{\varepsilon}^{[i]}(\sigma^{\mathcal{B}^{[i]}}(x), N)\,\mathrm{d}x \qquad (19a)$$
$$\left. + \int_{x=\bar{l}_{\mathrm{e}}^{[i]}(\sigma^\infty,N)/2}^{\bar{l}_{\mathrm{c}}^{[i+1]}(\sigma^\infty,N)/2} \hat{\varepsilon}^{[i]}(\sigma^\infty, N)\,\mathrm{d}x \right] .$$

Assuming that $\hat{\varepsilon}^{[i]}(\chi \cdot \sigma, N) \approx \chi \cdot \hat{\varepsilon}^{[i]}(\sigma, N)$ (which will be verified in Section 3.3), and approximating $\sigma^{\mathcal{B}^{[i]}}(x)$ within $|x| \in [a^{[i]}/2, \bar{l}_{\mathrm{e}}^{[i]}/2]$ by a linear field, this results into

$$\bar{\varepsilon}_{\mathcal{K}}^{[i+1]}(\sigma^\infty, N) \approx \left(1 + (k-1) \cdot \frac{a^{[i]}(\sigma^\infty, N) + \bar{l}_{\mathrm{e}}^{[i]}(\sigma^\infty, N)}{2 \cdot \bar{l}_{\mathrm{c}}^{[i+1]}(\sigma^\infty, N)} \right) \cdot \hat{\varepsilon}^{[i]}(\sigma^\infty, N) . \qquad (19b)$$

Combining Equations 18 and 19, we can now defined the full hierarchical scaling law for the expected strains in fibre bundles under longitudinal tensile fatigue:

$$\hat{\varepsilon}^{[i+1]}(\sigma^\infty, N) \overset{\text{Eq. 18,19}}{=} \left[\frac{S_{\mathcal{U}}^{[i+1]}(\sigma^\infty, N)}{S_{\mathrm{U,c}}^{[i+1]}(\sigma^\infty, N)} + \xi \cdot \frac{S_{\mathcal{K}}^{[i+1]}(\sigma^\infty, N)}{S_{\mathrm{U,c}}^{[i+1]}(\sigma^\infty, N)} \right] \cdot \hat{\varepsilon}^{[i]}(\sigma^\infty, N) ,$$
$$\text{where} \quad \xi \overset{\text{def}}{=} 1 + (k-1) \cdot \frac{a^{[i]}(\sigma^\infty, N) + \bar{l}_{\mathrm{e}}^{[i]}(\sigma^\infty, N)}{2 \cdot \bar{l}_{\mathrm{c}}^{[i+1]}(\sigma^\infty, N)} . \qquad (20)$$

Using the expected strains calculated in Equation 20, it is finally possible to predict degradation of the apparent longitudinal Young's modulus of a level–$[i+1]$ bundle as

$$\bar{E}^{[i+1]}(\sigma^\infty, N) = \frac{\sigma^\infty}{\hat{\varepsilon}^{[i+1]}(\sigma^\infty, N)} . \qquad (21)$$

3. Results

3.1. Fatigue life

Figure 3 compares the stochastic S–N curves predicted by the model against experimental results, for an unidirectional carbon/epoxy composite loaded under cyclic longitudinal tension. The input for the model (i.e. geometry of the specimens, properties of the carbon fibres, and properties of the matrices) are shown in Table 1.

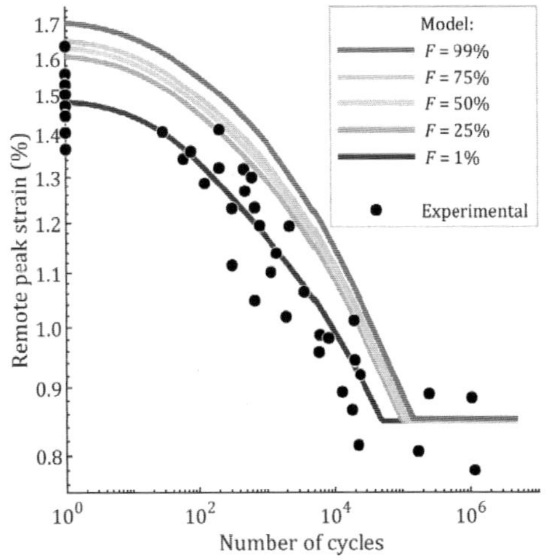

Figure 3: Stochastic S–N curves predicted by the model (for several failure probabilities F) compared against Gamstedt and Talreja's experiments [1].

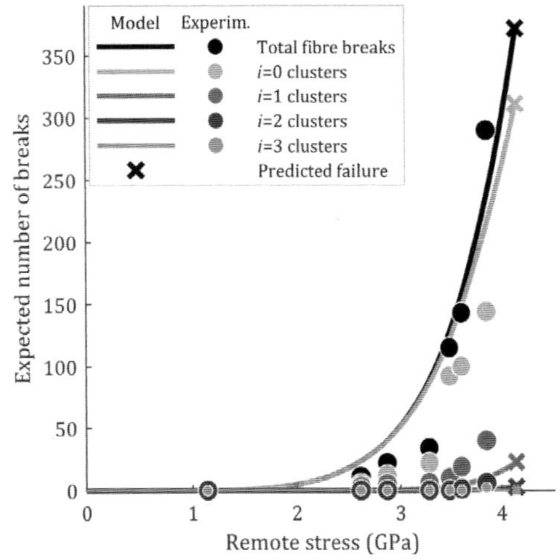

Figure 4: Static accumulation and clustering of fibre–breaks: expected values predicted by the model compared against Scott et al.'s experiments [5].

Table 1: Material properties used as model inputs for the result shown in this paper.

	Property	Units	Value: Gamstedt and Talreja [1]	Value: Scott et al. [5] / Garcea et al. [6]
	Fibre type		AS4 [14]	T700 [15]
ϕ^{f}	Fibre diameter	µm	6.85	7.1
E^{f}	Fibre Young's Modulus	GPa	222	240 [16]
m	SFTT Weibull modulus		4.8	5.86
σ_0^{f}	SFTT strength paramter	GPa	4.493	7.338
l_0^{f}	SFTT gauge length	mm	10	1.0
	Matrix type / Interface		PEEK APC–2 [9]	Epoxy M21 [8]
τ^{int}	Shear strength	MPa	110	105
$\mathcal{G}_{\mathrm{IIc}}^{\mathrm{int}}$	Mode–II fracture toughness	kJ/m^2	2.0	1.0
$C_{\mathrm{II}}^{\mathrm{int}}$	Mode–II Paris law scale constant	mm^2/cycle	2×10^{-3}	3.0×10^{-3}
$m_{\mathrm{II}}^{\mathrm{int}}$	Mode–II Paris law shape constant		2.0	2.2
$\mathcal{G}_{\mathrm{IIth}}^{\mathrm{int}}$	Mode–II fatigue threshold	kJ/m^2	0.025	0.025
$n^{[i]}$	No. fibres in cross–section		100,000	5444/7577
V^{f}	Fibre volume fraction	%	60	60
l_{s}	Specimen length	mm	127	1.8/3.6
R^{∞}	Loading ratio		0.1	n.a./0.1

IOP Conf. Series: Materials Science and Engineering **388** (2018) 012007 doi:10.1088/1757-899X/388/1/012007

3.2. Accumulation and clustering of fibre breaks

Figure 4 shows the accumulation and clustering of fibre breaks in a UD carbon/epoxy specimen under increasing applied static tension. Model predictions (using the inputs shown in Table 1) are compared against experimental results obtained by Scott et al. (through *in-situ* high–resolution computed tomography [5]).

Figure 5 shows the accumulation and clustering of fibre breaks in a UD carbon/epoxy specimen under cyclic longitudinal tension. The experimental results were obtained by Garcea et al. [6] through *ex-situ* high–resolution computed tomography of the same carbon/epoxy material as analysed in Figure 4, but with a slightly different specimen geometry (see Table 1).

The evolution of the size of the largest broken cluster with the accumulation of loading cycles is shown in Figure 6, for two different remote peak stresses, comparing model predictions against experiments.

3.3. Effect of cyclic loading on apparent stiffness

Figure 7 shows the evolution of the macroscopic longitudinal strain (Figure 7a) and apparent modulus (Figure 7b) of the same carbon/epoxy specimen analysed in Section 3.2, as predicted by the model.

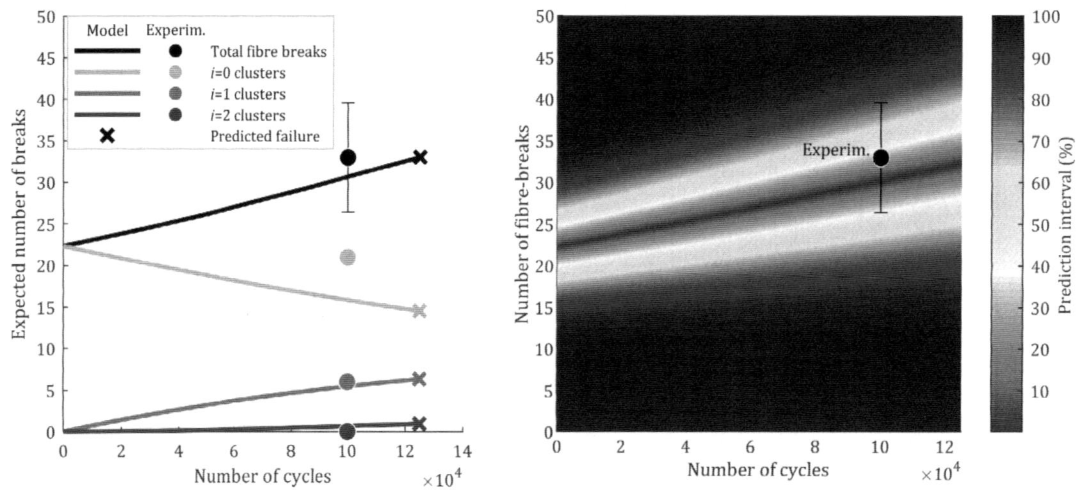

(a) Expected number of single–fibre and cluster breaks.

(b) Stochastic distribution of the total number of breaks.

Figure 5: Accumulation and clustering of fibre–breaks under cyclic loading with a remote peak load equal to 50% of the static strength. The experimental data shown was obtained by Garcea et al. [6].

IOP Conf. Series: Materials Science and Engineering **388** (2018) 012007 doi:10.1088/1757-899X/388/1/012007

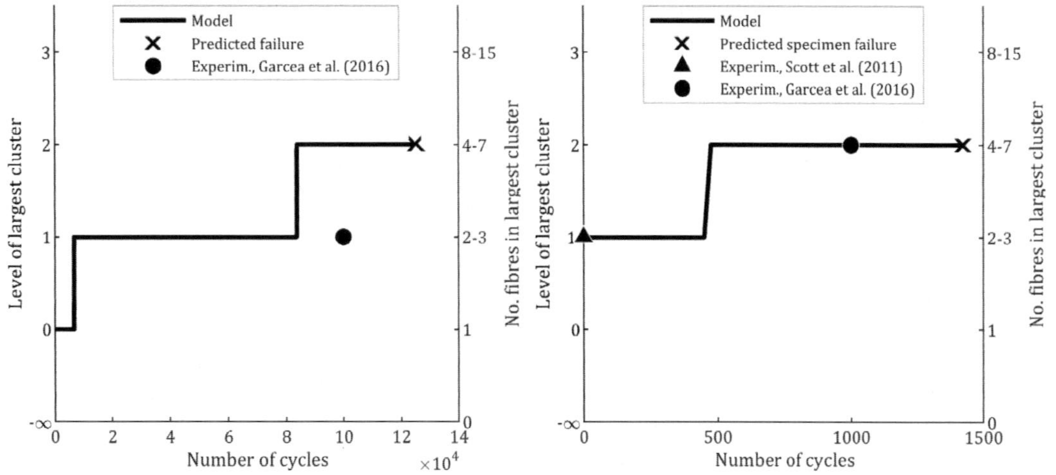

(a) Peak remote stress for cyclic load: 50% of static strength (see full distribution of fibre–breakage in Figure 5).

(b) Peak remote stress for cyclic load: 70% of static strength.

Figure 6: Evolution of the largest cluster size with cyclic loading: model predictions vs. experimental data [5, 6].

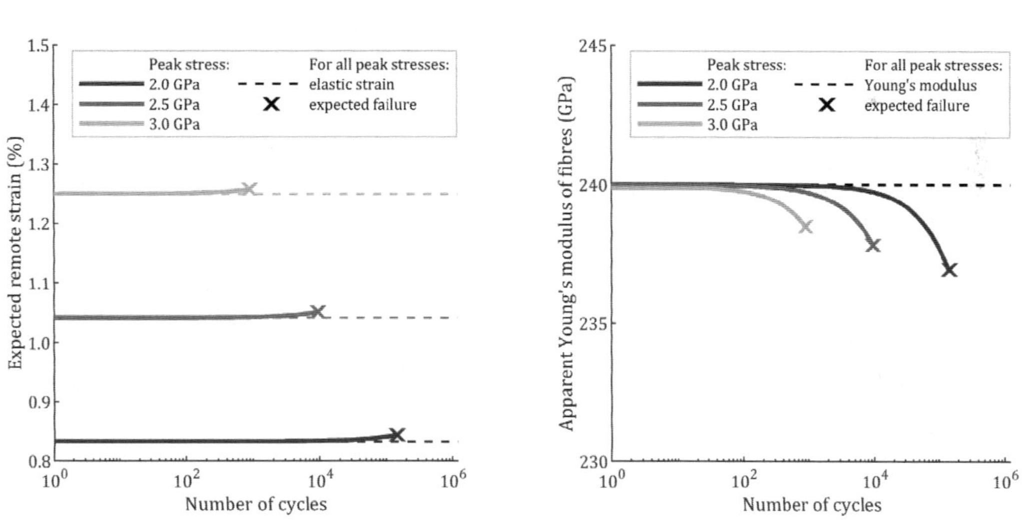

(a) Evolution of strain.

(b) Evolution of apparent modulus.

Figure 7: Effect of cyclic loading on the apparent stiffness of a carbon–epoxy composite system.

4. Discussion

4.1. Insight on the accumulation of damage in UD composites under cyclic longitudinal tension

The model proposed in this paper provides the following insight on the accumulation of damage in UD composites under cyclic longitudinal tension:

(i) The model predicts an increase of the overall density of fibre–breaks with progressive cyclic loading (Figure 5); this is due to the formation of new fibre–breaks near those previously formed during the first (i.e. static) loading cycle. Nevertheless, the total number of fibre–breaks at the moment of failure is predicted to be higher under static loading (Figure 4) than under high–cycle/low–stress fatigue (Figure 5);

(ii) The density of individual fibre–breaks decreases with progressive cyclic loading, but the density of clusters with 2–7 broken fibres increases (Figure 5a). This reveals that the kinetics of fibre–breakage under low–stress/high–cycle fatigue (where more than half of the total number of fibre–breaks is due clusters) is significantly different from that under static loading (where the large majority of the total number of fibre–breaks is due to individual/uncorrelated broken fibres, as shown in Figure 4);

(iii) The size of the largest cluster in the composite increases with progressive loading cycles (as shown in Figure 6). Interestingly, the largest cluster at the moment of failure is predicted to have the same number of fibres (4–7, i.e. level i = 2) under static conditions [8], low–cycle/high-stress fatigue (Figure 6b), and high–cycle/low–stress fatigue (Figure 6a).

(iv) Although extensive fibre–breakage occurs during cyclic longitudinal tension (Figure 5), this leads to a very limited reduction of the apparent modulus of the material (lower than 2%), as shown in Figure 7. This is because no large broken clusters are formed (as shown in Figure 6), and the increase of compliance due to individual fibre–breaks and small clusters is very local (i.e. limited to their effective recovery length);

(v) The loss of apparent stiffness at the moment of failure increases as one moves from low–cycle fatigue (see curve for σ^{∞} = 3.0 GPa in Figure 7b) to high–cycle fatigue (see curve for σ^{∞} = 2.0 GPa in Figure 7b). This suggests that unidirectional composites under longitudinal tension can accumulate larger amounts of overall damage under high–cycle/low–stress fatigue than under static loading and low–cycle/high–stress fatigue;

(vi) Combining the information from points (i), (iii), and (v) above, one can conclude that the main source of softening during fatigue loading cannot be the increase in the number of fibre–breaks or the size of broken–clusters; instead, the decrease

IOP Conf. Series: Materials Science and Engineering **388** (2018) 012007 doi:10.1088/1757-899X/388/1/012007

of apparent modulus during cyclic loading seems to be due to the increase of the effective recovery length which occurs due to mode–II fatigue of the interface near broken–fibres and clusters.

4.2. Validation of model predictions

Figures 5 and 6 show a good agreement between model predictions and experimental data. Nevertheless, the model predicts a large amount of variability for the total number of fibre breaks (see Figure 5b); this means that experimental data obtained with one specimen [5, 6] only is not sufficient to fully validate the model.

Moreover, while the model presented in this paper focuses on damage accumulation during cyclic loading, it is clear that the results will depend on the predictions of the model during the first loading cycle — i.e. under static conditions. For instance, Figure 4 suggests that the model overpredicts the density of fibre breaks under an applied stress between 60 and 80% of the static tensile strength; therefore, a similar overestimation can also been predicted for the number of fibre–breaks under cyclic loading at 70% of the static strength.

Regarding the effect of cyclic tension on the apparent Young's modulus of UD composites, it is widely accepted that UD composites under longitudinal tension present (*nearly*) no stiffness degradation under fatigue [1, 3]; this is corroborated by the predictions of the model shown in Figure 7b. A quantitative validation of the (small) stiffness degradation predicted by the model would require experimental data which, at the moment, is not available.

5. Conclusions

This paper proposed the first model in the literature to predict the kinetics of fibre–breakage and their effect on the macroscopic response of unidirectional composite materials under cyclic longitudinal tension. The following conclusions were reached:

- The model is based on the *analytical* evaluation of stress fields near fibre breaks, on a *hierarchical* statistical scaling law [7, 8], and on an *adaptive cycle jump* strategy for the fatigue analysis [9]. These features make the model capable of predicting damage accumulation in composite specimens with virtually any number of fibres and for their entire lifetime in less than 60 seconds;
- Model predictions for the accumulation and clustering of fibre–breaks under cyclic loading were successfully validated against experimental results from high–resolution computed tomography;
- The model suggests that cyclic loading leads to accumulation and clustering of fibre–breaks; however, the the critical cluster size does not vary significantly between

static, low–cycle/high–stress fatigue, and high–cycle/low–stress fatigue types of loading;

- The model suggests that unidirectional composites under longitudinal tension are able to undergo more softening under high–cycle/low–stress fatigue than under low–cycle/high–stress or static loading; this is mostly due to progressive degradation of the matrix/interface near fibre–fibres with cyclic loading.

Due to its computational efficiency and its easy–to–use analytical implementation, we expect that this model can have the following applications:

- Guiding the development of composite materials less sensitive to cyclic degradation (e.g. by predicting the effect that modifying fibre, matrix and/or interfacial properties would have on the accumulation of damage throughout the fatigue life);
- Developing macro–scale health–monitoring strategies for unidirectional composites under cyclic loading (e.g. by defining a critical value for the loss of longitudinal Young's modulus, after which the material may be prone to fatigue failure);
- Defining acceptable levels of micromechanical damage in composite structures under cyclic loading.

Acknowledgements

The research leading to these results has been done within the framework of the FiBreMoD project and has received funding from the European Union's Horizon 2020 research and innovation programme under the Marie Sklodowska–Curie grant agreement no. 722626. S. Pimenta also acknowledges the support from the Royal Academy of Engineering in the scope of her Research Fellowship on *Multiscale discontinuous composites for large scale and sustainable structural applications* (2015–2019).

The authors acknowledge Ian Sinclair from the University of Southampton and Serafina Garcea from the University of Manchester for discussions on the high–resolution computed tomography experiments.

Supporting data and the MATLAB® implementation can be requested from the corresponding author.

References

[1] E. K. Gamstedt and R. Talreja. Fatigue damage mechanisms in unidirectional carbon–fibre–reinforced plastics. *J Mater Sci*, 34(11):2535–2546, 1999.

[2] S. M. O. Tavares and P. M. S. T. de Castro. An overview of fatigue in aircraft structures. *Fatigue Fract Eng M*, 40:1510–1529, 2017.

[3] A. Somer and A. Bunsell. The tensile and fatigue behaviour of carbon fibres. *Plast Rub Compos Proc Appl*, 18:263–267, 1992.

[4] A. Bunsell et al. Benchmarking of strength models for unidirectional composites under longitudinal tension. *Compos Part A—Appl S*, 111:138–150, 2018.

[5] A. E. Scott et al. In situ fibre fracture measurement in carbon-epoxy laminates using high resolution computed tomography. *Compos Sci Technol*, 71:1471–1477, 2011.

[6] S. C. Garcea et al. Fibre failure assessment in carbon fibre reinforced polymers under fatigue loading by synchrotron X-ray computed tomography. *Compos Sci Technol*, 133:157 – 164, 2016.

[7] S. Pimenta and S. T. Pinho. Hierarchical scaling law for the strength of composite fibre bundles. *J Mech Phys Solids*, 61:1337–1356, 2013.

[8] S. Pimenta. A computationally efficient hierarchical scaling law to predict damage accumulation in composite fibre bundles. *Compos Sci Technol*, 146:210–225, 2017.

[9] M. Alves and S. Pimenta. A computationally–efficient micromechanical model for the fatigue life of unidirectional composites under tension–tension loading. *Int J Fatigue*, in press, 2018.

[10] B. W. Rosen. Tensile failure of fibrous composites. *AIAA Journal*, 2:1985–1991, 1964.

[11] G. W. Weibull. A statistical distribution function of wide applicability. *J Appl Math*, 18:293–297, 1951.

[12] A. Kelly and W. R. Tyson. Tensile properties of fibre–reinforced metals: copper/tungsten and copper/molybdenum. *J Mech Phys Solids*, 13:747–753, 2004.

[13] P. Paris and F. Erdogan. A critical analysis of crack propagation laws. *J Basic Eng, ASME*, December:528–534, 1963.

[14] I. J. Beyerlein and S. L. Phoenix. Stress concentrations around multiple fiber breaks in an elastic matrix with local yielding or debonding using quadratic influence superposition. *J Mech Phys Solids*, 44:1997–2039, 1996.

[15] S. Bai et al. Improving the adhesion between carbon fibres and an elastomer matrix using an acrylonitrile containing atmospheric plasma treatment. *Compos Interface*, 20:761–782, 2013.

[16] Toray Carbon Fibers America, Inc. Torayca T700G preliminary data sheet — Technical datasheet No. CFA 006, last accessed in June 2018.

IOP Conf. Series: Materials Science and Engineering **388** (2018) 012008 doi:10.1088/1757-899X/388/1/012008

High-speed bending-fatigue testing of composite materials

Y Ueki

Reliability Science Research Department, Research and Development Group, Hitachi, Ltd.

E-mail: yosuke.ueki.aq@hitachi.com

Abstract. A methodology for high-speed fatigue testing, especially for resin materials, including fibre-reinforced composites, was devised and evaluated. To exert periodic stress on a material at a frequency of more than 200 Hz, a specimen (made of glass fibre-epoxy laminate) was fixed as a cantilever to an electromagnetic vibrator and vibrated at its resonant frequency of the first bending mode by using a resonance tracking control. The shape of the specimen was designed with finite-element vibration analysis to obtain a resonant frequency of more than 200 Hz and a desired strain distribution for inducing fatigue damage under a certain stress level. The rise in temperature during the fatigue testing (due to damping loss) was estimated by applying heat-transfer theory and suppressed by external cooling to keep the specimen at room temperature. To confirm the validity of the devised high-speed-testing method, a completely reversed bending test at 1 Hz was also performed with identical specimens. The results from both testing methods (conducted at 230 Hz and 1 Hz, respectively) were plotted on a single power-law curve in an S-N plot. The good agreement between the two plots suggests that fatigue strength in the high- or giga-cycle region of resin and composites can be evaluated in a very short time if temperature is controlled appropriately.

1. Introduction

The range of applications of resin materials in industrial products is expanding year by year. Especially, fibre-reinforced composite (FRC)—combining carbon fibres and glass fibres as reinforcing materials—is excellent in terms of specific strength and specific stiffness, so it is increasingly used for transportation equipment such as cars and aircraft, which are required to be lightweight and highly reliable. Also, in the field of power production, FRC is applied for the main-load carrying member of rotor blades of wind turbines.

It is expected that the number of cases in which resin or FRC materials are adopted as structural members of industrial products will increase in the future. Resins and FRC materials to be evaluated are also becoming more diverse, and the choices for selecting FRC materials are increasing exponentially. On the other hand, it is known that the material properties of the resin depend on slight differences in composition, presence or absence of additives, curing conditions and the like. It is therefore necessary to sufficiently understand the characteristics of individual materials for designing reliable structures or products. If the reliability parameter is the static strength, it can be evaluated in a relatively short period of time. However, evaluating fatigue strength, which usually takes a very long period of time, is required when products are to be used under dynamic loading conditions. Therefore, fatigue testing sometimes becomes a bottleneck in the product-development process. For example, when a fatigue strength evaluation of 10^7 cycles is required, execution of a single test may require a long time (more than a month) since the test speed of a general fatigue-testing machine is at most about several 10 Hz. In recent

years, the number of products, such as rotor blades of wind turbines, requiring a fatigue design that assumes dynamic loading exceeding 10^7 cycles is increasing.
In light of those circumstances, a faster fatigue-test method applicable to resin and FRC materials is desired. Some fatigue-testing machines can actually attain a test speed exceeding 100 Hz; however, under the assumption large-size blades for mega-watt-class wind-power generators, etc., the repetition frequency of the load fluctuation achieved by the testing machine is actually in the order of several hertz [1]. When higher frequency is applied for shortening the testing time, it is important to verify the influence of the difference between the test frequency and the actual machine frequency on the result carefully. In particular, if the resin material is visco-elastic, heat generation due to viscous loss occurs as the test frequency is becoming higher, so it is considered that this will cause a difference from the test result under normal temperature. Regarding the relationship between fatigue-damage behaviour of the resin material and test frequency, although the frequency range is limited to several dozen hertz, several reports have been made [2-4]. However, a unified view has not been obtained, and there are many unknowns about fatigue-damage behaviour of the resin and FRC materials exposed to high-frequency loading.
One of the reasons for this state of affairs is that it is difficult to conduct a fatigue test under independently controlled specimen temperature and test frequency. Another reason is that it is practically difficult to attain both a higher test frequency together and a large deformation that are necessary for testing a resin material.
A carbon-fibre-reinforced (CFR) composite material was fatigue tested under a uniaxial tensile mode (stress ratio $R = 0.1$) at a test speed of 100 Hz by Hosoi et al. [5]. Under the assumption that the test did not cause significant heat generation, the test result is considered equivalent to a result obtained at 5 Hz. However, in their test, CRF resin was targeted because it has excellent heat conductivity, and a relatively thin test piece was used. Consequently, it was easy to suppress the temperature rise of the test piece even if the test is performed at high frequency. It was also easy to obtain a uniform temperature distribution in the cross-sectional direction. Despite those achievements, it is required to devise a novel test method applicable to a broader range of materials and specimen sizes and a higher testing frequency. In the present study, therefore, a test frequency of 200 Hz (at which a fatigue test of 10^7 cycles is completed in about half a day) is targeted, and a test method that can induce planar bending at high speed by using the resonance of the specimen is proposed [6]. The proposed method was applied to a general-purpose epoxy glass laminate, and the influence of test frequency and test-piece temperature on fatigue life was investigated.

2. Materials and Methods

2.1. Resonant bending-fatigue testing system

A schematic diagram of the test system developed for a resonance bending fatigue test (hereinafter referred to as "high-speed fatigue test") is shown in Figure 1. The fixed end of the cantilevered-beam specimen is attached to a vibrating head of an electromagnetic vibrator (i210/SA03, IMV, Japan). By subjecting the specimen to resonance at its natural frequency, large deformation can be generated in the specimen with comparatively little exciting force. Furthermore, control logic was applied to maintain the resonance condition of the specimen. That is, a small acceleration sensor (353C23, PCB, USA) for measuring response acceleration is attached to the free end of the specimen, and the phase difference between the excitation acceleration at the fixed end and the response acceleration at the free end is kept at 90° by this control logic. Since the stiffness of the test specimen fluctuates due to fatigue damage and temperature change during the fatigue test, the natural frequency changes moment by moment. However, this control logic maintains the resonant state of the specimen under stiffness fluctuation during the fatigue test. Bending strain and specimen temperature during the test were measured with a strain gauge and a T-thermocouple attached to the specimen, respectively. Amplitude of acceleration, strain, specimen temperature, and natural frequency during the test were recorded as time-history data by using a data logger (NR 500, Keyence, Japan) with a data acquisition rate of 1 Hz. The vibrating head of the electromagnetic exciter (including the test specimen) was installed in a thermostat chamber so that the ambient temperature could be arbitrarily controlled.

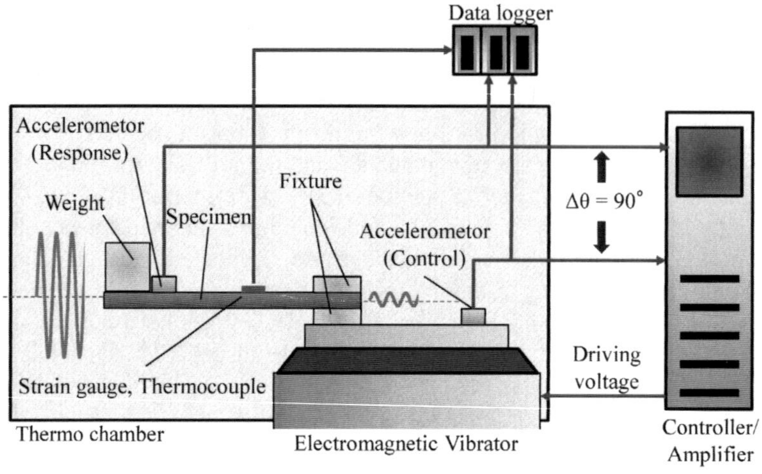

Figure 1. Schematic experimental setup for high-speed fatigue test [6].

2.2. Characteristics of materials

In the present study, which aimed to investigate the applicability of the high-speed fatigue test to composite materials, a general-purpose epoxy/glass-cloth laminate was used as a model material for evaluation. Its specifications and characteristics are listed in Table 1.

Table 1. Specifications of reference material [6]

Type	JIS-K6912, EL-GEM
Matrix resin	Epoxy
Reinforcement	Plain-woven glass fabric $[0/90]_{18}$
Thickness	3 mm
Specific weight	1.85 g/cm^3
Bending elastic modulus	22.6 GPa
Bending strength	420 MPa
Heat conductivity	0.471 W/m·K

2.3. Characteristics of test specimen

A test frequency of 200 Hz or higher was set as the target. The approximate specimen geometry was determined by assuming a cantilever beam with a uniform rectangular cross section, and the natural frequency was determined as

$$f = \frac{\lambda^2}{2\pi L^2}\sqrt{\left(\frac{EI}{\rho A}\right)} \tag{1}$$

where L is specimen length, E is elastic modulus, I is the second moment of area, ρ is density, and A is cross-sectional area. Parameter λ is 1.875 in the case of the first-order mode of the cantilever beam. Here, the dimensions of the test specimen (70× 15× 3 mm), with a natural frequency of 500 Hz or more (which is sufficiently higher than the target frequency of 200 Hz) were temporarily determined.

The optimum specimen shape, having the strain distribution shown in Figure 4, was determined by frequency-response analysis based on the finite element method in consideration of the attachment to the shaker, the method of adjusting the natural frequency, and the strain distribution in the specimen as follows. In the high-speed fatigue test, the specimen is gripped in a cantilevered manner and vibrated through its resonance state. Therefore, a fixed part is created on the fixed-end side, and a weight-attachment part (for adjusting natural frequency) is created on the free end. As shown in Figure 1, the specimen is attached to the electromagnetic vibrator via the fixing jig. Thus, in the fixed end, the local stress distribution will be spatially steep depending on the contact state between the fixing jig and the test piece. To overcome this non-ideal stress-distribution state, the test piece was formed as a fillet shape (namely, its width decreases sharply from the fixed end to the free end) so that maximum stress occurs at a position about 17 mm away from the fixed end. Adopting this shape has the advantage that mounting a sensor such as a strain gauge or a thermocouple becomes easy at the position where maximum stress is generated. For the determined specimen shape, bending-strain distribution was evaluated by frequency-response analysis based on the finite-element method. For the analysis, a general analysis environment, namely, ANSYS Workbench 14.0 (ANSYS), was used. In this analysis, it was aimed to evaluate the relative strain distribution in the test piece, so it was not necessary to acquire the absolute value. Therefore, the damping ratio was temporarily set as 1%. Displacement of the fixed part of the specimen was completely constrained, and the acceleration sensor and weight (including mounting bolt, washers, and nuts) on the free end were modelled as lumped masses at each barycentric position.
The bending-strain distribution in the longitudinal direction of the specimen obtained by the above-described analysis is shown in Figure 3 and Figure 4. The value of the bending strain was expressed as a relative value in regard to the strain at the position of the maximum generated bending stress. The natural frequency is an approximate number because it has an error of about 7% at the maximum between the measured value and the value predicted by the analysis. Since the deformation mode changes due to the mass of the weight mounted on the free end, it was a concern that the strain distribution would change as well. However, the strain distribution did not significantly change, and the deviation of the maximum-strain position to be evaluated was 1 mm or less. Even if a position 1 mm away from the maximum strain position was applied in the evaluation, the error of the strain under that condition would be 1% or less. Therefore, the difference between these deformation modes was considered to be negligible. Increasing the weight also raises the concern that bending stress caused by gravity may affect the test results; even so but the generated bending stress was at most 0.5 MPa. Static strength of the test material covered in this report is 400 MPa or more, which can be neglected as sufficiently small even when the expected fatigue strength is considered. Actual measured relative strain obtained by the strain gauge when the specimen was excited in the resonance state is shown in Figure 4. The values predicted by the analysis agree well with the measured values.

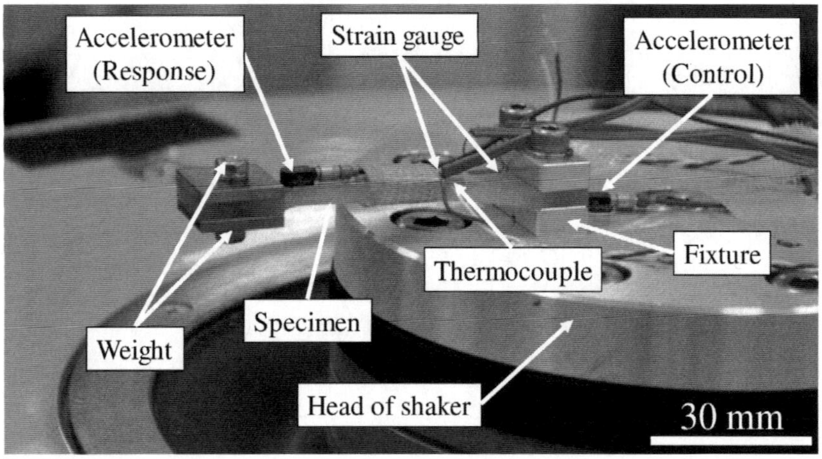

Figure 2. Photograph of experimental setup with specimen [6]

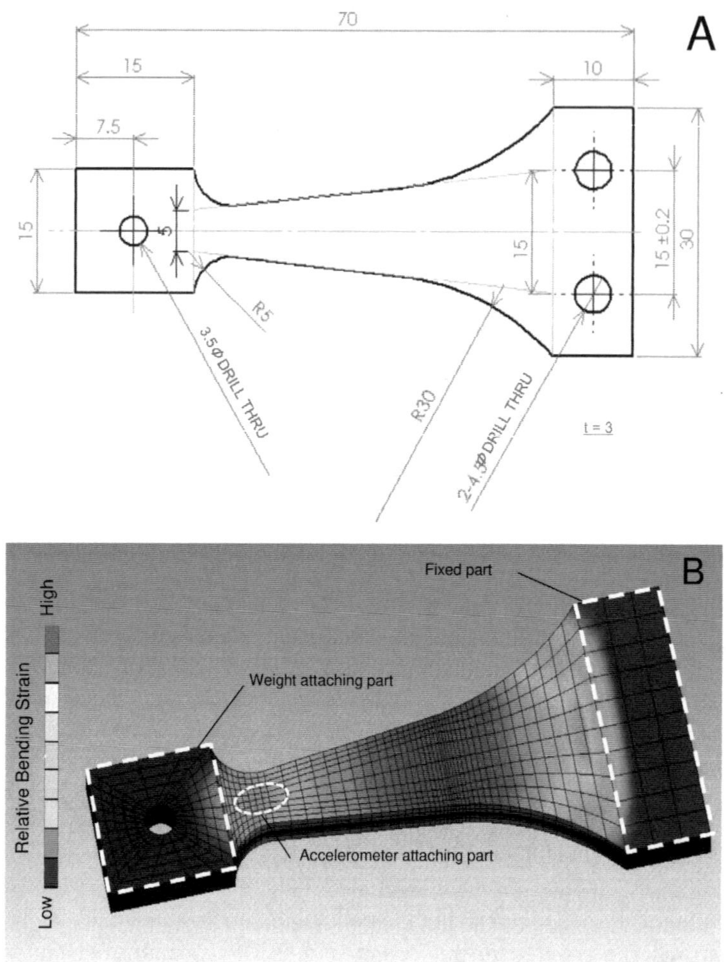

Figure 3. Geometry (A) and strain contours (B) of the specimen [6]

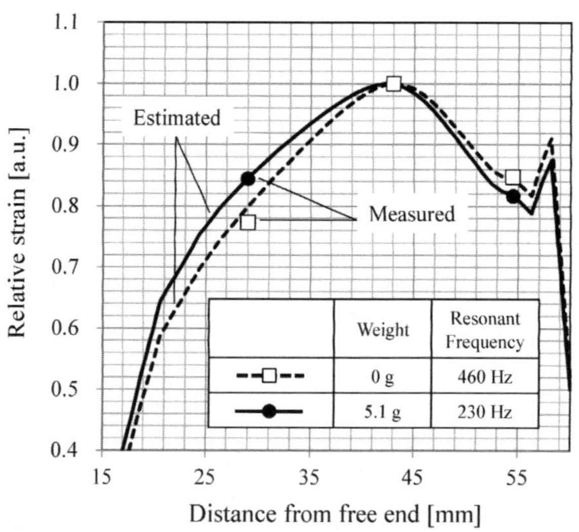

Figure 4. Bending-strain distribution in specimen along the longitudinal axis [6]

2.4. Regulation of stress amplitude

Since the high-speed fatigue test proposed in this report does not include a load-measuring device such as a load cell, it is difficult to directly measure and control the bending stress applied to the specimen. In addition, since the stiffness of the specimen changes from moment to moment (due to fatigue damage and temperature change) and the strain gauge reaches its fatigue life earlier than the target material does, a method based on strain measurement by strain gauge is difficult to adopt. However, in general, fatigue diagrams obtained under conditions of constant stress amplitude are often used in reliability design assuming a mechanical structure exposed to high-cycle fatigue loading. Therefore, it is important to keep a constant load amplitude during the fatigue test. In the present study, a method for defining the bending stress by combining strain measurement by strain gauge and the above-described displacement measurement at a response point (free end) based on response acceleration and vibration frequency was used.

The conditions under which the fatigue test was conducted with constant stress amplitude are described below. As shown in Equation (1), the natural frequency of the specimen is proportional to the square root of its elastic modulus. This condition also applies to the specimen geometry continuously changing cross-sectional shape. Therefore, Equation (1) can be transformed to

$$f^2 = C_1 E \tag{2}$$

where C_1 is a proportionality constant depending on the shape of the specimen and E is the bending sttiffness of the specimen. The response acceleration on the free end in resonance state is then taken as a. Accordingly, displacement d in the deflection direction at the response-acceleration measurement position can be expressed as

$$d = \frac{a}{(2\pi f)^2} \tag{3}$$

As fatigue damage progresses, the deformation mode may change due to the occurrence of cracks or the like. However, within a range where at least the deformation mode can be regarded as constant, the relationship between displacement d and strain ε at the evaluation point of the specimen can be assumed to be proportional and given as

$$\varepsilon = C_2 d \tag{4}$$

Stress σ at the evaluation point is the calculated from Equations (2) to (4) as

$$\sigma = \varepsilon E = C_2 \frac{a}{(2\pi f)^2} \cdot \frac{f^2}{C_1} = C_3 a \tag{5}$$

where C_3 is a proportionality constant expressed as

$$C_3 = \frac{C_2}{(4\pi^2 C_1)} \tag{6}$$

Based on Equation (5), the stress can be uniquely determined only by the response acceleration measured by the acceleration sensor independent of the stiffness of the specimen as far as no significant damage occurs. That is, even if the stiffness changes during the test, if it is possible to obtain C_3 beforehand and measure the response-acceleration amplitude during the test, the stress amplitude applied to the specimen can be determined from Equation (5). As shown in Figure 5, a vibration test was preliminarily performed at low excitation acceleration (10 m/s^2) to obtain a C_3. C_3 can be calculated from measurement data of the strain and the respose-acceleration and Equation (4) and (5). After that, the procedure was shifted to the main fatigue test. The fatigue test was carried out by controlling the amplitude of excitation acceleration as a constant value. Stress amplitude was evaluated from the measured response-acceleration amplitude after completion of the test.

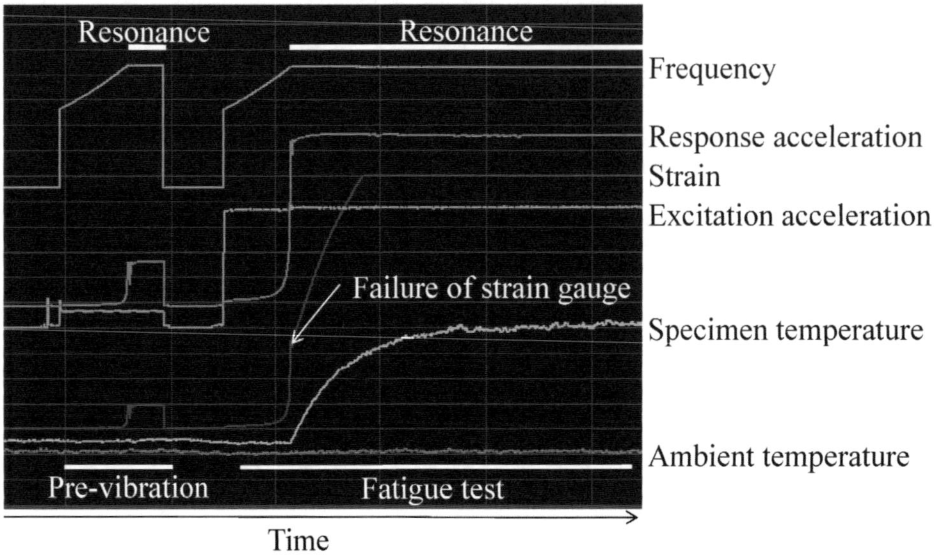

Figure 5. Temporal changes in parameters during the fatigue test. All parameters are shown as relative values.

2.5. Consideration of rise in specimen temperature.

If repeated strain is applied to a resin material at high speed, as shown in Figure 5, the specimen heats up remarkably due to viscous loss. Under the assumption of a high-speed fatigue test, this heat generation may raise the test temperature and affect the results of the fatigue test. Therefore, to determine appropriate test conditions for maintaining the specimen temperature in the normal temperature range, one-dimensional thermal conduction in the cross-section direction of the test piece was considered, and temperature distribution in the test piece was estimated. In the first case considered, the specimen resin (a viscoelastic body) was subjected to periodic repeated strain. Energy loss per unit time is expressed as

$$Q = fE"\varepsilon^2\pi \tag{7}$$

where f is repetition frequency, E" is loss elastic modulus of the material, and ε is strain amplitude [5]. As expressed by Equation (7), heat generation is proportional to repetition frequency and the square of strain amplitude. In the next case considered, strain distribution in the cross-section direction of the specimen is considered. In bending mode, axial strain increases linearly from the neutral axis toward the specimen surface, so strain distribution in the cross-section direction is expressed as

$$\varepsilon(x) = \frac{2}{t}\varepsilon_0 x \tag{8}$$

where x is distance from the neutral axis, $\varepsilon(x)$ is strain at position x, ε_0 is bending strain on the specimen surface, and t is thickness of the specimen. From Equations (7) and (8), heat distribution $Q(x)$ in the cross-sectional direction can be expressed as

$$Q(x) = \frac{4fE"\pi\varepsilon_0{}^2}{t^2}x^2 \tag{9}$$

Next, under the assumption of one-dimensional steady-state heat conduction in the cross-sectional direction, the heat conduction equation can be written as

$$\frac{\partial}{\partial x}\left(k\frac{\partial T}{\partial x}\right)+Q(x)=\frac{\partial}{\partial x}\left(k\frac{\partial T}{\partial x}\right)+\frac{4fE''\pi{\varepsilon_0}^2}{t^2}x^2=0 \tag{10}$$

where k is thermal conductivity of the specimen, and T is specimen temperature. In consideration of heat flux $q(x)$ at position x, the boundary condition can be expressed as follows. Since the heat flux is zero at the neutral-axis position, it follows that

$$q(x=0)=k\left.\frac{\partial T}{\partial x}\right|_{x=0}=0 \tag{11}$$

And since heat transfer to the surrounding atmosphere occurs on the surface of the test piece, it follows that

$$q\left(x=\frac{1}{2}t\right)=k\left.\frac{\partial T}{\partial x}\right|_{x=\frac{t}{2}}=h\left(T_b-T_{\frac{1}{2}t}\right) \tag{12}$$

where h is the heat-transfer coefficient, and $T_{\frac{1}{2}t}$ and T_b are surface temperature and environmental temperature, respectively. Solving the heat-conduction equation (Equation (10)) using the boundary conditions given by Equations (11) and (12), the temperature distribution in the cross-sectional direction is given as

$$T=-\frac{fE''\pi{\varepsilon_0}^2}{3t^2}\left\{\frac{x^4}{k}-\frac{t^3}{2k}-\frac{t^4}{16k}\right\}+T_b \tag{13}$$

Cross-sectional temperature distribution in the specimen given by Equation (13) is shown in Figure 6. Loss elastic modulus was calculated as 126 MPa from the measured acceleration frequency response, which was separately performed, and the heat-transfer coefficient was estimated as 44 W/m^{2}K from surface temperature measured during excitation. Thickness of the specimen was examined both in terms of the real value (3 mm) and twice of that value (6 mm). In addition, temperature distribution in the cross-sectional direction (similarly obtained by assuming uniaxial tensile-compression mode) is also shown in the figure. As for all results, test frequency of 200 Hz and stress amplitude of 100 MPa were assumed as test conditions.

In the case of test piece thickness t = 3 mm in bending mode, the temperature rise is less than 20°C, and temperature distribution is almost uniform. Therefore, if stress amplitude is within the range of about 100 MPa, it is considered that the test can be carried out by keeping ambient temperature at about 0°C in order to maintain the temperature of the evaluation point in the room-temperature range. In the test, surface temperature of the specimen is measured with a thermocouple attached to the surface of the test piece. Based on the analysis result, the measured temperature can be considered the specimen temperature. In the experimental system used in the present study, although it is impossible to directly control specimen temperature during the fatigue test, environmental temperature was set to either 0°C or 30°C (i.e., no cooling). On the other hand, the uniaxial tensile-compression mode induced temperature rise of more than 50°C. If the test is carried out under a lower-temperature environment, it is possible to keep the test-piece temperature in the normal temperature range even in this testing mode. However, practical problems in regard to cooling (such as frost formation) remain. If the temperature distribution of the specimen is focused on, it becomes clear that there is no noticeable difference in the slopes of the lines for bending mode and tensile-compression mode when specimen thickness was t = 3 mm. However, when thickness was t = 6 mm, not only absolute temperature rise exceeds 120°C but also the temperature distribution narrows by about 15°C. On the other hand, the temperature distribution in the bending mode is in the range of about 2°C. This result is due to the fact that the calorific value is smaller for a deeper position in the specimen (which is difficult to cool). Although the case of increased thickness was described here, the same tendency would be seen if a material having a smaller thermal conductivity were targeted, since thermal resistance between the neutral axis and the surface of the test piece would be increased. Therefore, the bending system adopted in the present study is considered to be able to cover a wider range of materials and specimen dimensions from the viewpoint of suppression of temperature rise and distribution.

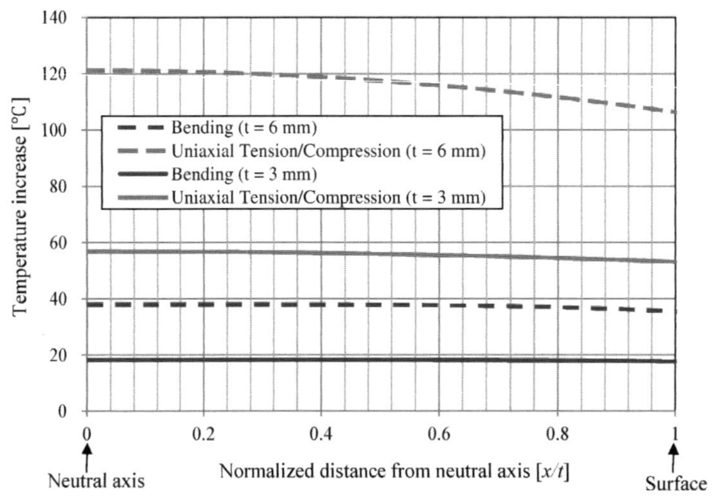

Figure 6. Cross-sectional distribution of temperature increase in specimens with thickness of 3 and 6 mm estimated under the assumption of one-dimensional heat transfer [6]. Testing conditions: frequency: 200 Hz; stress amplitude: 100 MPa.

2.6. Low-frequency alternating bending-fatigue test

In order to test the accuracy of the tests results obtained at high frequencies, it is necessary to evaluate the influence of the test frequency on the test result. Accordingly, a test at lower frequency with the same test specimen under the same strain-distribution condition as the high-speed fatigue test was carried out, and the results were compared with the results from the high-speed test. A general fatigue-testing machine was used for a fatigue test for that comparative study (hereinafter referred to as "low-speed fatigue test"). Bending deformation is obtained from the resonance of the specimen in the high-speed fatigue test. For that reason, it is possible to adjust the natural frequency by adjusting the weight mounted on the free end of the specimen. However, in consideration of a spring-mass system, since the natural frequency is inversely proportional to the square root of the mass, it is practically difficult to test the test frequency at several hertz when the same specimen is used. Accordingly, a low-frequency fatigue test (low-speed fatigue test) with the same specimen, complete strain distribution and same stress ratio was performed using a hydraulic-actuator fatigue-testing machine (Model 8871, Instron).

A photograph of the experimental setup is shown in Figure 7. To make the strain distribution in the specimen the same as that in the resonated specimen, bending deformation was applied by a load applied at the point of weight attachment. In the high-speed fatigue test, to attain bending test in resonance state, the stress ratio has to be R = -1 as alternate loading. To achieve the same stress ratio in the low-speed fatigue test, the specimen was sandwiched from above and below with a two-way indenter (having a wedge shape) connected to the hydraulic actuator. In addition, when the specimen was mounted, so that a bending moment is not generated at the loading point, a gap (about 0.3 mm) was set between the test piece and the tips of the indenters.

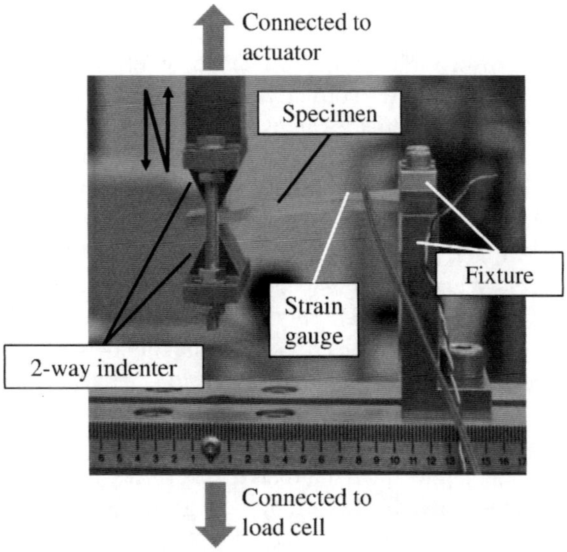

Figure 7. Photograph of experimental setup for low-speed fatigue testing [6].

Strain distributions based on the results of the static-load analysis and frequency-response analysis by the finite-element method under the assumptions that the low-speed fatigue test (with concentrated load applied) and the fast fatigue test (at about 230 Hz) were performed are compared in Figure 8. Both sets of results agree well; that is, the generated strain distributions are almost the same even though the way of exerting bending deformation was different.

The time histories of strain and load during the low-speed fatigue test are shown in Figure 9. The fatigue-testing machine was in the load-control mode, and load amplitude was 75 N with alternate loading (1 Hz). A sinusoidal waveform of the strain was observed, and it was confirmed that a fatigue test with stress ratio R = -1 is feasible. All tests were carried out in a normal temperature environment (25 to 30°C), and test frequency was 1 Hz. To record the change in stiffness of the specimen, the stroke of the actuator was recorded as a time series.

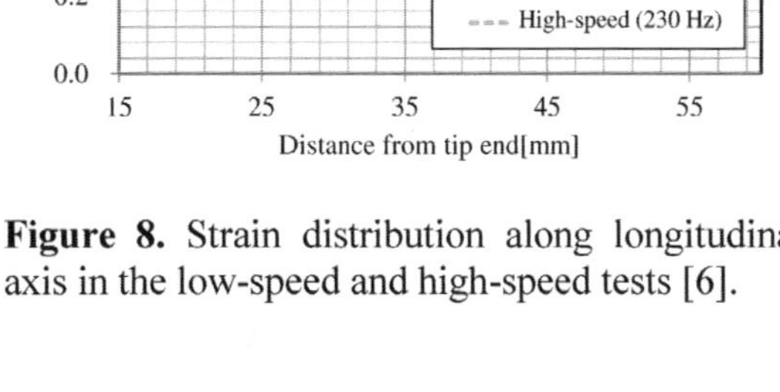

Figure 8. Strain distribution along longitudinal axis in the low-speed and high-speed tests [6].

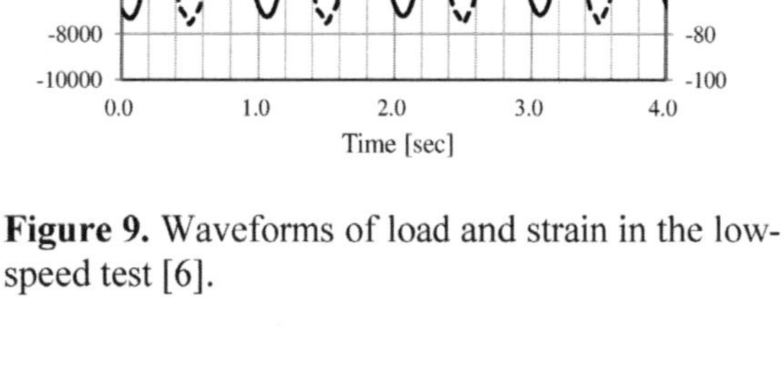

Figure 9. Waveforms of load and strain in the low-speed test [6].

3. Results and discussions

3.1. Evaluation of fatigue damaging based on stiffness change

An example of the change in stiffness retention ratio during the test is shown in Figure 10. In the case of the high-speed fatigue test, the stiffness of the specimen was obtained from the change of its natural frequency. Since the residual stiffness measured by the acceleration sensor was evaluated in the frequency domain, it is less susceptible to the influence of drift and noise, and relatively stable measurements were obtained as compared to directly measured strain by strain gauge or extensometer. In the case of the low-speed fatigue test, stiffness retention ratio was obtained from the stroke change of the testing machine. Stroke change is influenced not only by change in the stiffness of the specimen but also by the contact state change between the indenters and the specimen, deformation of the indenters or the testing machine, and so on. Therefore, the measurements from the low-speed test included a large noise component compared with the measurements from the high-speed test.

Both of the low-speed fatigue test and the high-speed fatigue test showed a tendency of rapid decrease of stiffness after the stiffness retention ratio fell below 95%. In the low-speed fatigue test, the specimen was broken immediately after this rapid decrease. In the high-speed fatigue test, the response-acceleration waveform distorted from the sinusoidal waveform when the damage progresses and a crack develops. Therefore, the resonance tracking could not work before the specimen reaches final failure; in other words, the stress amplitude sometimes decreases remarkably as described above. Therefore, in consideration of the above result, the moment that the stiffness retention ratio reached 95% was considered to be the life of the specimen. In the case of the high-speed fatigue test, the stiffness reduction due to the temperature rise was corrected by referring to the relationship between stiffness and temperature of the test specimen acquired in advance. In addition, stress amplitude could not be controlled directly. Therefore, if the damping ratio of the specimen changes during the test, the stress amplitude fluctuates. Indeed, at the end of the specimen's fatigue life, a decrease in stress amplitude, which is thought to be due to an increase in the specimen's damping characteristic (due to an increase in internal friction [8]), was confirmed. Therefore, as the value of the stress amplitude, an average value from the time immediately after the start of resonance until the stiffness retention ratio reached 95% was used. In addition, it was judged that the test results with fluctuation width exceeding 10% of the average value could not be regarded as a constant stress amplitude and excluded from the evaluation results.

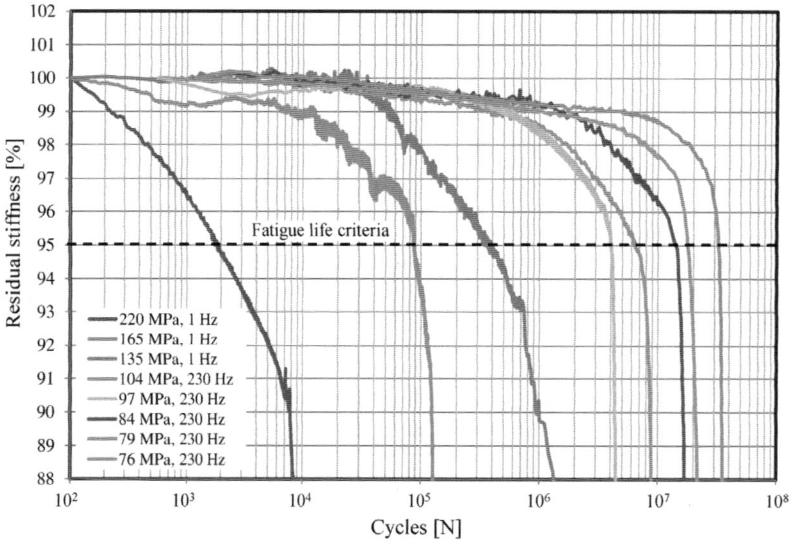

Figure 10. Residual stiffness of specimens during fatigue test. [6].

3.2. Observation of damaged specimen

Examples of the specimens subjected to the high-speed fatigue test and the low-speed fatigue test are shown in Figures 11(a) and (b), respectively. Visible cracks are observed in the test piece subjected to the high-speed fatigue test. The positions where the cracks occurred are within the range of approximately ± 2 mm from the maximum stress position. It was judged that the error of the stress amplitude within this range was within -3% of the maximum value, so its influence on the test result was small. In addition, when the crack was observed in detail, there was a case where it was slightly discoloured to brown. It was considered to be a discoloration due to frictional heat generated by sliding fracture surfaces in the cracks. In fact, in some cases, a sharp rise in the specimen temperature was observed after the stiffness retention ratio fell below 95%. In addition, in the vicinity of the crack position, a spot-like whitening (which is thought to be caused by separation of the matrix resin and the glass cloth) was observed. In the low-speed-fatigue-test specimen [Figure 11(b)), whitening was also observed around the fracture surface of the specimen. However, the influence of the frequency on damaging-process is not sufficiently studied at present because in the case of high-speed fatigue test, it is impossible to continue the test until final failure (as described above), and heat generation due to sliding between fracture surfaces of cracks may affect the failure mode.

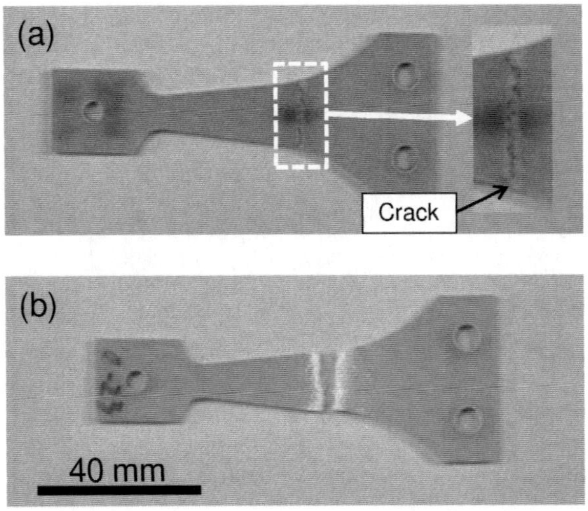

Figure 11. Photographs of specimens of glass fibre-epoxy laminate after fatigue tests: (a) high-speed test and (b) low-speed test [6].

3.3. S-N curve

Based on the fatigue-life criteria described in Section 3.1 (stiffness retention ratio of 95%), S-N curves for both tests were derived. S-N curves derived from the results of the low-speed test at room temperature (ambient: 20-25°C), high-speed test at room temperature (specimen: 8-15°C; ambient: 0°C.), and high-temperature conditions (specimen: 40-60°C; ambient: 30°C) are shown in Figure 12. It is clear that fatigue strength decreases by about 10% to 20% under the high-temperature condition compared to the room-temperature condition in regard to the high-speed test. It can be speculated that fatigue strength of this material varied under the influence of temperature. This result agrees with the previous report that if test frequency is raised, the specimen lifetime will decrease due to temperature rise (Rotem, 1993). The results from the low-speed test and the high-speed test obtained under room temperature conditions are focused on hereafter. For glass-fibre-reinforced resin and carbon-fibre-reinforced resin, it is known that a high cycle fatigue curve obtained under the load-control condition follows the power law described as

$$S = CN^{-\frac{1}{m}} \tag{14}$$

where S is stress amplitude, N is number of repeating cycles, and m and C are constants [8]. For fibre-reinforced resins using epoxy resin as a matrix, it has been reported that m (which represents the slope of the S-N curve) is within the range of about 6 to 15 [9]. Approximated curves obtained from Eq. (14) are shown for the low-speed test and the high-speed test in Figure 12. A high-cycle fatigue range (10^4 times or more) was chosen for the approximation range. Both of sets of results agree well, and it is considered that the high-speed test provides comparable results to the test at several hertz if cooling was carried out properly to suppress excessive temperature rise. However, comparable fatigue lives were not obtained in the high-speed test and the low-speed test due to difficulties such as test time (for the low-speed test) at low stress levels and frosting (for the high-speed test) at high stress levels. In the future, the author plans to expand the stress-amplitude range of each test and verify the equivalence of both tests. In the present study, evaluation was limited only to a single material/laminated structure. In the future, we plan to conduct tests on different resin materials or composite materials with different laminated constructions and confirm the applicable range of the devised test method.

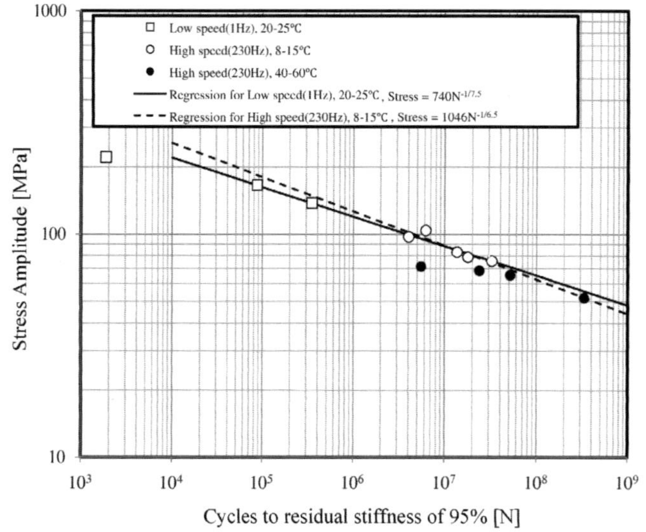

Figure 12. S-N plot of glass fibre-epoxy laminate obtained from high-speed and low-speed fatigue tests. (Stress ratio: R = -1)

4. Conclusions

A high-speed fatigue test method utilizing resonance applicable to resins including fibre-reinforced materials was devised and to a general-purpose glass-epoxy laminate to evaluate its effectiveness. The results of the evaluation are summarized below

- When the fatigue test was conducted at a test frequency of about 230 Hz, it was found that the results plotted on the S-N curve were approximately the same as the test results performed at test frequency of 1 Hz if the specimen temperature was appropriately controlled.
- Specimen temperature during the fatigue test was estimated by assuming one-dimensional thermal conduction. It was found that the specimen temperature can be kept sufficiently in the normal temperature region by external cooling within the assumed stress amplitude range even if the test is performed at loading frequency of more than 200 Hz.

These results indicate that the devised test method has is suitable for obtaining fatigue strength of resins in a short period of time.

5. References

[1] Ronald K O, Wedel-Heinen J and Christensen C J 1992 Reliability-based fatigue design of wind turbine rotor blades, *Eng. Struct.*, **21** 12

[2] Rotem A 1993 Load frequency effect on the fatigue strength of isotropic laminates, *Compos. Sci. Technol.* **46** 2

[3] Sun C T and Chan W S 1979 Frequency effect on the fatigue life of a laminated composite, *Composite Materials: Testing and Design(Fifth conference) ASTM.*, pp. 418-430

[4] Mishnaevsky L J and Brøndsted P 2006 Cyclic loading frequency effect on the damage growth in the low-frequency range, *Proceedings of the 27th Risø International Symposium on Materials Science*, pp. 239-248.

[5] Hosoi A, Arao Y, Karasawa H and Kawada H 2007 High-cycle fatigue characteristics of quasi-isotropic CFRP laminates, *Adv. Compos. Mater.*, **16** 12

[6] Ueki Y, 2015, A high-speed bending fatigue testing method for resin materials (An attempt of application to a glass fiber-reinforced plastic), *Trans. JSME,* **81** 821, (in Japanese)

[7] Zangenberg J, Brøndsted P and Gillespie J W J 2013 Fatigue damage propagation in unidirectional glass fibre reinforced composites made of a non-crimp fabric, *J. Compos. Mater.*, **48** 22

[8] Nijssen R P L, 2010 Phenomenological fatigue analysis and life modeling, *Fatigue life prediction of composites and composite structures* ed Vassilopoulos A P (Cambridge :Woodhead Publishing) pp. 47-78

[9] Mandell J F, Samorsky D D and Miller D A 2013 Effects of resin and reinforcement variations on fatigue resistance of wind turbine blades, *Advances in wind turbine blade design and materials* ed Brøndsted P and Nijssen R (Cambridge :Woodhead Publishing) pp. 210-250

IOP Conf. Series: Materials Science and Engineering **388** (2018) 012009 doi:10.1088/1757-899X/388/1/012009

Structural behavior of secondary-bonded composite joints subjected to Mode II fatigue induced delamination

F P Garpelli[1,2], F M G Ramírez[1], H B Resende[2] and M V Donadon[1]

[1] Divisão de Engenharia Mecânica, ITA – Instituto Tecnológico de Aeronáutica, Praça Marechal Eduardo Gomes, 50, 12228-900, São José dos Campos/SP, Brasil
[2] Laboratório de Estruturas Leves, IPT – Instituto de Pesquisas Tecnológicas, Estrada Dr. Altino Bondensan, 500 – Distrito Eugênio de Melo, 12210-131, São José dos Campos/SP, Brasil

E-mail: felipegarpelli92@yahoo.com.br

Abstract. This paper analyses the Mode II fatigue delamination growth onset for secondary bonded joints compared to co-cured joints. The materials used were composed by two carbon fiber reinforced sub-laminates joined by the secondary bonded and co-cured (without adhesive) methods. Mode II fatigue tests were performed using three-point bending *End Notched Flexure* test setup. The tests were performed under displacement control and a sinusoidal displacement applied at a frequency of 5 Hz with a $R_d = 0.1$, that mean, $\delta_{min} = 0.1\ \delta_{max}$. The main objective of this study was to obtain the strain energy release rate (*SEER*) versus number of cycles (N_f) in order to evaluate the effect of the adhesive on fatigue life of bonded joints. A Closed form solution for 3-ENF setup was used in order to define the displacement amplitudes that were applied in the fatigue test. The results show that the use of adhesive causes a reduction on Mode II fatigue delamination growth onset *SERR* (G_{th}) for secondary-bonded compared to co-cured joints. Finally, a scanning electron microscopy (*SEM*) was used to analyze the fracture surfaces of the Mode II secondary bonded specimens. The fractography images show no crack growth for the specimens tested below or at G_{th} loading levels, which indicated that the value corresponds to the threshold.

1. Introduction

Composite materials been widely employed in the aeronautical industry for the construction of structural components and the aircraft. This trend can be seen through an assessment of the new aircraft produced by today's largest manufacturers, Boeing and Airbus. Boeing's new 787-9 Dreamliner features a set of new technologies of which 50% of the weight is fabricated with composite materials, including fuselage and wings [1]. The Airbus A350XWB was built using 70% of advanced materials, which includes 53% of composites materials and the rest in aluminum and titanium alloys [2]. The properties that can be highlighted are high performance materials with improved mechanical and physical properties. However, composite materials are prone to delamination under cyclic loading, resulting in loss of properties and even component failure. Therefore, it is first necessary to evaluate delamination toughness of the material for the development of damage tolerant structural components [3, 4]. Interlaminar fracture toughness characterization is usually done through interlaminar crack propagations in Mode I (the opening mode), Mode II (the sliding mode) and Mode III (the tearing mode) [3, 4].

In recent years, adhesively bonded joints are being widely used to join composite structures, a method concurrent with traditional mechanical joining [5]. Its advantages over mechanical fastening are: better load transfer, improved fatigue life, damage tolerance, less weight, flexibility and others [6]. For such a methodology to be widely used, in the aeronautical industry, it is necessary further studies to evaluate the behavior of the union, selecting a process that results in good structural performance and reliability in order to obtain parameters that measure the influence of the joint: adhesive, geometry, surface preparation, operating environment, among others. This technology is an alternative that can contribute to reducing the number of parts and final costs, improve the production process and reduced delamination problems [7]. The most commonly used techniques for composite assembly are: co-cured (CC), co-bonded (CB) and secondary-bonded (SB). Secondary-bonding corresponds to the process of joining two cured structural parts with an adhesive [8]. Fernández [9] investigated the 3-ENF procedure in order to evaluate if the results obtained presented reliability for Mode II fatigue test in composite materials using joint technology. It was found that the 3-ENF test is valid and reliable for Mode II fatigue tests on bonded composite joints. So far there is no Mode II standard, therefore in this research these tests will use the 3-ENF configuration, as recommended by Fernández [9].

A deep understanding is needed on the failure modes of bonded joints, since there are several failure modes in adhesive joints, the main ones being: a) delamination of the laminate; b) failure of the interface between the adhesive and the laminate; c) cohesive failure; and d) the combination of them [10, 11]. In this context, the present work aims to evaluate the structural behavior of composite secondary bonded joints subjected to Mode II fatigue induced delamination tested at room temperature ambient in order to obtain the G-N_f curve, following the standard [12], and compare the obtained results with results reported for co-cured joints [13]. A fractography analysis is performed in order to verify the failure mechanism and whether or not there was crack propagation.

2. Material and Methodology

The 3-ENF fatigue tests were performed on unidirectional Torayca® prepreg made of T800 carbon fiber embedded in a 3900-2C epoxy resin. This prepreg also contain thermoplastic polyamide particles on the epoxy resin in order to improve the toughness and consequently reduce the crack propagation [7, 13-16]. The laminates had a $[0°]_{26}$ layup with a Teflon® film insert in the mid-plane, in order to produce an initial crack for the subsequent interlaminar fatigue tests. The adhesive used for the secondary-bonded (SB) was Loctite® EA 9695 (produced by Henkel®) with a pre-impregnated polyester peel ply to act as thickness controller. The specimens dimensions are: length (L) 175.50 mm, width (B) 20.00 mm, thickness (h) 4.99 mm and initial crack (a_0) 57.50mm. The Flexural Young's modulus (E) was measured in a quasi-static test and the value was 124.12 GPa, this result was provided by Brito [7].

Figure 1. 3-ENF setup fatigue test.

2.1. Mode II fatigue tests

The Mode II fatigue tests were performed on a servo-hydraulic test machine (MTS370.10) with a 15*kN* load cell, at the Lightweight Structures Laboratory (LEL/IPT). The specimens were cycled under displacement control, respecting a sinusoidal function with a displacement ratio of $R_d = 0.1$, that mean, $\delta_{min} = 0.1\ \delta_{max}$, at a 5 *Hz* frequency. A preload of 10 *N* was applied to hold the sample in the test device, as shown in Figure 1.

The criteria used to stop the test was $N_{5\%}$ a (5 % decrease in the maximum cyclic load) as an evidence of crack propagation or N_f equal to 106 cycles that represents the threshold Strain Energy Release Rate (SERR) according to Doker and Marci [17]. The experimental procedure was performed in order to identify the number of cycles to delamination and determining the G - N_f curve.

The displacement values, maximum displacement (δ_{max}), minimum displacement (δ_{min}) and mean displacement (δ_{mean}) required to define the amplitude of the displacement in fatigue cycle tests were determined based on the beam theory. The following equations represent the displacement as function of the load in the middle of the beam, firstly for the linear part, before Eq. (1)) and after (Eq. (2)) the delamination starts. The critical load and displacement can be found from the intersection of Equation (1) and Equation (2) [18].

$$\delta = \frac{P\left(2L^3 + 3a_0^{\ 3}\right)}{12EI} \tag{1}$$

$$\delta = \frac{P}{12EI}\left(2L^3 + \frac{\left(8G_{II}BEI\right)^{\frac{3}{2}}}{\sqrt{3}P^3}\right) \tag{2}$$

where *I* the moment of inertia of the cross-sectional area of the specimen, which is represented by Eq. (3), and G_{IIc} is the Mode II critical strain energy release rate (*SERR*).

$$I = \frac{Bh^3}{12} \tag{3}$$

According to the standard [12], the fatigue Mode I test must be performed cycling between a minimum and maximum displacement at a set frequency. Due to the absence of the standard to 3-ENF fatigue test, the same criteria were adopted to use in Mode II fatigue tests. The maximum cyclic displacement is obtained from an expression which relates the δ_c with an Energy ratio (R_G) to be used in each test, as shown in Equation 4:

$$\left(\frac{\delta_{max}}{\delta_c}\right)^2 = \frac{G_{IIMax}}{G_{IIc}} \Rightarrow \delta_{max} = \delta_c\sqrt{\frac{G_{IIMax}}{G_{IIc}}} \Rightarrow \delta_{max} = \delta_c\sqrt{R_G} \tag{4}$$

Knowing the maximum cyclic load (P_0) acquired at N = 1, it is possible to calculate the maximum *SEER* (G_{IImax}) for different energy (R_G), using Equation 5:

$$G_{IIMax} = \frac{3P_c^{\ 2}a_0^{\ 2}}{8BEI} \tag{5}$$

3. Results and discussion

A typical load displacement curve obtained from quasi-static test using 3-ENF is shown in Figure 2, it is possible to notice that a fairly good correlation between the experimental and analytical results. The critical load and displacement are 2700 *N* and 2.931 *mm*, respectively. The Mode II fracture toughness obtained by Brito [7] were considered, $G_{IImax} = 5.941$ *N/mm*.

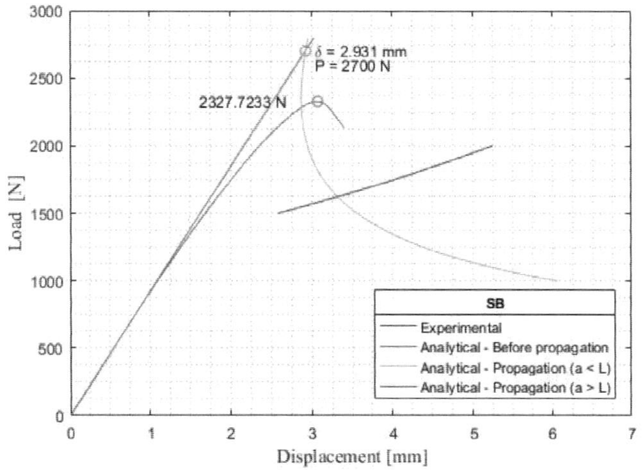

Figure 2. Analytical and experimental load versus displacement curves obtained for Mode II loading.

The results for the fatigue tests of the SB are presented in Figure 3. In all, 14 specimens are tested for energy ratios from 0.95 % to 3.66 %, in order to development the G-N_f curve. A trend line was adjusted to obtain an excellent correlation with the experimental points and verify the behavior each sample subjected to different R_G.

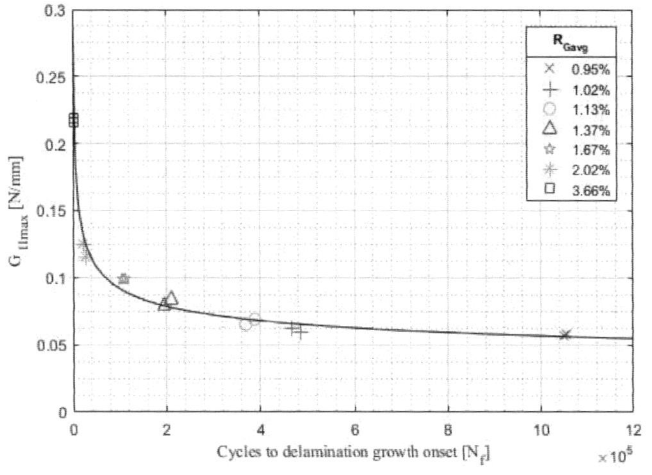

Figure 3. Analytical load versus displacement curve obtained for Mode II loading.

The delamination onset threshold value (G_{th}) under Mode II fatigue tests was found to be 0.95 %, which represent an average value of 0.0567 *N/mm*. Figure 4 represent the comparison of Mode II fatigue curves between SB and CC joint. The fatigue limit under 3-ENF for the CC joint was 8.12 %, corresponding to an average value of G_{th} = 0.0648 *N/mm* for 10^6 cycles [13]. It can be observed that the G_{th} value was higher for the CC joints, followed by the SB.

4. Fractographic analysis

A scanning electron microscope (SEM), model VEGA3 TESCAN, was used in order to verify if there was no crack propagation when reaching 10^6 cycles (threshold) and elucidate the failure modes. Figure 5 presents the fractographic aspects of SB sample tested in RG of 0.95 %. It was observed that no propagation occurred, only microcracks due to the difference in physical properties between the matrix and the adhesive. The mesh carrier (thickness controller) contributed to crack barrier and act as an obstacle to the crack front. It was confirmed that the sample supported 10^6 cycles without any crack nucleation and the threshold was certified.

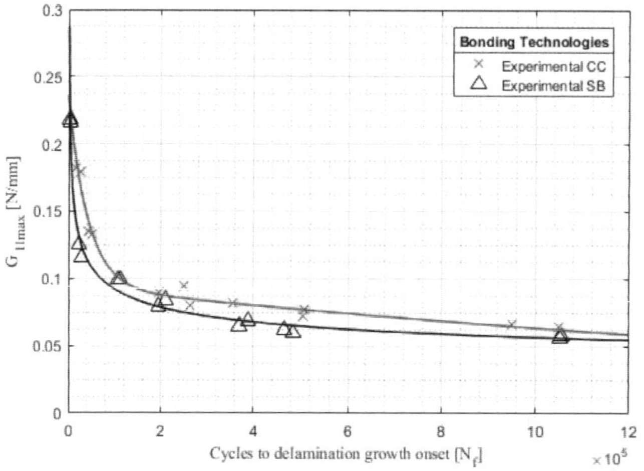

Figure 4. Comparison of Mode II fatigue curves delamination onset.

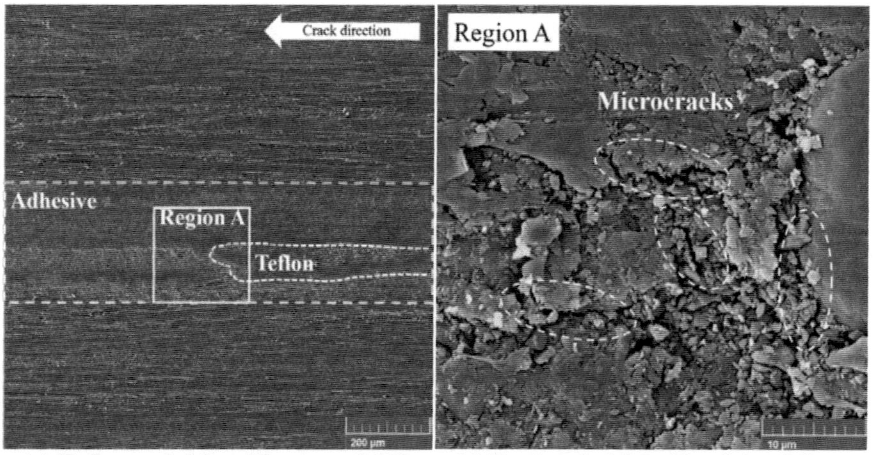

(a) Teflon end. Magnification of 500x. (b) Region A. Magnification of 5kx.

Figure 5: Fracture surface of SB samples tested under Mode II fatigue test.

Figure 6 shows the fracture surface of samples tested under fatigue test. Figure 6a shows that the failure mode was mixed because the initial crack propagation was in the adhesive (cohesive) and propagate to the region between adhesive and matrix (interlaminar failure). The polyamide particles did not act as a crack barrier to the SB sample. Figure 6b shows the polyamide particles in the CC sample. It is possible to note that the polyamide particles have carried out their aim, that accommodates the crack propagation and caused a cohesive failure.

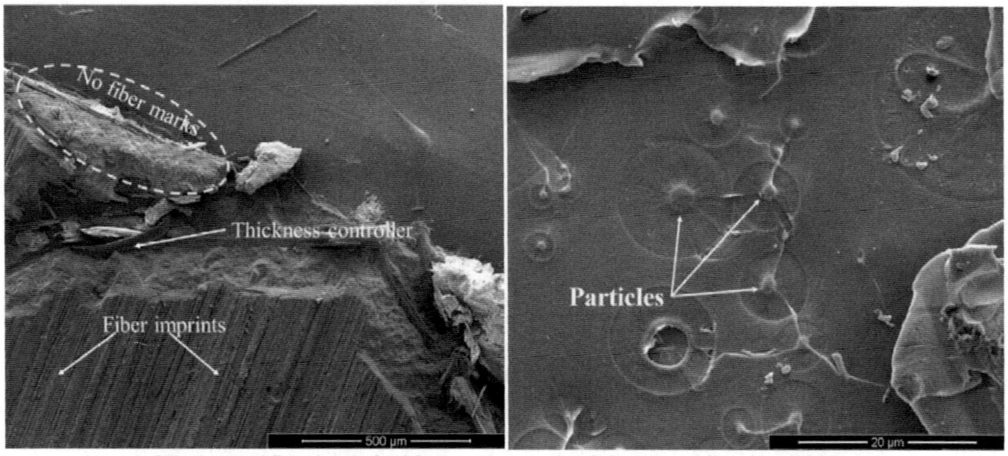

(a) SB. Magnification of 100x. (b) CC. Magnification of 1kx.

Figure 6: Fracture surface of samples tested under fatigue test.

5. Conclusions

The fatigue tests based on Mode II delamination growth onset in secondary-bonded joint at a room temperature condition was explained. The results show that the use of adhesive causes a reduction on Mode II fatigue delamination growth onset SERR (G_{th}) for secondary-bonded because the polyamide particles are not as efficient as in the CC sample. The fatigue threshold value found for Mode II was 0.0567 *N/mm*. Finally, a scanning electron microscopy (SEM) was used to analyze the fracture surfaces. The fractography images show no crack growth for the specimens tested below or at G_{th} loading levels, which indicated that the value corresponds to the threshold.

Acknowledgements

This project is partially supported by the CNPq processes 126346/2016-0, 142761/2016-8, 800319/2016-8, 800392/2016-7 and 300893/2015-9, Capes 1543373, FINEP 0114018300, IPT/FIPT 991914T and FAPESP 2015/16733-2. The authors also acknowledge the LEL-IPT, the Laboratory of Aerospace Structures and Laboratory of Materials and Processes at ITA for providing the infrastructure and equipment needed for the development of this work.

References

[1] BOEING.787 dreamliner advanced composite use, 2017. Available at: www.boeing.com/commercial/787/by-design/advanced-composite-use. Acessed on : Dec, 2017.

[2] AIRBUS.A350, 2017. Available at: www.aircraft.airbus.com/aircraftfamilies/passengeraircraft/- Acessed on : Dec, 2017.

[3] O Al-Khudairi, H Hadavinia, A Waggott, E Lewis, and C Little. Characterising Mode I/Mode II fatigue delamination growth in unidirectional fibre reinforced polymer laminates. *Materials & Design*, **66**:93-102, 2015.

[4] Iréne Maillet, Laurent Michel, Frédéric Souric, and Yves Gourinat. Mode II fatigue delamination growth characterization of a carbon/epoxy laminate at high frequency under vibration loading. *Engineering Fracture Mechanics*, **149**:298-312, 2015.

[5] M D Banea and Lucas F M da Silva. Adhesively bonded joints in composite materials: an overview. *Proceedings of the Institution of Mechanical Engineers, Part L: Journal of Materials: Design and Applications*, **223(1)**:1-18, 2009.

[6] Min-Gyu Song, Jin-Hwe Kweon, Jin-Ho Choi, Jai-Hyun Byun, Min-Hwan Song, Sang-Joon Shin, and Tae-Joo Lee. Effect of manufacturing methods on the shear strength of composite single-lap bonded joints. *Composite Structures*, **92(9)**:2194-2202, 2010.

[7] C B G Brito. *Hygrothermal effects on interlaminar fracture toughness of composite joints*. Master's thesis, Instituto Tecnológico de Aeronáutica, ITA. São José dos Campos /SP. 2017.

[8] A A Baker. *Composite materials for aircraft structures*. AIAA, 2004.

[9] M V C Fernández. *Fracture characterization of composite bonded joints under fatigue loading*. PhD thesis, Universidade do Porto (Portugal), 2013.

[10] S Budhe, M D Banea, S de Barros, and L F M da Silva. An updated review of adhesively bonded joints in composite materials. *International Journal of Adhesion and Adhesives*, **72**:30-42, 2017.

[11] AMERICAN SOCIETY FOR TESTING AND MATERIALS. **D5573-99**: Standard Practice for Classifying Failure Modes in Fiber-Reinforced Plastic FRPM Joints. West Conshohocken, PA: ASTM International, 2012.

[12] AMERICAN SOCIETY FOR TESTING AND MATERIALS. **D6115**: Standard Test Method for Mode I Fatigue Delamination Growth Onset of Unidirectional Fiber-Reinforced Polymer Matrix Composites. West Conshohocken, PA: ASTM, 2011.

[13] F P Garpelli, F M Gonzalez, R Sales, M A Arbelo, M Y Shiino, H B Resende, and M V Donadon. Experimental characterization of Mode II fatigue delamination growth onset in composite joints tested at RTA condition. *Composite Structures (Submitted)*, 2018.

[14] Nobuyuki Odagiri, Hajime Kishi, and Masaki Yamashita. Development of TORAYCA prepreg P2302 carbon fiber reinforced plastic for aircraft primary structural materials. *Advanced Composite Materials*, **5(3)**:249-254, 1996.

[15] Jin Zhang and Bronwyn L Fox. Manufacturing influence on the delamination fracture behavior of the T800H/3900-2 carbon fiber reinforced polymer composites. *Materials and manufacturing processes*, **22(6)**:768-772, 2007.

[16] Jin Zhang, Bronwyn Fox, and Qipeng Guo. Consistent model predictions for isothermal cure kinetics investigation of high performance epoxy prepregs. *Journal of applied polymer science*, **107(4)**:2231-2237, 2008.

[17] H Doker and G Marci. Threshold range and opening stress intensity factor in fatigue. *International journal of fatigue*, **5(4)**:187-191, 1983.

[18] M.V. Donadon and A. Faria. Analysis and design of composite structures – Fracture mechanics applied to composite laminates – Note Class. Technical report, Instituto Tecnógico de Aeronáutica – ITA. São José dos Campos, SP. 02/2016.

IOP Conf. Series: Materials Science and Engineering **388** (2018) 012010 doi:10.1088/1757-899X/388/1/012010

Specimen for cyclic delamination crack growth rates from ply-drops

Stergios Goutianos and Bent F. Sørensen

Department of Wind Energy, Section of Composite Mechanics and Structures, Technical University of Denmark, Risø Campus, DK-4000 Roskilde, Denmark

E-mail: gout@dtu.dk; bsqr@dtu.dk

Abstract. The cyclic growth rate of delaminations initiated from ply-drops under tensile cyclic load was experimentally measured using a test specimen geometry, with internal ply-drops, that allows stable crack growth. It is found that the fatigue delaminations grow at a constant rate.

1. Introduction

In many applications, the external geometrical shape of lightweight composite structures (e.g. aerofoil shape for a wind turbine rotor blade) is specified by aerodynamic considerations. An additional requirement is usually to minimize the weight of the structure but without significant decrease in the structural stiffness and strength or fatigue life [1, 2]. To achieve these requirements, the thickness of the load-carrying composite laminates is varied along the length of the structure by terminating or dropping off plies (ply-drops) at specific locations. Ply-drops result in material and geometrical discontinuities which induces stress concentrations. As a result, delaminations can potentially start at these locations and propagate [3], reducing the load carrying capability of the composite laminates/structures.

Due to the significant effect on the load carrying capability of composite structures, the initiation and propagation of delaminations in composites with ply-drops (tapered composites) has been been the subject of research for many years [4–10]. Despite, however, the large amount of research on the effect of ply-drops on the structural strength, challenges still exist a) in accurate structural predictions of *e.g.* the safe lifetime of composite structures with ply-drops and b) in designing more damage tolerant tapered composites by investigating the parameters that influence the onset and growth of the delaminations.

The present work, is an ongoing effort towards the development of a fracture mechanics based methodology to predict the initiation and propagation of delaminations starting from ply-drop locations. Such an approach can potentially allow the development of composites with ply-drops that possess high damage tolerance by engineering the interface properties between the different plies. A high damage tolerance is a prerequisite in damage tolerance design of composite structures where damage is accepted as long it can be detected and its growth will be stable and can be accurately predicted [11]. To identify the failure mechanisms, simplified composite specimens with ply-drops are manufactured and tested under cyclic loading. The present work focuses on the measurement of the delamination growth rates under cyclic loading at different applied stress levels. The initiation of delaminations and the delamination growth rate change when the thickness of the specimen changes (due to the presence of ply-drops) are not discussed.

2. Experimental Details

2.1. Materials

A nearly unidirectional glass fibre fabric (2400 tex) with area weight equal to 1134 g/m^2 was obtained from Saertex. The fibre diameter was 17 μm. The area weight of the 90° fibres was 54 g/m^2 and the tex value was 68. The fibre diameter was 10 μm. The sewing threads were made of polyester fibres (110 dtex) with an area weight equal to 12 g/m^2. The stitching type was tricot-warp. A biaxial fabric ($\pm 45^o$) with a total area weight equal to 720 g/m^2 was provided by Ahlstrom. The area weight of the $\pm 45^o$ fibres (600 tex) was 592 g/m^2, of the 90° fibres (200) 19 g/m^2, of chopped strand mat ($<$ 30 tex) 592 g/m^2 and of the stitch yarns (7.6 tex) 10 g/m^2. The diameter of the glass fibres in the $\pm 45^o$ and 90° layers was equal to 16 μm. In the chopped strand mat, the fibres had a diameter equal to 12 μm. The stitching type was tricot-warp. The epoxy matrix material, Araldite LY 1568, and the hardener, Aradur 3489 CH, were provided by Huntsman.

2.2. Test specimen

Ten unidirectional fabrics/plies were placed on the vacuum infusion table as shown in Fig. 1a. Then, four unidirectional plies, each with different length, were placed one by one on top of the ten unidirectional plies. The distance between the ply-drops is shown in Fig. 2. It should be mentioned that all the unidirectional plies were placed with the 90° fibres face down. A continuous biax fabric, with a length equal to the length of the ten unidirectional plies, was afterwards placed on top of the four unidirectional plies and the epoxy resin was infused (see Fig. 1b). The curing cycle, after vacuum infusion, was 19 hrs at 40^oC followed by 5 hrs at 75^oC. After curing, the thickness of each unidirectional (UD) layer was approximately 0.9 mm and the thickness of the biax layer was approximately 0.6 mm.

Figure 1. Ply layup to manufacture the composite specimens with ply-drops by vacuum infusion.

The ply-drop specimens were cut-out from the vacuum infused plate (Fig. 1b). The specimen has had a total length equal to 630 mm and a width equal to 30 mm (see Fig. 2). The distance between ply-drops 1 and 2 was approximately 10 mm, the distance between ply-drops 2 and 3 was equal to 290 mm and between ply-drops 3 and 4 equal to 50 mm. Glass fibre epoxy composite tabs (height equal to 4 mm) were glued on the right-end of the ply-drop specimens. The dimensions of the ply-drop specimens *e.g.* length, height, distance between ply-drop 1 and tab were designed to minimise bending of the specimens during testing. The bending ratio in the thin section of the ply-drop specimens (at x_1 =60 mm, see Fig. 2) was less than 0.03 under static tensile load. The bending ratio was calculated from the strains measured by two strain gauges on opposite faces of the ply-drop specimens.

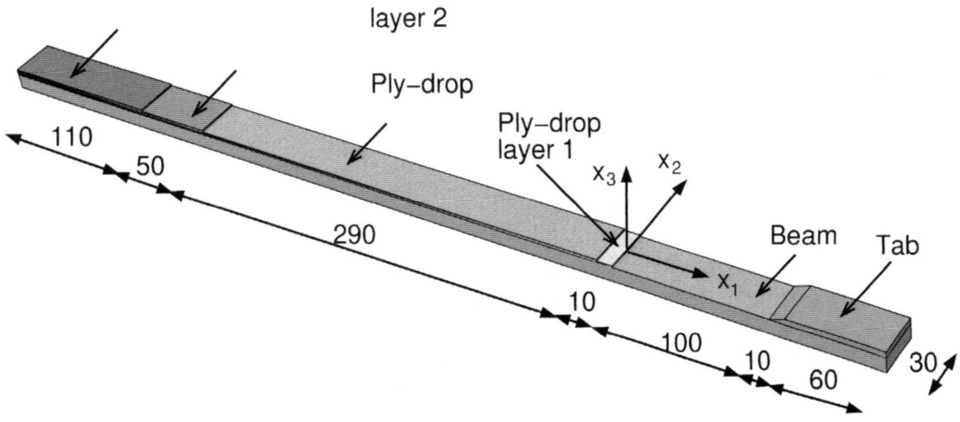

Figure 2. Specimen dimensions and position of the ply-drops. The biax layer that is on top is not shown (all dimensions in mm).

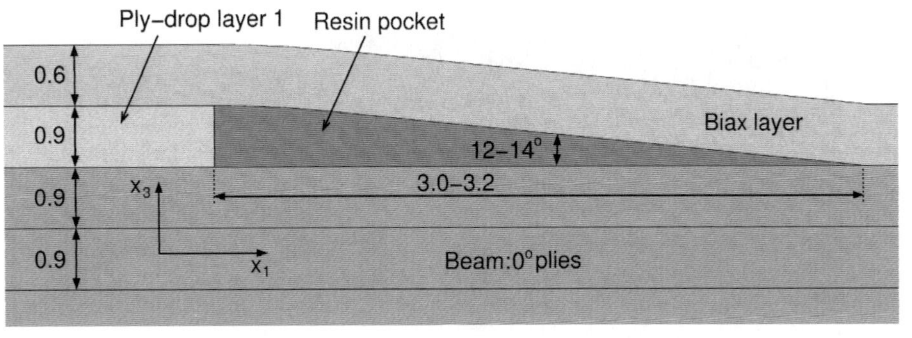

Figure 3. Schematic illustration of the material and geometric discontinuity at the ply-drop location (all dimensions in mm).

As mentioned above, a continuous biax layer was placed on top of the four unidirectional plies. Fig. 3 shows schematically the geometry at the location of ply-drop 1. As can be seen at the ply-drop location there exist both a material and a geometric discontinuity. It should be noted, that no effort was made to control the geometry of the ply-drop *e.g.* length of the resin pocket, and steepness of the biax layer. The shape of the geometric discontinuities was dictated by the vacuum infusion process with the variation from specimen to specimen not being significant based on SEM observations.

2.3. Cyclic loading

Two ply-drop specimens were first tested under monotonic static tensile loading in order to measure the stiffness of the specimens, measure the bending ratio and find the stress level at which a delamination initiates at the location of the ply-drop 1. The tests were conducted on an Instron servo-hydraulic testing machine with a 250 KN load cell at displacement rate of 0.0166 mm/min. The load, the position of the cross-head and the strain-gauge signals were acquired at a sampling rate of 10 Hz.

The cyclic tests were performed between fixed stress values (lower than the stress level at which delamination initiates under static loading) at a frequency of 3 Hz. The R-ratio was equal to 0.1. The cyclic tests were performed on the same machine as the static tests. The position of the cross-head and maximum and minimum loads for each cycle were recorded. Photographs of the specimen were obtained at a predefined number of cycles in order to measure the fatigue delamination crack length. By default 1 photograph per cycle was acquired for the first 100 cycles, 1 photograph for every 10 cycles in the range 100 to 1000 cycles, 1 photograph for every 100 cycles in the range 1000 to 10000 cycles etc. However, it was possible to change the acquisition rate during the experiments if necessary. It was find that it was easier to identify the delamination front by taking photographs at the $x_1 - x_2$ plane (see Fig. 2) instead at the $x_1 - x_3$ plane. Prior to each cyclic test, a photograph with a scale bar on the surface of the specimen (x_1 − x_2 plane) was taken. The photographs from the cyclic tests were then calibrated using the ImageJ software [12] and the delamination lengths were measured. Several crack growth measurements at different maximum applied load could be obtained from a single specimen when the delamination was in the region between ply-drops 2 and 3 (see Fig. 2).

3. Results and Discussion

Fig. 4 shows a series of photographs taken from a ply-drop specimen tested under tensile cyclic loading.

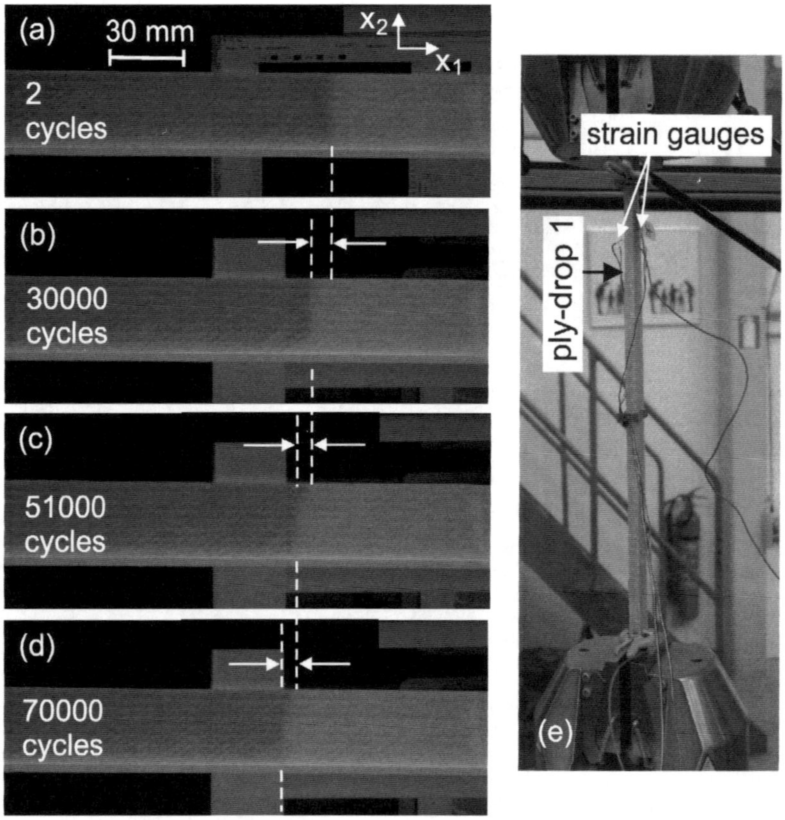

Figure 4. Measurement of fatigue delamination crack growth: a) 2 cycles, b) 30000 cycles, c) 51000 cycles, d) 70000 cycles. e) Specimen mounted on the hydraulic testing machine. The maximum applied stress, σ_{max}, is 142.9 MPa, R=0.1 and frequency, v, is 3 Hz.

As mentioned above, the current work focuses on the measurement on the fatigue delamination growth rates when the delamination is between ply-drops 2 and 3 (see Fig. 2). The maximum applied stress, σ_{max}, is 0.75 of the stress, σ_u, where the delamination initiates between the first ply-drop layer and the 0° plies (beam, see Fig. 3) under static tensile loading. σ_u is in the range of 205 to 250 MPa. As Fig. 4a shows, the delamination has already grown (under cyclic loading) to the constant thickness region between ply-drops 2 and 3 and delamination growth rate is measured from this point. Fig. 4b-d shows the delamination extension at 30000, 51000, and 70000 cycles, respectively. It can be seen that the crack front is nearly constant across the specimen width (x_2 direction) and clearly visible allowing accurate measurement of the delamination growth with the number of cycles. It should be mentioned that the delamination had initiated between the first ply-drop layer and the 0° plies (beam) and had remained at this plane when it reached the region between ply-drops 2 and 3 (Fig. 4) and it stays in this plane Fig. 4b-d.

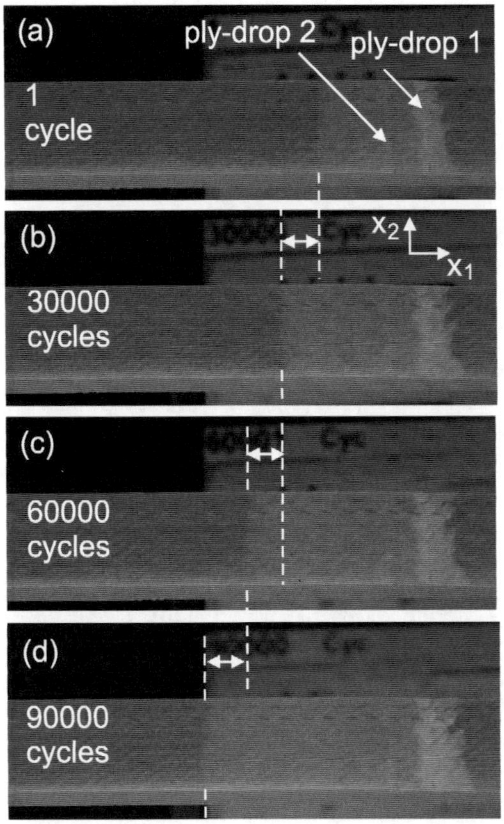

Figure 5. Measurement of fatigue delamination crack growth: a) 1 cycle, b) 30000 cycles, c) 60000 cycles, d) 90000 cycles. The maximum applied stress, σ_{max}, is 146.4 MPa, R=0.1 and frequency, v, is 3 Hz.

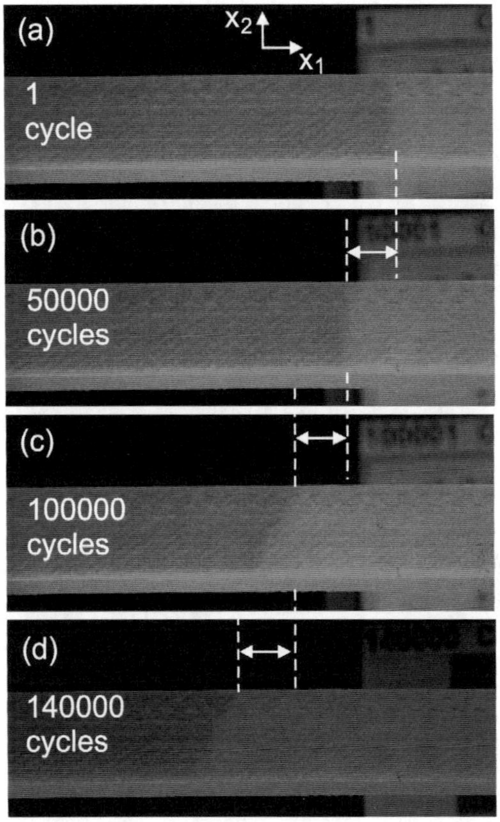

Figure 6. Measurement of fatigue delamination crack growth: a) 1 cycle, b) 50000 cycles, c) 100000 cycles, d) 140000 cycles. The maximum applied stress, σ_{max}, is 136.7 MPa, R=0.1 and frequency, v, is 3 Hz.

Fig. 5 shows a series of photographs, at different number of cycles, from a another specimen. As in Fig. 4, the delamination has already grown in the region between ply-drops 2 and 3. The maximum applied stress is equal to 0.77 σ_u. As mentioned above, several measurements of the delamination fatigue growth rates could be made in the region between ply-drops 2 and 3

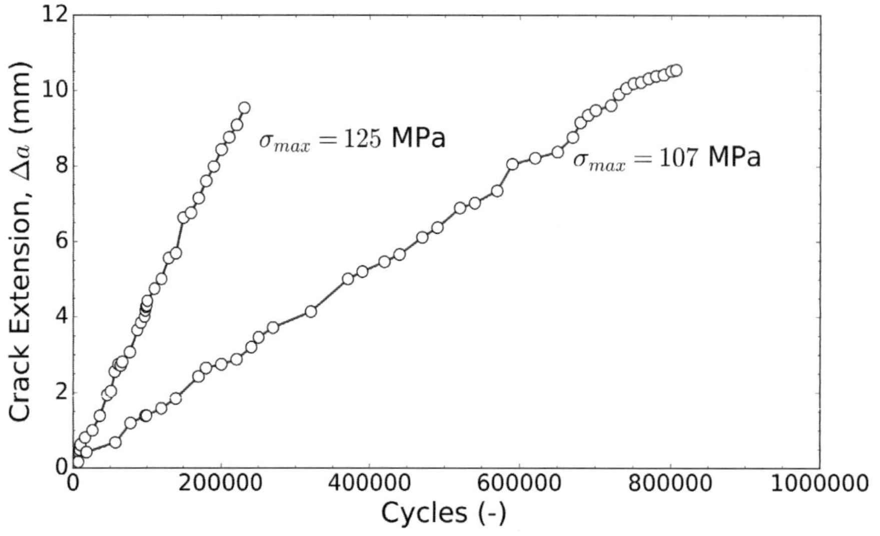

Figure 7. Measured cyclic delamination (crack) growth as a function of the number of cycles for two different maximum applied stress levels.

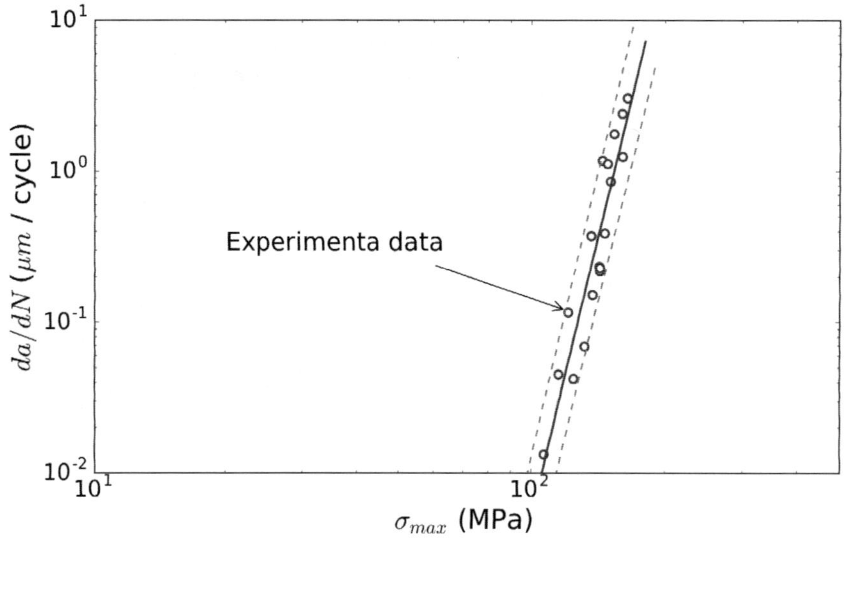

Figure 8. Cyclic delamination (crack) growth rate as function of the maximum stress level, σ_{max}, normalised with the Young's modulus of the unidirectional plies, E_{UD}.

since this region is relatively long. After 90000 cycles (Fig. 5d), the test was interrupted, the maximum applied stress was set equal to 0.71 σ_u and thereby a new cyclic test was started. A series of photographs are shown in Fig. 6 at 1 cycle, 50000 cycles, 100000 cycles and 140000 cycles, respectively. Unlike the specimen shown in Figs. 4 and 5, for this specimen the crack front is not perpendicular to the x_1 direction. Such a behaviour was observed in few cases but usually the crack front after a number of cycles became again perpendicular to x_1.

From measurements such as shown in Figs. 4, 5 and 6, the fatigue crack growth rate can be derived. Two examples are shown in Fig. 7 for σ_{max} equal to 0.66 and 0.56 of σ_u, respectively. It can be seen that the delamination (crack) growth rate is constant and as expected it increases with increasing the maximum applied stress. It should be noted that delamination growth rate is constant for large number of cycles (larger than 800000 cycles) showing that the ply-drop specimen developed is a steady-state specimen allowing accurate measurement of the cyclic delamination growth rate.

The cyclic delamination growth rate can be calculated from a linear fit of the experimental data shown in Fig. 7. The results are shown in Fig. 8 and follow a straight line in a log-log plot. Upper and lower bounds estimates are drawn.

As mentioned in Section 2, during manufacturing the unidirectional plies were placed with the 90° fibres on the bottom face. This probably creates a weak interface by preventing fibre bridging between the plies. Examination of the crack surfaces after the fatigue tests showed that there were no fibres bridging the crack faces and thus the cyclic crack growth can be treated using Linear Elastic Fracture Mechanics (LEFM) and Paris Law to predict the cyclic delamination growth rate.

4. Conclusions

A test specimen with multiple ply-drops was designed to allow stable cyclic delamination crack growth. The specimen was used to measure the tension-tension delamination fatigue crack growth rate initiated from a ply-drop. It was shown that delamination growth rate is constant when the delamination grows in a region of constant thickness and can be described by Paris Law.

Acknowledgments

This work has received funding from the European Unions Horizon 2020 Research and Innovation Programme under Grant Agreement No 761072 (DACOMAT). Leonardo Di Crescenzo is acknowledged for manufacturing the test specimens and Hans Christian Jensen for technical assistance in the mechanical testing.

References

[1] Gan K W, Allegri G and Hallett S R 2016 *Materials and Design* **108** 570-80
[2] Helmy S and Hoa S V 2014 *Composites Science and Technology* **102** 10-19
[3] He K, Hoa S V and Ganesan R 2000 *Composites Science and Technology* **60** 2643-57
[4] Fish J C and Lee S W 1989 *Engineering Fracture Mechanics* **34** 43-54
[5] Murri G B, OBrien T K and Salpekar S A 1993 *Journal of the American Helicopter Society* **38** 29-37
[6] Wisnom M R, Jones M I and Cui, W 1995 *AIAA Journal* **33** 911-18
[7] Murri G B, O'Brien T K and Rousseau C Q 1998 *Journal of the American Helicopter Society* **43** 146-155
[8] Thomsen O T, Rits W and Eaton D C G 1996 *Composites Science and Technology* **56** 407-22
[9] Thomsen O T and Mortensen F 1999 *Composites Science and Technology* **59** 1213-26
[10] Murri G B and Schaff J R 2006 *Composites Science and Technology* **66** 499-508
[11] McGugan M, Pereira M, Sørensen B F, Toftegaard, H and Branner, K. 2015 *Philosophical Transactions A* **373** 20140077
[12] Rueden C T, Schindelin J, Hiner M C, DeZonia B E, Walter A E, Arena E T and Eliceiri K W 2017 *BMC Bioinformatics* **18** 529

IOP Conf. Series: Materials Science and Engineering **388** (2018) 012011 doi:10.1088/1757-899X/388/1/012011

Damage mechanisms of directly bonded carbon fibre reinforced thermoplastics and aluminium with nanostructured surface

K M Jespersen[1,2], J C Chung[2], K Okamoto[2], H Abe[2], A Hosoi[2,*], and H Kawada[2,3]

[1]Kanagawa Institute of Industrial Science and Technology (KISTEC), 3-2-1 Sakado, Takatsu-ku, Kawasaki-shi, Kanagawa, 213-0012, Japan

[2]Department of Applied Mechanics and Aerospace Engineering, Waseda University, 3-4-1 Okubo, Shinjuku-ku, Tokyo 169-8555, Japan

[3]Kagami Memorial Research Institute for Materials Science and Technology, 2-8-26 Nishiwaseda, Shinjuku-ku, Tokyo 169-0051, Japan

E-mail: `hosoi@waseda.jp`*

Abstract. The current study presents a direct bonding method making it possible to obtain a high interface strength of aluminium joined to carbon fibre reinforced thermoplastic (CFRTP) plates by hot pressing. This is achieved by subjecting the aluminium to a combination of anodising, etching, and silane-coupling treatments prior to bonding. Different types of aluminium are subjected to different treatments and bonded to different types of CFRTP laminates. The effect of the surface structure on the static bonding strength and fatigue life measured by single-lap testing is compared and discussed. The bonding strength is found to be highly dependent on the anodisation conditions along with the type of thermoplastic resin.

1. Introduction

In recent years, fibre reinforced plastics are increasingly replacing conventional materials such as steel and aluminium due to their high specific strength and stiffness along with their great fatigue resistance. Particularly in the aerospace and automotive industries, carbon fibre reinforced thermoplastics (CFRTPs) are of interest due to their great formability and productivity, making them suitable for mass production. Furthermore, the possibility of recycling CFRTPs, which is difficult for the thermoset-type counterparts, makes them a suitable material choice for a sustainable future.

Nevertheless, it is rarely feasible to build an entire structure of CFRTPs alone, and thus it is necessary to join CFRTPs to other materials. However, bonding between dissimilar materials contains several challenges. Mechanical joining e.g. by bolts increases the weight of the structure and hole drilling causes stress concentrations. Dissimilar materials can be joined by adhesives, however the adhesive has to be compatible with both materials and generally do not bond well to thermoplastic materials. Furthermore, there are issues such as long curing times and degradation of the bond strength over time. Therefore, alternative bonding methods to join CFRTPs to other materials have been proposed. Some examples are ultrasonic welding [1, 2], friction spot joining [3, 4], laser direct joining [5], induction heating [6], friction lap joining

[7] methods among others. However, these methods require expensive equipment, there is a limitation to the thickness, and they can generally only be applied to small areas.

In response to these challenges, a novel method to directly bond carbon fibre reinforced thermoplastics and aluminium was proposed by the authors [8, 9] (patent: Japanese patent application no. 2016-197812 and 2018-53610). The proposed method is based on carrying out an initial surface treatment by anodising and etching of the aluminium surface in a way that results in either a nano-porous or nano-spike structured surface. The method to establish a nano-spike structure on the surface of A1051 aluminium was proposed by Yu et al [10] for light absorption purposes. This principle was adopted for enhancing the bonding strength with the idea of mimicking easy-to-bond structures found in nature such as pollen, virus molecules, and the feet of a gecko. By performing anodisation of the aluminium surface and subsequently subjecting it to a silane coupling treatment, it was possible to obtain a high bonding strength between the modified aluminium and CFRTPs by hot-pressing [9].

The current study applies the above mentioned method to fabricate nano-porous and nano-spike microstructures on two types of aluminium alloys (A1050 and A5052) and to bond them to CFRTP by hot pressing. The different microstructures are compared to one another and discussed in relation to the bonding strengths measured from single-lap testing. Furthermore, single-lap fatigue testing is carried out for two chosen test cases to compare the effect of the nano-spike structure on the fatigue life.

2. Experimental method

This section will outline the materials and method used to manufacture the nanostructures on the aluminium surface along with the method to join the CFRTP and aluminium.

2.1. Materials

Toray T300B-3K woven fabrics with an average fibre diameter of $7\mu m$ and a $[(0/90)_9]_T$ layup were used for the CFRTPs. Two types of CFRTPs were made using Polyamide-6 (PA6) and Polyamide-MXD6 (MXD6) as matrix (also known as Nylon-6 and Nylon-MXD6) having a melting point of 225°C and 243°C, respectively. The fibre volume fraction of the final composites was $V_f \approx 50\%$. Two types of aluminium (A1050 and A5052) were considered and subjected to different types of surface treatments resulting in a range of different nanostructures on the aluminium surfaces.

2.2. Aluminium surface treatment

Fig. 1 shows a schematic of the process to fabricate a nanostructure on the aluminium surface. Initially the aluminium specimen is degreased by ultrasonic cleaning while submerged in acetone for 10 mins. The aluminium specimen is then anodised in an acidic solution (Fig. 1a) using a carbon cathode and then etched in an acid (Fig. 1b). This process is generally carried out twice. Depending on the anodising and etching parameters along with the chemical solution type, different surface structures can be obtained. Fig. 1c shows an example of a fabricated nano-spike structure on the A5052 aluminium alloy surface.

Considerable experimental work was carried out on finding the optimal anodisation and etching conditions, however the current study will only include the conditions that resulted in the highest interface strengths. The chemical solutions used for the surface treatments are listed in Table 1. The two different aluminium types (A1050 and A5052) were each subjected to two different series of chemical treatments, as listed by a-d below. For each aluminium type, a nano-spike and a nano-porous type of structure was obtained resulting in four different aluminium surface structures as shown in Fig. 2.

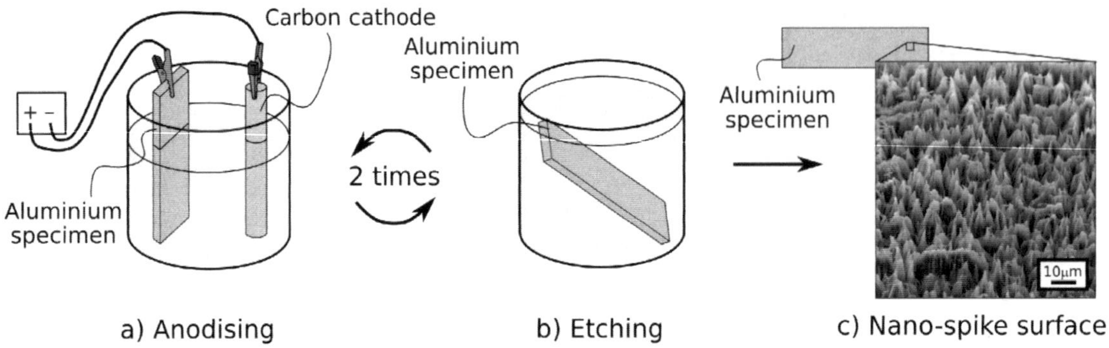

Figure 1. Schematic of the (a) anodising and (b) etching surface treatment approach to generate a (c) nano-spike structure on the aluminium. The shown example is on the A5052 aluminium surface.

Table 1. Overview of chemical surface treatments

ID	Chemical Solution	Current [V]	T [°C]
Anod-A	0.5M Sulfuric Acid	25	3
Anod-B	0.1M Phosphoric Acid	160	5
Anod-C	2wt% Citric Acid + 2wt% Ethylene Glycol (2:1)	400	10
Etching-A	6wt% Phosphoric Acid + 1.5wt% Chromic Acid	-	63
Etching-B	6wt% Phosphoric Acid + 1.8wt% Chromic Acid	-	63
Etching-C	0.5M Phosphoric Acid	-	RT

(a) **A1050 Nano-spike:**
Anod-C (9h) ⟶ Etching-A (48min)

(b) **A1050 Nano-porous:**
Anod-B (60min) ⟶ Etching-B (60min) ⟶ Anod-B (60min) ⟶ Etching-C (30min)

(c) **A5052 Nano-spike:**
Anod-A (60min) ⟶ Etching-A (60min) ⟶ Anod-A (15min) ⟶ Etching-C (50min)

(d) **A5052 Nano-porous:**
Anod-C (9h) ⟶ Etching-B (60min) ⟶ Anod-C (9h) ⟶ Etching-B (20min)

As seen from Fig. 2 varying the anodising treatment conditions and even the type of aluminium has a great influence on the nanostructure developed on the surface of the aluminium. For A1050, a nano-spike structure (Fig. 2a) can be produced by 400V anodising (Table 1) as also reported by Yu et al [10], however in the case of A5051 the 400V treatment results in a less spiky structure (Fig. 2c) and instead nano-spikes (Fig. 2d) can be produced at 25V anodisation (Table 1). After the anodisation and etching treatments, some of the specimens were subjected to a silane coupling treatment using the coupling agent KBE-9007 by Shi-Etsu Silicones. Other types of silane coupling agents were also tested, but KBE-9007 was found to provide the highest bonding strength.

2.3. Test specimens

The aluminium with different surface structures were bonded to the CFRTPs by hot pressing to form single-lap specimens in accordance to the Japanese standard JIS K6850. The specimen geometry is shown in Fig. 3. Hot pressing was carried out at 290°C for the A1050 cases and at 300°C for the A5052 under a constant load of ~0.13 MPa for 3 minutes. This was followed

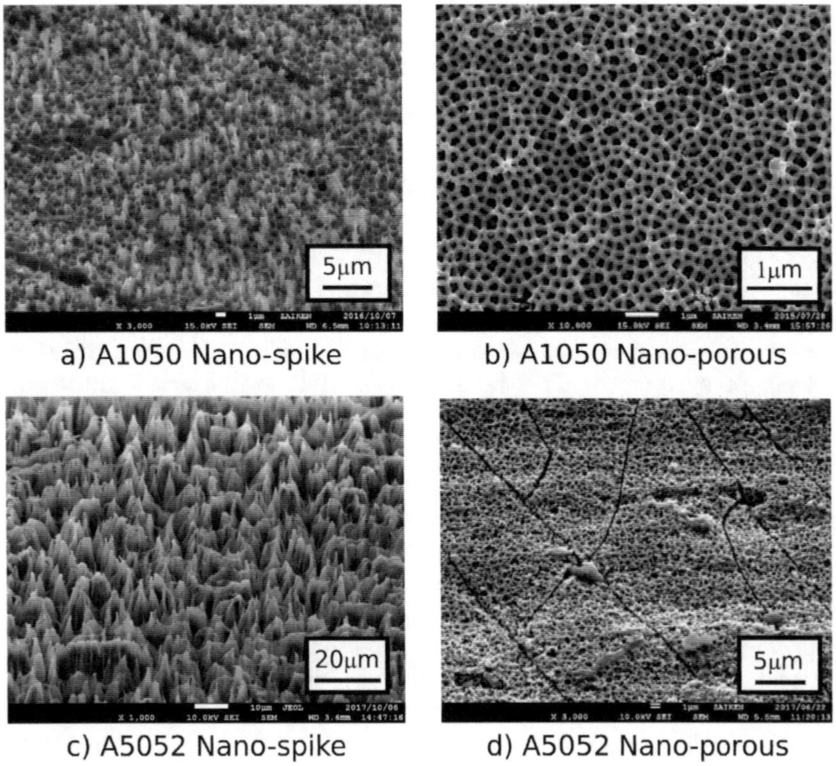

Figure 2. Images of the fabricated nanostructures used for the single-lap tests.

by natural cooling while the load was kept constant until the temperature decreased to below 70°C to make sure that the CFRTP matrix had solidified.

Two types of aluminium with 5 different surface structures (nano-porous, nano-spike, silane + as-rolled, silane + nano-porous, silane + nano-spike) were bonded to CFRTPs with PA6 matrices. Furthermore, A5052 aluminium alloys subjected to three different surface treatments (silane + as-rolled, silane + nano-porous, silane + nano-spike) were bonded to CFRTPs with MXD6 matrices, giving a total of 13 different combinations (see also table 2 later).

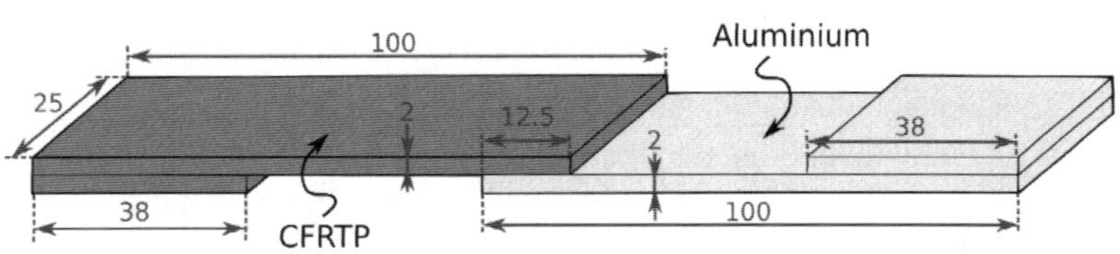

Figure 3. Single lap test geometry in accordance to the Japanese JIS K6850 standard.

2.4. Single-lap static tensile tests

Single-lap specimens were tested in static tension on a hydraulic testing machine and the average bonding strength was evaluated from the fracture load and the overlap area by $\tau = \frac{P}{A}$ according to the Japanese standard JIS K6850. Three tests were carried out for each condition and all tests were carried out until failure. For A5052/PA6 and A5052/MDX6 the tests were carried out with an overlap length of 12.5mm. However, for the A1050/PA6 the specimens subjected

IOP Conf. Series: Materials Science and Engineering **388** (2018) 012011 doi:10.1088/1757-899X/388/1/012011

to silane treatments broke in the aluminium part instead of at the interface and therefore the overlap was modified to 5mm to be able to obtain interface fracture. This should be taken into account when comparing this data series to the other two.

2.5. Single-lap fatigue tests

Fatigue tests were carried out for two cases both subjected to a silane coupling treatment namely as-rolled A5052/PA6 (Si-AR) and nano-spike A5052/PA6 (Si-NS). The same geometry as for the static tests (Fig. 3) with an overlap length of 12.5mm were used for the fatigue tests. The tests were carried out in load control with a load ratio of $R = 0.1$, and at a test frequency of 5Hz. A Total of 12 A5052/PA6 (Si-AR) specimens and 2 A5052/PA6 (Si-NS) were tested. For the Si-AR case, tests were carried out at loads corresponding to 80%, 70%, 60%, 50%, 40%, 30% of the static shear strength (τ_s^{AR}). For the Si-NS case tests were carried out at 40% of τ_s^{AR}.

3. Results and discussion

3.1. Bonding strength comparison

The average shear strengths obtained from single-lap testing are summarised in Table 2. Tests were also carried out for as-rolled aluminium only subjected to degreasing and then joined to CFRTPs, however the interface strength was practically zero and has therefore not been included in Table 2. It is seen that both the nano-porous and nano-spike structures increase the bonding strength, which has previously been discussed to be a result of the anchoring effect of the nanostructure penetrating into the CFRTP surface [9]. All the specimens subsequently subjected to a silane coupling treatment show significantly improved interface strengths as a result of the improved chemical bonding between the surfaces. The A5052/PA6 case generally showed lower bonding strengths than for A1051/PA6, however as previously mentioned it should be noted that the a shorter overlap length was used for A1051/PA6 might affect the results. Nevertheless, the nano-spike structured A1050 type aluminium resulted in higher bonding strength than for nano-porous structure whereas the A5052/PA6 nano-spike case showed a significantly lower bonding strength than the nano-porous structure.

Table 2. Comparison of average bonding strength for different conditions (*carried out with an overlap of 5mm). The numbers in parenthesis are the standard deviations.

Bonded materials	Nano-porous	Nano-spikes	Silane + As-rolled	Silane + Nano-porous	Silane + Nano-spikes
A1050/PA6*	5.7 (2.0)	11.0 (4.1)	24.4 (2.8)	23.9 (1.2)	24.9 (1.5)
A5052/PA6	17.8	13.8	17.4	20.6	14.7
A5052/MXD6	-	-	23.8	24.8	21.7

A detailed discussion on the effect of the nano-spike structure on the A1051 aluminium alloy bonded to PA6 is given in [8] where nano-spike structures of two different heights were also compared. Here it was found that the higher nano-spikes gave higher interface strength. This was discussed to be a result of the increased anchoring effect from the higher spikes. However, for the nano-spikes fabricated on the A5052 aluminium a significantly lower interface strength was obtained despite the spikes being high. Fig. 4 shows high resolution FE-SEM images of the nano-spikes on the A1050 (Fig. 4a) and A5052 (Fig. 4b) aluminium surfaces where it is seen that the overall shape of the spikes is significantly different. Hence, it is likely that the spikes in Fig. 4a simply are stronger than those in Fig. 4b. For the MXD6 matrix case, the overall bonding strengths were higher than for to PA6, which is believed to mainly be due to the higher strength of the MXD6 matrix. Furthermore, the difference between the bonding strength for the nano-porous and nano-spike cases is less visible.

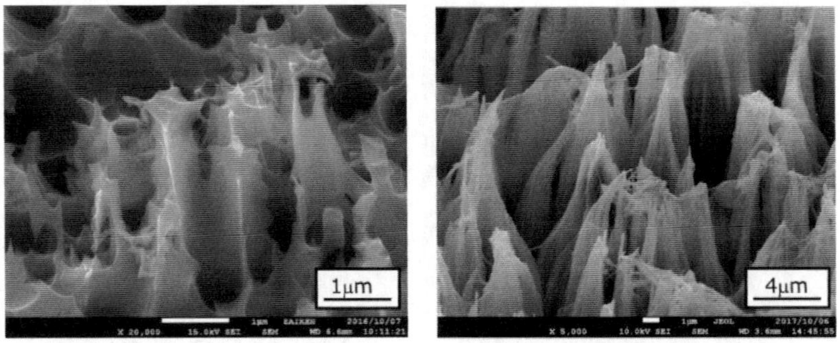

Figure 4. High resolution FE-SEM images of the two types of nano-spike structures

Fig. 5 shows a comparison between high resolution FE-SEM images of the nano-porous structured A1050 and A5052 aluminium cases. Although both up till now have been referred to as nano-porous, it is seen that the structure for A5052 is slightly spiky and might actually be more of a mixture between the A1051 nano-spike (Fig. 4a) and the nano-porous (Fig. 5a) structures. It is possible to be one of the reasons for the difference between the interface strengths of the A1050 and A5052 nano-porous cases.

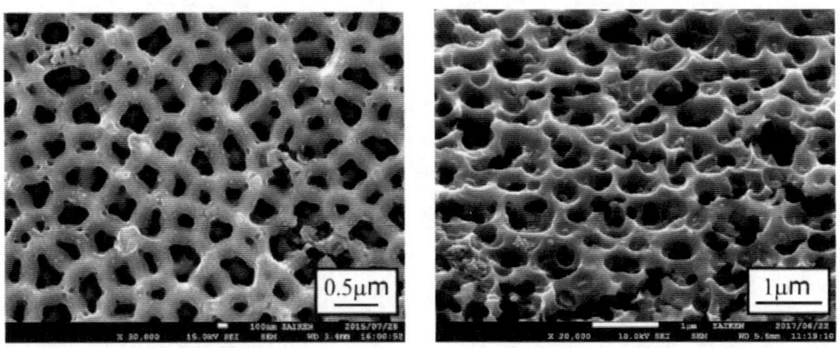

Figure 5. High resolution FE-SEM images of the two types of nano-porous structures

3.2. Experimentally observed damage mechanisms
Fig. 6 shows examples of the fracture surfaces observed by FE-SEM of the A1051/PA6 case on the A1051 side (top images) and the corresponding PA6 side (bottom images) for the silane coupling treated as-rolled and nano-porous cases. Although similar bonding strengths were observed for these cases (Table 2), the fracture surfaces look quite different. It is seen that the nanostructure results in a more ductile fracture with a high amount of matrix remaining on the fracture surface.

3.3. Fatigue testing results
Fig. 7 shows the S-N curve obtained from the fatigue tests for the single-lap specimens. Fig. 7a shows the absolute applied maximum load during the fatigue tests as a function of the cycles and Fig. 7b shows the applied average shear load normalised with respect to the static failure shear load of the silane treated as-rolled case (τ_s^{AR}) to include the variation of the bonding

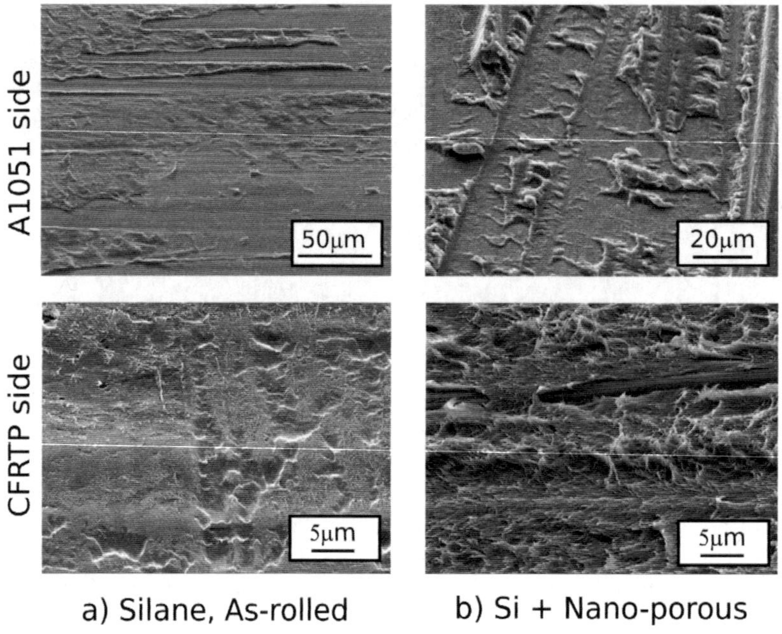

Figure 6. Example of fracture surfaces for A1051/PA6

surface area present in the specimens. As is often the case for fatigue testing, a high degree of scatter is observed in the fatigue lifetime. Comparing Fig. 7a and 7b, it seems that including the variation in the bonding area does not affect the scatter of the data. The grey point is an Si-AR specimen, which showed significant bending in the bonding region. Reverse bending of single-lap joints has been found to improve the bond strength up to a certain angle [11] and furthermore non-flat surfaces have been found to lead to longer fatigue lifetimes. This could explain the significantly longer fatigue life of this particular specimen. Furthermore, as even a few degrees of reverse bending in the specimen can significantly affect the strength [11], it could be one of the causes of the general scatter in the data. However, the degree of bending in the specimens would have to be measured to confirm this.

Nevertheless, it is seen from Fig. 7 that the fatigue lifetime of the A5052/PA6 (Si-NS) case is in the upper part of the scatter band of A5052/PA6 (Si-AR) despite the static strength of A5052/PA6 (Si-NS) actually being lower than A5052/PA6 (Si-AR) (see Table 2 previously). This tendency might be related to the ductile fracture as was observed for the A1051/PA6 nano-porous case (Fig. 6). However, further data and additional observations of the fracture surfaces are necessary to confirm this matter.

4. Conclusion

In the current study a direct bonding method to join aluminium and CFRTP by fabricating a nano-porous or nano-spike structure on the aluminium surface was presented. Two types of aluminium subjected to various surface treatment conditions were bonded to two types of CFRTPs and tested by single-lap testing. The presence of the microstructure was found to improve the bonding properties compared to as-rolled aluminium. By applying a silane coupling treatment, the interface strength was improved but the effect of the microstructure became less visible on the measured bond strength. However, it was seen from fracture surface observations that the surface modified aluminium case experienced ductility to a significantly higher extent than the as-rolled case. Fatigue tests were carried out for the A5052/PA6 for silane coupling

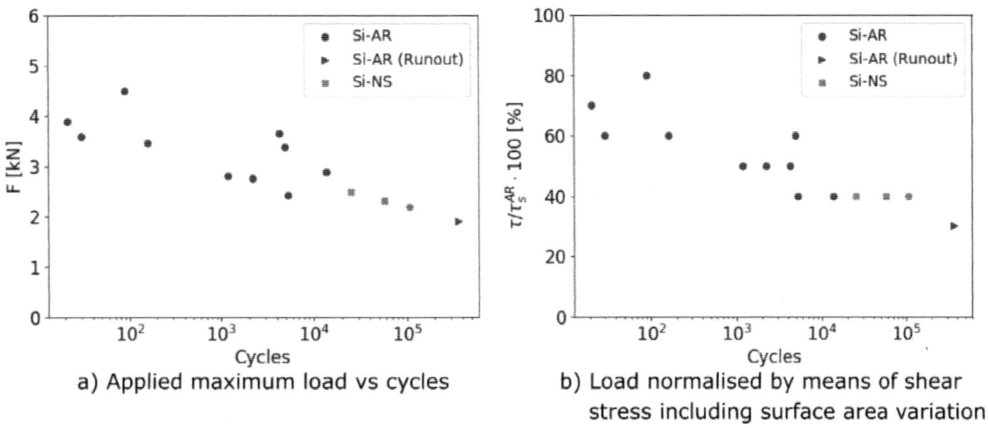

Figure 7. S-N Curve for A5051/PA6 single lap test specimens tested in fatigue (a) for the maximum fatigue load and (b) the normalised average shear load including the effect of variation in the surface area.

treated as-rolled and nano-spike structured A5052 aluminium bonded to a CFRTP with a PA6 matrix. Despite large scatter in the data, the results showed a tendency towards longer lifetimes of the nano-spike case. However, additional data is required to confirm this matter.

Acknowledgments

This paper is based on results obtained from a project of Kanagawa Institute of Industrial Science and Technology (KISTEC), and was supported by JSPS KAKENHI Grant Numbers 18H01342. We would also like to show our gratitude to Mr. Arai from SPIC Corporation and Mr. Maruo from Mitsubishi Engineering-Plastics Corporation for supplying the MXD6 CFRTP used in this study.

References

[1] Balle F, Wagner G and Eifler D 2009 *Advanced Engineering Materials* **11** 35–39 ISSN 15272648
[2] Balle F and Eifler D 2012 *Materialwissenschaft und Werkstofftechnik* **43** 286–292 ISSN 09335137
[3] Goushegir S M, dos Santos J F and Amancio-Filho S T 2015 *Materials and Design* **83** 431–442 ISSN 18734197 URL `http://dx.doi.org/10.1016/j.matdes.2015.06.044`
[4] André N M, Goushegir S M, Dos Santos J F, Canto L B and Amancio-Filho S T 2016 *Composites Part B: Engineering* **94** 197–208 ISSN 13598368 (*Preprint* `arXiv:1011.1669v3`)
[5] Jung K W, Kawahito Y, Takahashi M and Katayama S 2013 *Materials and Design* **47** 179–188 ISSN 18734197 URL `http://dx.doi.org/10.1016/j.matdes.2012.12.015`
[6] Mitschang P, Velthuis R, Emrich S and Kopnarski M 2009 *Journal of Thermoplastic Composite Materials* **22** 767–801 ISSN 08927057
[7] Nagatsuka K, Yoshida S, Tsuchiya A and Nakata K 2015 *Composites Part B: Engineering* **73** 82–88 ISSN 13598368 URL `http://dx.doi.org/10.1016/j.compositesb.2014.12.029`
[8] Abe H, Chung J C, Mori T, Hosoi A and Kawada H 2018 Effect of nanospike structures in direct bonding between aluminum plates and CFRTP composites on adhesion strength properties (Submitted)
[9] Chung J C, Okamoto K, Abe H, Hosoi A and Kawada H 2017 Effects of interfacial nanostructures on adhesive strength properties between a5052 aluminum alloys and cfrtp laminates 9th Japanese Conference on Composite Materials
[10] Yu R, Ching K L, Lin Q, Leung S F, Arcrossito D and Fan Z 2011 *ACS Nano* **5** 9291–9298 ISSN 19360851
[11] You M, Li Z, ling Zheng X, Yu S, yan Li G and xin Sun D 2009 *International Journal of Adhesion and Adhesives* **29** 280–285 ISSN 01437496

Manufacturing curing cycles influence on local fracture toughness properties at micro and macro scale

F Lahuerta and S Raijmaekers

Knowledge Centre WMC. Kluisgat 5, 1771 MV Wieringerwerf, The Netherlands

E-mail: f.lahuerta@wmc.eu

Abstract. Ultimate strengths and fatigue properties (S-N curves) of thick laminates vary through the thickness showing unexpected values in comparison with thin laminate. This is related to curing cycles, residual strains and manufacturing thermal history. These observations were reported using the sub-laminate technique where thick laminates through-thickness mechanical properties can be described according to a prescribed curing cycle. A similar behaviour might be also expected at a micro level for the fibre/matrix interface properties. In the present work, the curing cycles influence is studied for fracture toughness properties at a macro and micro level. For this purpose, standard test coupons are obtained with the sub-laminate technique and single fibre fragmentation tests are carried out at different curing cycles.

1. Introduction

Epoxy glass fibre reinforced plastics (GFRP) are widely used in the manufacturing of large-scale structures such as wind turbine blades. The manufacturing process consists of an open mould vacuum assisted resin transfer moulding (VARTM). Once the resin infusion is finalized, a thermal curing cycle is applied to solidify the resin and fibre forming the composite structure. During this process, the epoxy resin is transformed into a thermoset plastic by a cross-linking reaction which shows an exothermal behaviour.

Curing cycle design aims to achieve the shortest manufacturing times and avoid excessive manufacturing temperatures driven by the exothermic reaction above the resin degradation temperature. In particular, GFRP present a big challenge from the manufacturing process design point of view. Due to the small thermal conductivities of GFRP and the resin exothermal reaction, large temperature gradients can be observed during the thick laminate manufacturing through the thickness [1, 2, 3, 4]. This can lead to strong temperature peaks above the resin degradation temperatures hampering the resin and composite mechanical properties. Below the resin degradation temperatures, the variability of the GFRP mechanical properties with the manufacturing history has been reported due to the curing influence [5, 6, 7, 8] or the thermal residual strains [9, 10, 11, 12]. The sub-laminate technique consists on extracting sub-laminates plates from a thick laminate with the help of peel ply layers at different thickness positions. The sub-laminates plates are extracted with a smaller thickness than the thick laminate. This allows manufacturing coupons with standard dimensions which have equal properties as the ones of a thick laminate at a given thickness position [13]. The use of the sub-laminate technique to study thick laminates showed a dependence of static and fatigue properties with the curing rates thermal history. Compression, tension and shear properties were studied. [13, 14, 15, 16].

In this regard, two complementary causes have been hypothesized to describe the variation of the mechanical properties with the manufacturing thermal history and thermal gradients. On one hand, the residual strains which drive a micro-damage mechanism by which interface fractures diminish the macro-mechanical properties. And on the other, the resin and interface adhesion properties depend on the curing path, which leads to a gradient of mechanical strengths related to the thermal history [17].

The sub-laminate technique offers the possibility to study the static and fatigue mechanical properties at a coupon level and relate those properties to the curing thermal history and residual strains measurements. At a micro-mechanical level single fibre fragmentation tests allow studying the mechanical properties of the fibre / matrix interface for composite materials as well as residual strains measurements when combined with the shear lag model [18, 19]. In the present work, the curing cycle influence on fracture toughness properties is studied for a GFRP at the macro and micro levels. For this purpose, the sub-laminate technique is used with standard coupon tests and the single fibre fragmentation tests are used to study the fibre/matrix interface properties.

2. Materials and methods

As reinforcements, unidirectional (UD) glass fiber type E 948 $[g/m^2]$ non-crimp fabric and biaxial (biax) ($\mp 45°$) glass fiber type E 812 $[g/m^2]$ fabric were used as base material for this study. A commonly used wind energy epoxy resin Hexion RIM 135 epoxy with Hexion RIM 137 hardener was used. The research was divided into four steps:

- The manufacturing of a thick laminate and extraction of sub-laminates
- Static mechanical tests of the fracture toughness of sub-laminates coupons
- Manufacturing of the single fragmentation test coupon at two different curing cycles
- Single fibre fragmentation tests

2.1. Standard coupons, sub-laminates technique

An 84 layer $600x600$ $[mm]$ thick laminate was manufactured. Peel ply layers were used to define sub-laminates to study the lamina properties at different thickness positions (see reference [15] for more details). Sub-laminates for interlaminar fracture toughness (G_{Ic}) tests consisted of plates of 6 plies UD fabric with a thin (62 $[\mu m]$) Teflon strip placed between each third and fourth layer, which served as delamination initiator for fracture toughness tests. A polytetrafluoroethylene (PTFE) peel ply was used to allow the extraction of the sub-laminates after the laminate was cured. Eight thermocouples (type K) and strain gauges (5 $[mm]$ gauge section TML FLA-10-11) were embedded in the laminates at different thickness positions to monitor the temperatures and strains through the thickness during the infusion and curing process. Both were placed in the middle of the $600x600$ $[mm]$ laminate at different Z positions between the fabrics and oriented with the UD fibre direction, as figure 1 shows. Temperature deviations were corrected on strain gauges based on thermocouples temperatures and the temperature deviation curve supplied by the strain gauges manufacturer. The sensors were connected to a data acquisition system and data was recorded every second during the manufacturing process. Heating rates were computed from the thermocouples temperature recordings as $max(dT/dt)$.

The laminate was infused using a vacuum bagging technique on a single-sided mould. The mould and resin were heated to 25 $[°C]$ before infusion. A two-stage cure cycle was used after two hours of infusion. The hardening of the resin took place in soak 1 (380 minutes) with a mould set temperature of 35 $[°C]$. Soak 2 had a duration of 800 minutes with a temperature of 80 $[°C]$. The long post-curing duration promoted the temperatures stabilization through the thickness, with each layer experiencing an 80 [°C] uniform post curing cycle temperature. The mould was setup with temperature ramps of 1 [°C/min]. Full vacuum (0.9 [$bars$]) was applied during the infusion and curing process.

The fibre weight ratio and void content were calculated by a calcination method according to ISO 1172. Glass transition temperatures were determined with a differential scanning calorimeter (DSC) according to ISO 1357. Moreover, UD G_{Ic} coupons were milled from the sub-laminates and tested in static loading. Six coupons per sub-laminate were tested. Interlaminar fracture toughness was tested in Mode I using DCB coupons according to ASTM D5528-13. One side of the coupons was painted with brittle white paint and the crack initiation distances were marked. Static tests were carried out using an MTS 10 [kN] hydraulic test frame with a 1 [kN] load cell (see figure 2). A black background was used and illumination was controlled to avoid shadows and glare to maximize contrast between the coupon and background. The coupons were pre-cracked with an initial cycle and then loaded with 3 [mm/min] until maximum machine displacement was reached. An HD camera was placed perpendicular to the coupon and using a video image processing technique [20], crack propagation was determined. G_{Ic} fracture toughness values were computed by means of deviation from linearity (NL) and 5% offset according to the three calculation methods contained in the standard: modified beam theory (MTB), compliance calibration (CC) and modified compliance calibration (MCC). For each DCB sub-laminate tests, batches of six coupons were carried out.

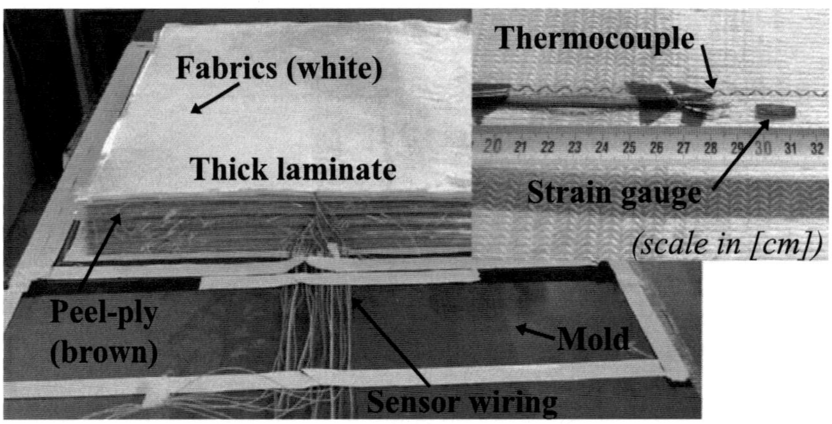

Figure 1. Laminate before infusion (left) and detail of embedding thermocouple and strain gauge (right).

2.2. Single fiber fragmentation tests

Eleven fibre fragmentation coupons were tested with an MTS 10 kN test frame equipped with a 1 kN load cell. The coupons were manufactured using type E glass fibres obtained from the fabrics described in section 2.1 and the same epoxy resin. Coupons dimensions are shown in figure 8. For this purpose, latex moulds were used to manufacture the single fibre fragmentation coupons. The moulds were instrumented with type K thermocouples to measure the coupons manufacturing curing cycles temperatures (see figure 3). Due to the coupons small size in comparison with the thermocouples and the latex mould large thermal inertia, the exothermal peaks were not visible in this setup. Half of the coupons were manufactured at a low curing cycle divided in a first soak at 50 ° for 500 min and the second soak at 70 ° for 700 min. The other half of coupons was manufactured at a high curing cycle with a single curing soak at 90 ° during 800 min. In both cases, the heating rates of the mould were setup at 1 [°$/min$].

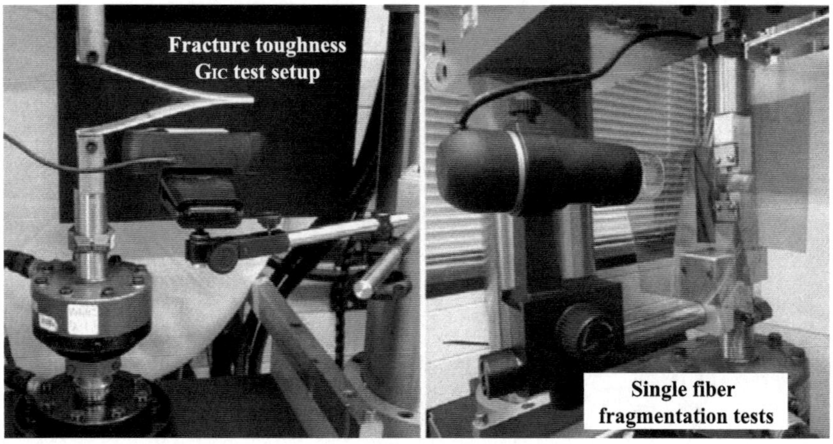

Figure 2. Test setup interlaminar fracture toughness test (left) and single fiber fragmentation (right).

Fibre diameters were measured with a 100x magnification microscope prior the manufacturing with average values of 18 $[\mu m]$. The fibre fragmentation coupons were positioned in the test frame between two polarized films and recorded with a digital USB microscope with an HD resolution (see figure 2 right). The videos were image processed to obtain the strain σ_c/E_c at the gauge section and the fibre debonded length l_d temporal signal for each fragment appeared during the test (see figure 4). Figure 4 shows an example of fibre fragments that were discarded due to having an unclear broken fibre zone or because some fragments were too close to others. In addition, fragments which did not show a linear dependence between the strain σ_c/E_c and the debonded length l_d were also discarded. Since it was not possible to directly measure the broken fibre gap $2 \cdot \delta$ with the image processing automated system, the dimension $2 \cdot \delta$ was measured at the beginning of the fibre failure and the end of the test. It was then assumed to show a linear behaviour with the measured strain. Based on the measured average values of $2 \cdot \delta$ a value $\delta \approx 0.8 \cdot r$ was obtained. The residual strain $\Delta\epsilon^T$, the sliding shear stress τ_s and the interfacial fracture energy G_c^i were computed from the temporal signals [18]. For these computations, a matrix modulus $E_m = 3.13\,[\text{GPa}]$, fibre modulus $E_f = 73\,[\text{GPa}]$ and a fibre radius $r = 9\,[\mu m]$ were considered. The fibre fragmentation tests were run at a speed of $1\,[\text{mm/min}]$.

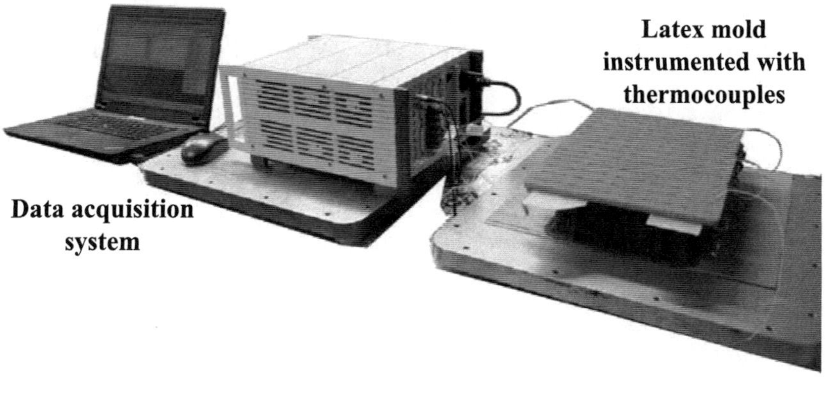

Figure 3. Single fiber fragmentation coupons manufacturing setup.

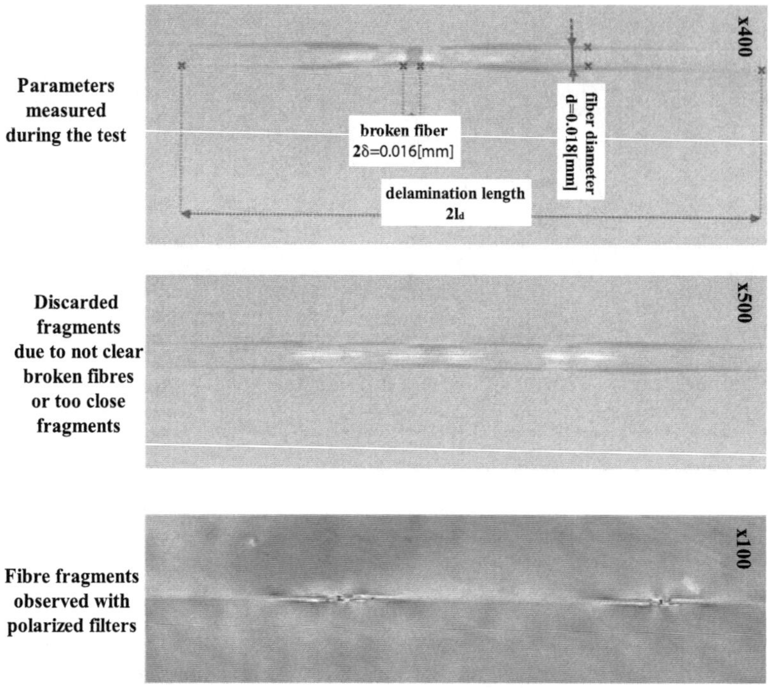

Figure 4. Single fiber fragments observed during the tests and schematic of the measured values.

3. Results and discussion

The sub-laminate technique was successfully applied and allowed to sub-divide the thick laminates into plates of reduced thicknesses which could be tested as standard fracture toughness coupons. The fibre weight ratio (FWR), void content and glass transition temperatures of each sub-laminate were measured to study the evolution of those manufacturing indicators through the thick laminate thickness. FWRs varied from 72.4% to 74.7% through the thickness showing no correlation with the thickness position. Sub-laminate void contents were below 0.1%, for sub-laminates 01 and 02 void contents were 0.4% and 0.2% respectively. The DSC glass transition temperatures varied from 91 [°C] in the lower sub-laminate to 85 [°C] in the upper sub-laminate. Fibre weight ratios, void contents and glass transition temperatures showed uniform values through the laminate thickness, indicating uniform levels of fabrics impregnation and curing degrees.

Large temperature gradients were observed through the thickness (Z positions) during the curing process of the laminate. Figure 6 shows the evolution of temperature of the embedded thermocouples. During thick laminate soak 1, the mould set temperature of 35 [°C] was exceeded due to the exothermic reaction of the resin to maximum temperatures of 85 [°C] in the middle sub-laminates. Temperature differences were up to 35 [°C] within the laminates and up to 50 [°C] with the mould set temperature. Based on these temperature measurements it can be stated that each thickness position followed a local curing cycle, and each sub-laminate was cured according to those local curing cycles. For each local curing cycle, the maximum heating rates $max(dT/dt)$ were computed.

The maximum strains recorded through the thickness during the process were up to 600 $[\mu\epsilon]$. However, once the plate cooled down, the residual strains between plates reduced to up to 200 $[\mu\epsilon]$ (0.02% elongation). Since each corresponding sub-laminate experienced a different local

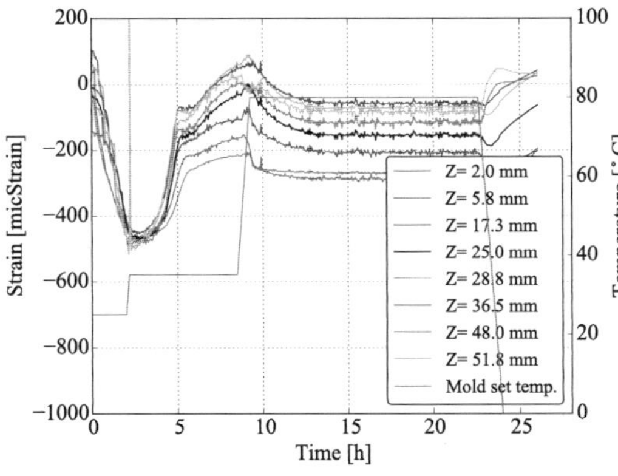

Figure 5. Evolution of strains during infusion and curing through the thickness, thick laminate. Maximum residual strains 200 [$\mu\epsilon$].

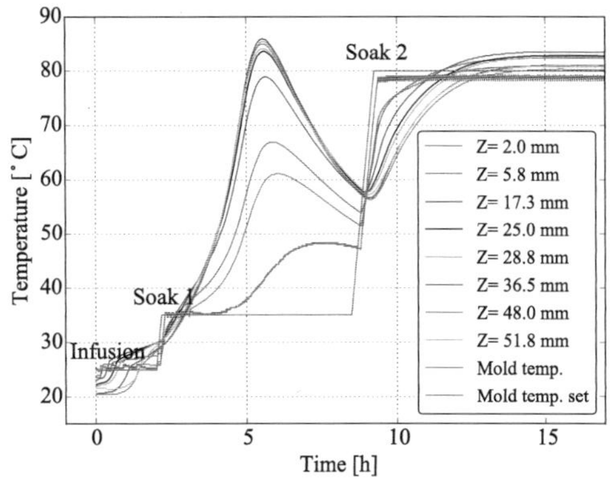

Figure 6. Evolution of temperatures during infusion and curing. Thick laminate.

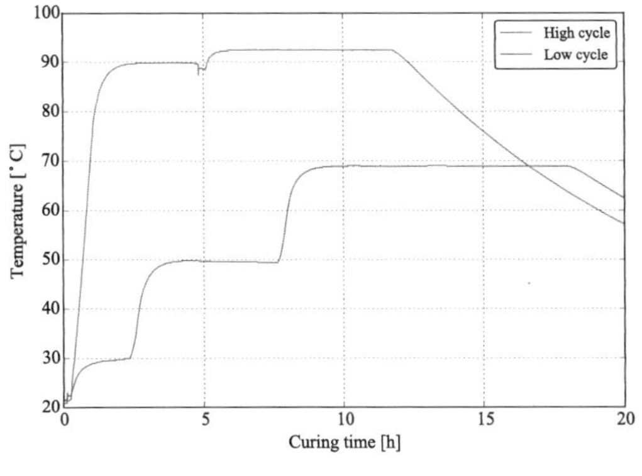

Figure 7. Evolution of temperatures during the single fibre fragmentation coupons curing. Two different cycles were used in the manufacturing of the coupons.

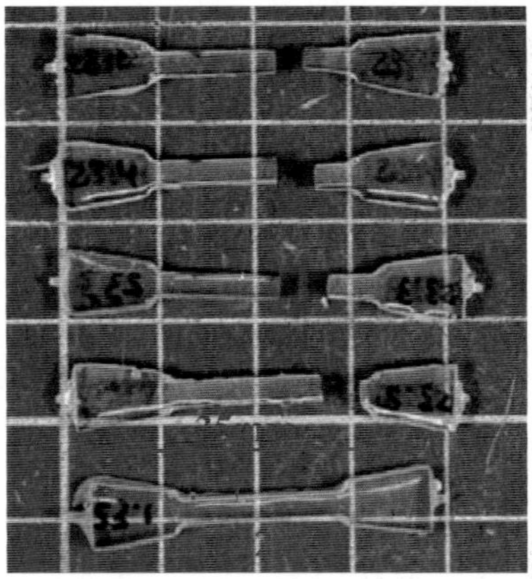

Figure 8. Single fibre fragmentation coupons after the test.

curing cycle, the sub-laminates which experienced the highest curing temperature peaks showed the highest contractions during the manufacturing process. A residual strain gradient could be observed through the thickness between the outer layers and the inner layers (see figure 5).

The influence of the manufacturing process on the G_{Ic} fracture toughness was evaluated for sub-laminates with UD lay-up using DCB tests. Fracture toughness was evaluated according to the MCC and MBT methods described in the standard ASTM D5528-13. The results showed an evolution of the fracture toughness for both analytical methods within the thickness position, where larger values can be observed in the positions with the larger maximum temperatures and heating rates. The average fracture toughnesses varied from 718 to 1200 [J/m^2] through the thickness. Figure 9 shows a cross-correlation between sub-laminate fracture toughnesses and maximum heating rates monitored for each sub-laminate with the embedded thermocouples. Cross-correlation R coefficients vary with the G_{Ic} calculation methods chosen between 0.84 and 0.87. In the case of the residual strains measurements, it was not possible to state a similar correlation between the measured fracture toughnesses and the residual strains. This fact and

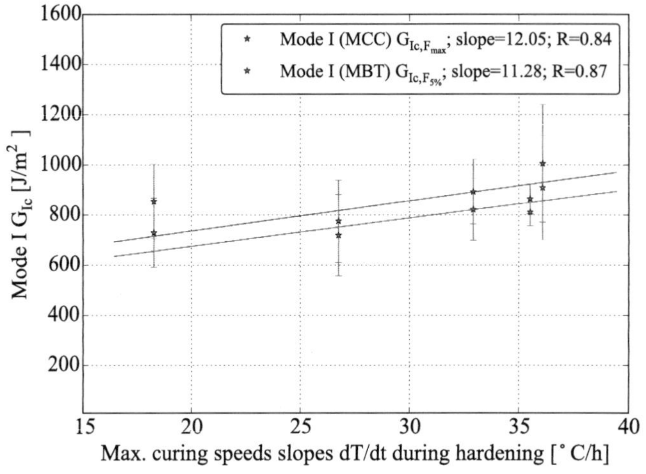

Figure 9. Fracture toughness versus maximum heating rates during the hardening. Each point represents the average of six coupons.

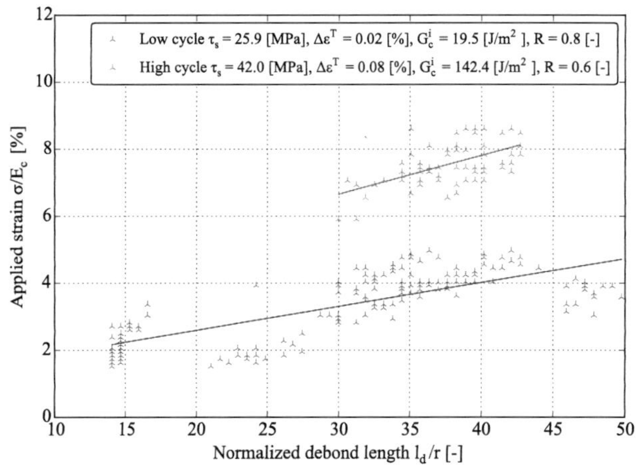

Figure 10. Single fibre fragmentation test results at two different curing cycles. Residual strain $\Delta\epsilon^T$, sliding shear stress τ_s and interfacial fracture energy G_c^i measurements. Aggregated values of fragments of 5 coupons per curing cycle.

the lower relative residual strains magnitudes in comparison with the maximum resin strain strengths (around 5 to 7%) suggest a weaker dependence of the residual strains in comparison with the maximum heating rate indicators.

Single fibre fragmentation tests were carried for coupons manufactured at two different curing cycles. Due to the small dimensions of the single fragmentation coupons (see figure 8) and the low sensitivity of the thermocouples it was not possible to record the exothermic peaks during manufacturing. However, as shown in figure 7 the coupons were manufactured with two different curing cycles, a low one and high one, resulting in two different curing states. Temporal signals of normalized debonded length l_d/r and applied strain σ/E_c for different coupons were aggregated and grouped in a low curing cycle and high curing cycle group. Figure 10 shows the single fragmentation test results divided into two different groups by curing cycle conditions. Single fragmentation coupons manufactured at low curing cycles show lower residual strain $\Delta\epsilon^T$ sliding shear stress τ_s and interfacial fracture energy G_c^i when compared to coupons manufactured at a higher curing cycle. This trend agrees with ones observed in the fracture toughness sub-laminates test. The residual strain $\Delta\epsilon^T$ showed a difference of 0.06 [%] between both curing cycles, which implies that the resin gelification and solidification temperatures occurred at different temperatures. It should be mentioned that the magnitude of the residual strains $\Delta\epsilon^T$ is rather low in comparison with the maximum strains, and of the same order of magnitude of the residual strains measured during the thick laminate manufacturing. In addition, an increase of the interfacial fracture energy G_c^i with the curing cycles was observed. Despite the thick laminates tests G_{Ic} (in mode I), fracture toughness is not comparable with the measured interfacial fracture energy G_c^i (mainly in mode II). Both show a dependence with the curing conditions, suggesting a strength gradient dependent on the thermal history.

One can hypothesize on which are the mechanisms behind the relation between the thermal history and the changes in the residual strain, shear stress and interface fracture energy. In the case of the residual strains, the variation with the curing cycles is related with the gel and solidification matrix temperatures. To vary the curing cycles modifies at which temperature occurs the matrix solidification or hardening. Therefore at which temperature are the resultant residual strains locked. On the other hand, the shear stress and interface fracture energy are related to the adhesion mechanisms of the fibre to the matrix. The adhesion mechanisms also appear to be dependent on the thermal history according to the experimental data reported in the present work. This dependence of the interface cohesive properties with the thermal gradient is also observable at a macro level. Where G_{Ic} fracture toughness shows a relation with the thickness position of a thick laminate and the curing speeds during or temperature profiles through the thickness.

4. Conclusions

The influence of curing cycle in fracture toughness properties was studied at a macro and micro levels. Using the sub-laminate technique, a dependence on the mode I fracture toughness G_{Ic} with the heating curing rate was observed. In addition, a weak variation on the residual strains through the thickness was observed. In a similar manner, single fragmentation tests were performed for coupons manufactured at two different curing cycles. High curing cycles coupons showed higher residual strain $\Delta\epsilon^T$ in comparison with lower curing cycles. The order of magnitude of the residual strains was similar to the one measured during the sub-laminate test, showing in both cases values below 1000 [$\mu\epsilon$]. As observed for sub-laminate coupons, single fibre fragmentation test showed higher interfacial fracture energy G_c^i values for the high curing cycles than for the low curing cycles.

Acknowledgments

The authors acknowledge the contribution of the EU 7th framework programs FP7-ENERGY-2013 IRPWind Project (under Grant Agreement 609795) and the Dutch TKI-WoZ research SLOWIND project to motivate and partly fund this research.

References

[1] Oh J H and Lee D G 2002 *Journal of Composite Materials* **36** 19–45 URL `http://dx.doi.org/10.1177/0021998302036001300`
[2] Sorrentino L, Esposito L and Bellini C 2016 *Composites Part B* URL `http://dx.doi.org/10.1016/j.compositesb.2016.10.064`

[3] Wieland B and Ropte S 2017 *Materials* **10** 1157 ISSN 1996-1944 URL `http://dx.doi.org/10.3390/ma10101157`

[4] Grama S, Nagaraja A, Gururaja S and Mohan R 2018 *Machining Science and Technology* **0344** ISSN 15322483 URL `https://doi.org/10.1080/10910344.2017.1415934`

[5] Lee S y and Springer G S 1988 *Journal of Composite Materials* **22** 15–29 URL `http://dx.doi.org/10.1177/002199838802200102`

[6] Singh S, Vummadisetti S and Chawla H 2017 *Journal of Building Engineering* ISSN 23527102 URL `http://linkinghub.elsevier.com/retrieve/pii/S2352710217304977`

[7] Manalo A, Maranan G, Sharma S, Karunasena W and Bai Y 2017 *Composite Structures* **173** 255–267 ISSN 02638223 URL `http://dx.doi.org/10.1016/j.compstruct.2017.04.040`

[8] Djabali A, Toubal L, Zitoune R and Rechak S 2018 *Composite Structures* **184** 178–190 ISSN 02638223 URL `https://doi.org/10.1016/j.compstruct.2017.09.069`

[9] Russell J D, Madhukar M S, Genidy M S and Lee A Y 2000 *Journal of Composite Materials* **34** 1926–1947 URL `http://dx.doi.org/10.1106/UY9U-F2QW-2FKK-91KG`

[10] Nawab Y, Tardif X, Boyard N, Sobotka V, Casari P and Jacquemin F 2012 *Composites Science and Technology* **73** 81–87 ISSN 02663538 URL `http://dx.doi.org/10.1016/j.compscitech.2012.09.018`

[11] Esposito L, Sorrentino L, Penta F and Bellini C 2016 *The International Journal of Advanced Manufacturing Technology* ISSN 0268-3768 URL `http://link.springer.com/10.1007/s00170-016-8613-5`

[12] Hther J and Brøndsted P 2016 *Proc. from the 37th Risø International Symposium on Materials Science. IOP Conf. Series: Materials Science and Engineering* **139** ISSN 1757-8981 URL `http://dx.doi.org/10.1088/1757-899X/139/1/012023`

[13] Lahuerta F, Nijssen R, van der Meer F and Sluys L 2016 *Journal of Composite Materials* **51** 563–575 URL `http://dx.doi.org/10.1177/0021998316648758`

[14] Lahuerta F and Nijssen R 2016 *37th Risø International Symposium on Materials Science. IOP Conf. Series: Materials Science and Engineering* **139** Submitted ISSN 1757-8981 URL `http://dx.doi.org/10.1088/1757-899X/139/1/012028`

[15] Smit D, Lahuerta F, Borst M and Nijssen R 2017 *Materials Performance and Characterization (ASTM)* **7**

[16] Mikkelsen L P 2017 Influence of curing profile and fibre architecture on the fatigue resistance of composite materials for wind turbine blades *Wind Energy Science Conference* (Lyngby, DK) p 1

[17] Lahuerta F, Nijssen R, van der Meer F and Sluys L 2016 *Composites Science and Technology* **123** 115–124 ISSN 02663538 URL `http://www.sciencedirect.com/science/article/pii/S0266353815301615`

[18] Sørensen B F and Lilholt H 2016 *IOP Conference Series: Materials Science and Engineering* **139** ISSN 1757-8981 URL `http://dx.doi.org/10.1088/1757-899X/139/1/012009`

[19] Sørensen B F 2017 *Mechanics of Materials* **104** 38–48 ISSN 01676636 URL `http://dx.doi.org/10.1016/j.mechmat.2016.10.002`

[20] Lahuerta F, Westphal T, Nijssen R P L, van der Meer F P and Sluys L J 2014 *Composites Part B: Engineering* **63** 1–7 ISSN 13598368 URL `http://dx.doi.org/10.1016/j.compositesb.2014.03.003`

Process-property-performance relationships in CFRP composites using fractal dimension

Amjed Saleh Mahmood [1,2], M Neil James [1,3] and John Summerscales [1]

[1]Advanced Composites Manufacturing Centre, School of Marine Science and Engineering, Reynolds Building, University of Plymouth, Plymouth PL4 8AA, England

[2]Head of Electromechanical Engineering Department, College of Engineering, Samarra University, Samarra, Iraq

[3]Department of Mechanical Engineering, Nelson Mandela Metropolitan University, Summerstrand Campus (North), Port Elizabeth, 6031, South Africa

E-mail: dramjed78@gmail.com

Abstract. The present study considers the dependence of the mechanical properties on the fibre architecture for composite laminates. Woven carbon/epoxy fabric composites were manufactured by the resin transfer moulding (RTM) process. Image analysis and fractal dimension (D) were used to quantify fibre distribution and resin-rich volumes (RRV) then correlate these data with the mechanical properties of the fibre-reinforced composites. The static strength of these composite laminates showed a clear dependence on the fibre distribution. Four-point flexural fatigue tests were conducted under load control, at a sinusoidal frequency of 10 Hz under load amplitude control with a stress ratio (R=σmin/σmax) of 0.1. Specimens were subjected to maximum fatigue stresses of 95% to 82.5% step 2.5% of the ultimate flexural strength (UFS). The fatigue properties of the composite laminates are shown to have statistically significant correlations with the fibre distribution and the static properties of the laminates. A loss of 5-6 % in the flexural modulus of the composite laminates predicted an increasing probability of failure of the composite laminates under fatigue loading. This occurs at about 90% of the fatigue life at a particular stress level.

1. Introduction

The increasing demand for high performance materials, (i.e. those which have high modulus- and/or high strength-to-weight ratio combined with improved fatigue performance and high toughness) has led to rapid advances in the field of fibre–reinforced composites. The estimated global demand could reach 146,000 tonnes of carbon composites produced annually by 2020 across the various applications [1]. High-performance composites often use autoclave consolidation to improve their mechanical properties by increasing the fibre volume fraction, and hence reducing the defect population. However, autoclaves are both an expensive capital equipment cost and a significant constraint on the rate of production. There is therefore increasing interest in composite manufacturing processes which do not require an autoclave stage, e.g. the resin transfer moulding process (RTM). Whilst out-of autoclave (OOA) manufacturing can increase the production rate of composite components and significantly reduce the energy consumption, it can also lead to lower fibre volume fractions (Vf) due to the compressibility characteristics of the fibres [2-5]. The lower Vf, obtained at reduced consolidation pressure, may lead to resin-rich volumes (RRV) in the composite and a consequent disproportionate reduction in the mechanical properties of the composite laminates. The level of RRV is highest in spray-up and hand-lamination techniques (no consolidation pressure), less common in OOA and liquid composite moulding (LCM) procedures, and lowest in autoclave and compression moulding processes. Increasing the pressure during manufacturing reduces the RRV, especially during autoclave consolidation [6, 7]. The improved nesting of fibres leads to improved mechanical properties (both static and fatigue).

Studying the effect of fibre volume fraction on the performance of composite laminates, without taking full account of the effect of changes in the fibre architectures and the consequent RRV for the reinforcement system in use, may lead to erroneous estimates of the service performance. Fractal dimension (D) looks promising for the quantification of the fibre distribution. Researchers [8-25] have already used D to correlate some mechanical properties and processing parameters. One of the techniques used to calculate the fractal dimension is the box-counting method which has been widely used in D research [17, 26-33].

The mechanical properties of composites have been extensively considered in analytical and experimental investigations for over four decades for both static properties [34-45] and fatigue properties [46-57]. These studies indicate that the fibre distribution is the most common parameter that links other parameters such as the RRV and internal defects, the stacking sequence, the stiffness and the anisotropic properties of the composite laminates. Modelling and prediction of the fatigue life requires significant experimental S-N data for each type of composites [58-60]. This study is believed to be the first systematic endeavour reported in the literature aimed at correlating fractal dimension with the fatigue life for continuous woven reinforcement fabric laminates.

2. Experimental methodology

RTM is a process for producing composite laminates by clamping the reinforcement between the upper and lower parts of a mould tool, then infusing resin to fill the inter fibre porespace. A schematic of the RTM process is given in Figure 1. Two 6K carbon fibre fabric architectures were woven by Carr Reinforcements using the same batches of fibre for each warp and weft. The modulus of elasticity of the carbon fibres was 235 GPa. The areal densities of the two fabrics tested in this work were: plain weave 300 g/m^2 and twill weave 320 g/m^2. The resin was Sicomin SR8100 epoxy with Sicomin SD8824 hardener and a weight mix ratio of 100:22. Six different composite laminate plates were manufactured for this work, comprising:

- 3 plain weave plates, utilising either 4 layers, 5 layers or 6 layers of fabric.
- 3 twill weave plates, utilising either 4 layers, 5 layers or 6 layers of fabric.

Images of the microstructure were captured from polished sections using an Olympus BX60 microscope with Olympus Stream image analysis software (SM04733) and analysed using the ImageJ software package with the FracLac add-in.

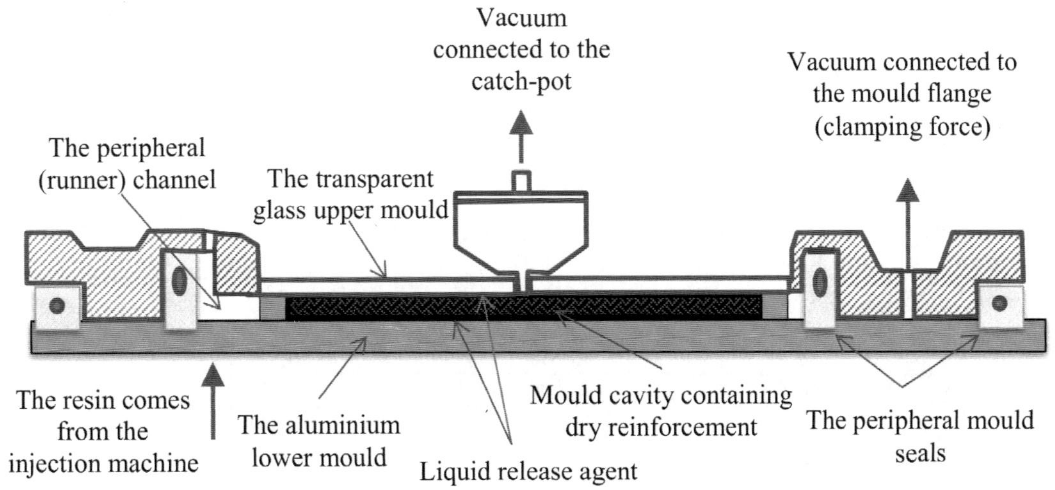

Figure 1: The RTM process as implemented in this study.

Four-point bend (static and fatigue) tests were conducted according to the British Standard EN ISO 14125:1998. An Instron Electropuls E300 (serial number: 5527-103) with a ±5 kN load cell (serial number: 107190) and test speed 1 mm/min was used. Tests were carried out in the warp and the weft direction with specimen dimensions of 100*15*2 mm (L*W*T). Five rectangular samples were cut in both the warp and weft directions from each of the six composite laminate plates: plain or twill and 4, 5 or 6 layers (giving 60 samples in total). The interlaminar shear strength was assessed in this work using the ISO 14130:1998. Five samples were cut from each direction of 3 different plates for each weave style (again giving 60 samples in total). Tests were conducted using an Instron 5582 universal testing machine (serial number: 5582J7466) with a ±100 kN load cell (serial number UK195) and a test speed of 1 mm/min. The four point sinusoidal bend S-N fatigue tests were performed with a stress ratio ($R=\sigma_{min}/\sigma_{max}$) of 0.1. The tests were carried out on specimens machined with their longitudinal axis parallel with either the warp or the weft directions in the as-manufactured laminate plates. The load amplitudes were chosen on the basis of using a set percentage of the ultimate flexural strength UFS.

3. Fractal dimension (D)

ImageJ with FracLac software creates an imaginary network of boxes over the binary image of the sampling region of the laminate. At every step i, the program counts the number of boxes that contain fibre. A ln-ln plot of number of fibre-containing boxes, against box size is produced as the output from this process. Each iteration gives a point in this plot. The slope of the regression line equals the negative estimated fractal dimension of the laminate. For different grid positions, the mean fractal dimension has been calculated. For consistent image analysis, all images should have (a) the same pixel size, (b) uniform assignment of colors to different parts of the microstructure (e.g. resin as black background), and (c) the ImageJ software options set to obtain the same size of sampling grid area in each case.

4. Results and discussion

Table 1 presents the mean mechanical properties for both fibre architectures (plain and twill weave) with 4, 5 and 6 layers of fabric. Figure 2 illustrates the correlation between UFS and D for all laminates, a strong second order correlation is apparent, with $R^2 = 0.9034$ and p=0.000027. Figure 3 illustrates the correlation between ILSS and microstructure, characterised by the fractal dimension. There is a reasonable correlation between the two parameters as $R^2 = 0.5579$ and p=0.025 as the ILSS is strongly influenced by resin properties and interfacial adhesion rather than depending solely on the fibre distribution. Figure 4 presents the correlation between E_f and microstructure, where $R^2 = 0.6822$ and p=0.005. From the data contained in Figures 2 to 4, the regression equations linking mechanical properties with fibre distribution via D were found to be:

UFS (MPa) = -21339D2 + 82155D - 78329 (1)
ILSS (MPa) = -550D2 + 2131D – 2014 (2)
E_f (GPa) = - 1278D2 + 4881D – 4616 (3)

Table 1 : The mean mechanical property and image analysis data for woven carbon fibre laminates.

Laminate	Direction	Notation	D	UFS(MPa)	ILSS(MPa)	E_f(GPa)
4 layers (Plain)	Warp	P4wp	1.835	561	43.8	38.9
	Weft	P4wt	1.809	463	40.5	31.2
4 layers (Twill)	Warp	T4wp	1.826	548	46.8	30.9
	Weft	T4wt	1.862	608	45.9	34.6
5 layers (Plain)	Warp	P5wp	1.859	670	47	40.8
	Weft	P5wt	1.841	567	44.1	39.2
5 layers (Twill)	Warp	T5wp	1.844	642	51.3	36.8
	Weft	T5wt	1.893	719	49.3	42.4
6 layers (Plain)	Warp	P6wp	1.892	725	48.9	45.6
	Weft	P6wt	1.850	623	44.9	41.4
6 layers (Twill)	Warp	T6wp	1.863	711	50.1	40.6
	Weft	T6wt	1.912	743	51.7	42.4

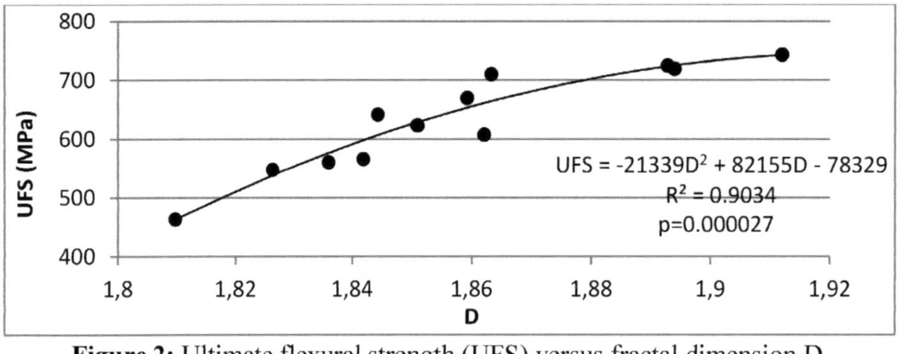

Figure 2: Ultimate flexural strength (UFS) versus fractal dimension D.

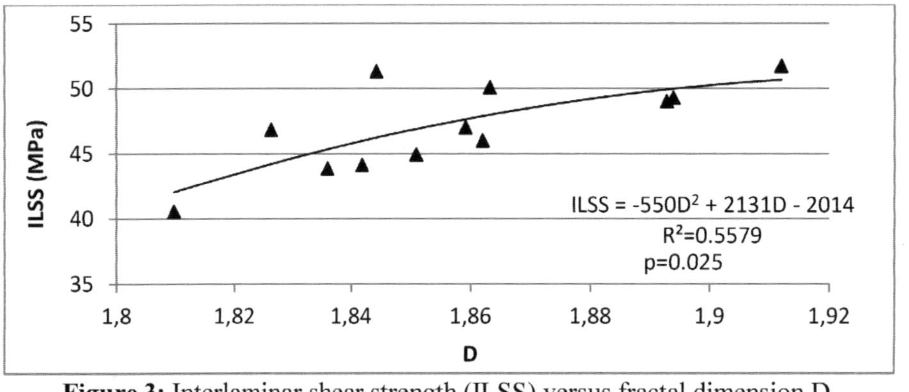

Figure 3: Interlaminar shear strength (ILSS) versus fractal dimension D.

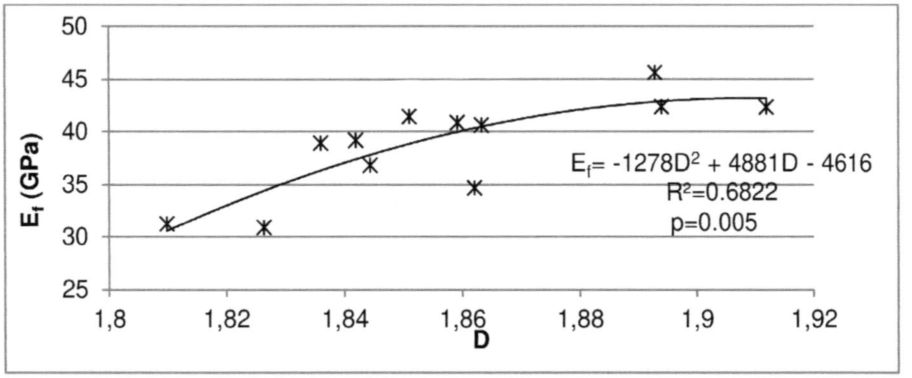

Figure 4: Flexural modulus of elasticity (E_f) versus fractal dimension D.

Figures 5-7 present the S-N fatigue data for specimens with 4, 5 and 6 layers of fabric. Figure 5 demonstrates that the T4wt laminate shows a high fatigue strength deriving from its high UFS and high D, while the P4wt laminate shows a lower fatigue strength due to its low UFS and low D (see Table 1). Fatigue life ranks in the sequence T4wt >P4wp >T4wp > P4wt, i.e. when the laminates are subjected to the same flexural stress, it is expected that, on average. This observation can be related to the static properties, where $UFS_{T4wt} > UFS_{P4wp} > UFS_{T4wp} > UFS_{P4wt}$. Furthermore, the ranking is the same for the fibre distribution, which is measured by the D, where $D_{T4wt} > D_{P4wp} > D_{T4wp} > D_{P4wt}$ as shown in Table 1. Similar behaviour is observed, for the same reasons, with the five and six fabric layer laminates, as shown in Figures 6 and 7 respectively. These observations lead to the conclusion that in these laminates, the fatigue life depends on the fibre distribution (as characterised by the D) and that laminates with high flexural strength will show high fatigue life, as has been observed by other workers [56]. This leads to general hypothesis that fatigue life is proportional to the D and to the UFS.

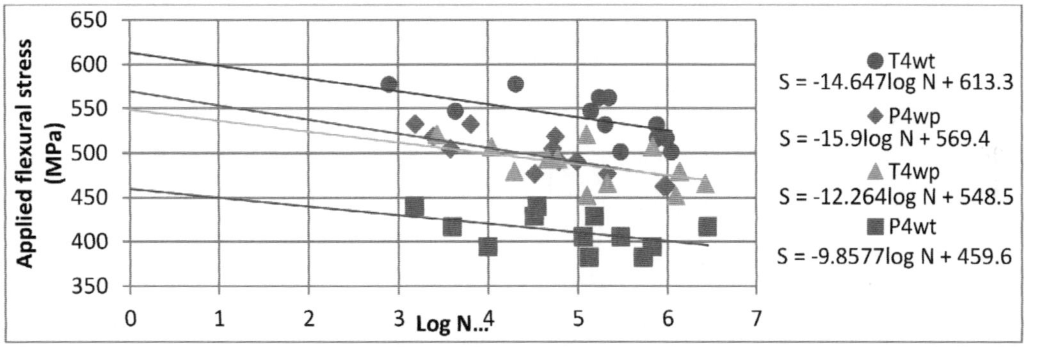

Figure 5: S-N curves for 4 layer plain and twill weave laminates in both the warp and weft directions.

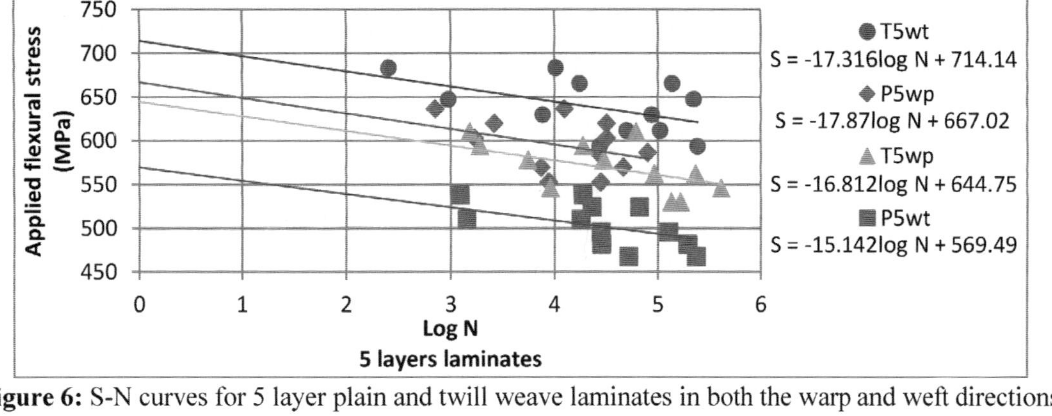

Figure 6: S-N curves for 5 layer plain and twill weave laminates in both the warp and weft directions.

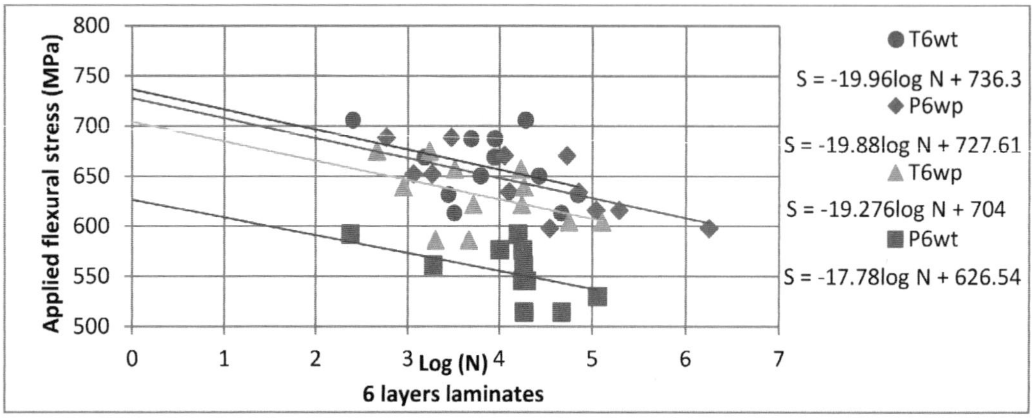

Figure 7: S-N curves for 6 layer plain and twill weave laminates in both the warp and weft directions.

The applied fatigue loads were recorded automatically by the Electropuls testing machine, allowing the actual flexural modulus, E_{ff}, to be calculated at any point in the fatigue test. Figure 8 presents normalised flexural modulus data, i.e. the ratio of E_{ff} to the original flexural modulus, E_o, as a function of life fraction, N/N_f, for all five layer laminates at an applied load of 85%UFS. As expected, it demonstrates that the change in flexural modulus can be divided into three regions, and that the data for all the 5-layer laminate types except for the plain weave fabric in the warp direction, group into a single scatter band of data. This particular case shows a higher value of E_{ff}/E_o for approximately the first 60% of the fatigue life than observed with the other laminate types.

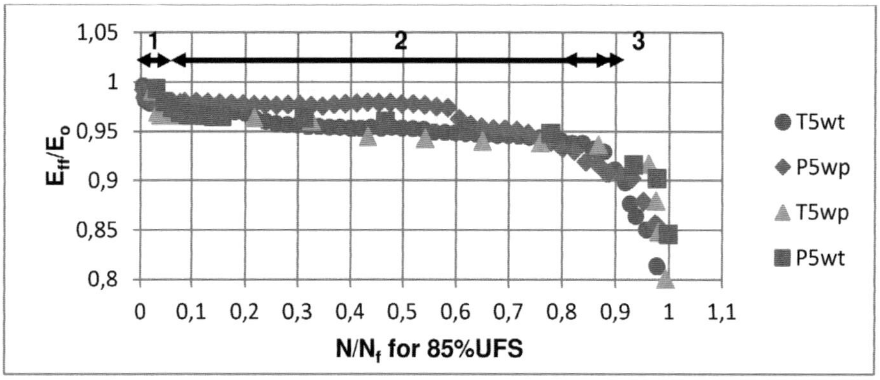

Figure 8: Normalised flexural modulus as a function of fatigue life fraction for the 5-layer laminates tested at 85% of the UFS.

In the specimens tested in this research, a 5-6% decrease in the flexural modulus indicates the beginning of the region of final failure in the composite laminates under fatigue loading. This decrease could be related to the rise of temperature that usually accompanies increasing deformation during fatigue testing [54]. It was observed that most of the fatigue specimens failed in the expected central region (middle of the inner loading roller span). However, some specimens failed near to one of the loading points, particularly at higher applied load levels, i.e. 90-95%UFS. In these cases cracking did not begin exactly below the loading point, but rather in close proximity to it. This could be due to the local stress concentration becoming more important at higher applied loads, as a result of the compression stresses induced at the loading point (leading to a localised multiaxial stress state) as well as a potential thermal influence from friction and fretting between the loading point and the specimen

5. Conclusions

The following conclusions can be drawn from the work reported in this study:

a. Fibre distribution, characterised by the fractal dimension (D), can be used as a parameter to characterise the static and fatigue properties of fibre-reinforced composites tested in four-point bend.

b. The corollary to this conclusion is that the ultimate flexural strength (UFS) of the composite laminates shows a clear dependence on the fibre distribution.

c. Interlaminar shear strength, ILSS, can also be correlated with the fibre distribution of the composite laminates, although the correlation is less significant than that identified for the UFS. d. In these composite specimens, a decrease of 5-6 % in their flexural modulus indicates the occurrence of damage zones in the plates and the onset of fatigue failure. This occurs at about 90% of the fatigue life at a particular stress level.

References

[1] Holmes, M., Global carbon fibre market remains on upward trend. *Reinforced Plastics*, 2014. 58(6): p. 38-45.

[2] Toll, S. and J.A.E. Manson, Elastic Compression of a Fiber Network. *Journal of Applied Mechanics*, 1995. 62(1): p. 223-226.

[3] Toll, S. and J.-A. Månson. An analysis of the compressibility of fibre assemblies. in *6th Int. Conf. Fibre Reinf. Compos*. 1994.

[4] Pearce, N. and J. Summerscales, The compressibility of a reinforcement fabric. *Composites Manufacturing*, 1995. 6(1): p. 15-21.

[5] Toll, S., Packing mechanics of fiber reinforcements. *Polymer Engineering & Science*, 1998. 38(8): p. 1337-1350.

[6] Abraham, D., S. Matthews, and R. McIlhagger, A comparison of physical properties of glass fibre epoxy composites produced by wet lay-up with autoclave consolidation and resin transfer moulding. *Composites Part A: Applied Science and Manufacturing*, 1998. 29(7): p. 795-801.

[7] Botelho, E., et al., Evaluation of fatigue behavior on repaired carbon fiber/epoxy composites. *Journal of Materials Science*, 2008. 43(9): p. 3166-3172.

[8] Worrall, C. and G. Wells. Fiber distribution in discontinuous fiber reinforced plastics: characterization and effect of material performance. in ECCM-7: *Seventh European Conference on Composite Materials. Realising Their Commercial Potential*. 1996.

[9] Pearce, N.R.L., F.J. Guild, and J. Summerscales, An investigation into the effects of fabric architecture on the processing and properties of fibre reinforced composites produced by resin transfer moulding. *Composites Part A: Applied Science and Manufacturing*, 1998. 29(1-2): p. 19-27.

[10] Pearce, N.R.L., J. Summerscales, and F.J. Guild, Improving the resin transfer moulding process for fabric-reinforced composites by modification of the fabric architecture. *Composites Part A: Applied Science and Manufacturing*, 2000. 31(12): p. 1433-1441.

[11] Summerscales, J., et al., Voronoi cells, fractal dimensions and fibre composites. *Journal of Microscopy*, 2001. 201(2): p. 153-162.

[12] Summerscales, J., et al., The fractal dimension of X-ray tomographic sections of a woven composite. *Advanced composites letters*, 2004. 13(2): p. 113-121.

[13] Picu, R., et al., Composites with fractal microstructure: The effect of long range correlations on elastic–plastic and damping behavior. Mechanics of Materials, 2014. 69(1): p. 251-261.

[14] Gao, S.-L., E. Mäder, and S.F. Zhandarov, Carbon fibers and composites with epoxy resins: Topography, fractography and interphases. *Carbon*, 2004. 42(3): p. 515-529.

[15] Allen, M., G.J. Brown, and N.J. Miles, Measurement of boundary fractal dimensions: review of current techniques. *Powder Technology*, 1995. 84(1): p. 1-14.

[16] Celli, A., et al., Fractal analysis of cracks in alumina–zirconia composites. *Journal of the European Ceramic Society*, 2003. 23(3): p. 469-479.

[17] Biancolini, M.E., et al., Fatigue cracks nucleation on steel, acoustic emission and fractal analysis. *International Journal of Fatigue*, 2006. 28(12): p. 1820-1825.
[18] Kuznetsov, P.V., V.E. Panin, and J. Schreiber, Fractal dimension as a characteristic of deformation stages of austenite stainless steel under tensile load. *Theoretical and Applied Fracture Mechanics*, 2001. 35(2): p. 171-177.
[19] Venkatesh, B., D.L. Chen, and S.D. Bhole, Three-dimensional fractal analysis of fracture surfaces in a titanium alloy for biomedical applications. *Scripta materialia*, 2008. 59(4): p. 391-394.
[20] Wang, L., L.-d. Wang, and W.-d. Fei, Fractal analysis of fracture surfaces in aluminum borate whisker-reinforced aluminum alloy 6061 composite. *Transactions of Nonferrous Metals Society of China*, 2011. 21(3): p. 461-466.
[21] Yuan, C.Q., et al., The use of the fractal description to characterize engineering surfaces and wear particles. *Wear*, 2003. 255(1–6): p. 315-326.
[22] Mishnaevsky, L., Hierarchical composites: Analysis of damage evolution based on fiber bundle model. *Composites Science and Technology*, 2011. 71(4): p. 450-460.
[23] Aniszewska, D. and M. Rybaczuk, Fractal characteristics of defects evolution in parallel fibre reinforced composite in quasi-static process of fracture. *Theoretical and Applied Fracture Mechanics*, 2009. 52(2): p. 91-95.
[24] Campos, K.A.d., et al., 3-D reconstruction by extended depth-of-field in failure analysis – Case study II: Fractal analysis of interlaminar fracture in carbon/epoxy composites. *Engineering Failure Analysis*, 2012. 25(0): p. 271-279.
[25] Pimenta, S. and S.T. Pinho, An analytical model for the translaminar fracture toughness of fibre composites with stochastic quasi-fractal fracture surfaces. *Journal of the Mechanics and Physics of Solids*, 2014. 66: p. 78-102.
[26] Mishnaevsky Jr, L., Hierarchical composites: Analysis of damage evolution based on fiber bundle model. *Composites Science and Technology*, 2011. 71(4): p. 450-460.
[27] Foroutan-Pour, K., P. Dutilleul, and D. Smith, Advances in the implementation of the box-counting method of fractal dimension estimation. *Applied Mathematics and Computation*, 1999. 105(2): p. 195-210.
[28] Pitchumani, R. and B. Ramakrishnan, A fractal geometry model for evaluating permeabilities of porous preforms used in liquid composite molding. *International journal of heat and mass transfer*, 1999. 42(12): p. 2219-2232.
[29] Buczkowski, S., P. Hildgen, and L. Cartilier, Measurements of fractal dimension by box-counting: a critical analysis of data scatter. *Physica A: Statistical Mechanics and its Applications*, 1998. 252(1–2): p. 23-34.
[30] Li, J., Q. Du, and C. Sun, An improved box-counting method for image fractal dimension estimation. *Pattern Recognition*, 2009. 42(11): p. 2460-2469.
[31] Buczkowski, S., et al., The modified box-counting method: analysis of some characteristic parameters. *Pattern Recognition*, 1998. 31(4): p. 411-418.
[32] Chen, S.S., J.M. Keller, and R.M. Crownover, On the calculation of fractal features from images. Pattern Analysis and Machine Intelligence, *IEEE Transactions* on, 1993. 15(10): p. 1087-1090.
[33] Jin, X. and S. Ong. Scale limits in estimating fractal dimensions of image surfaces. in Networks, 1993. International Conference on Information Engineering'93.'Communications and Networks for the Year 2000', *Proceedings of IEEE Singapore International Conference* on. 1993. IEEE.
[34] Basford, D.M., et al., Relationship between mechanical performance and microstructure in composites fabricated with flow-enhancing fabrics. Composites, 1995. 26(9): p. 675-679.
[35] Hale, R.D., An experimental investigation into strain distribution in 2D and 3D textile composites. *Composites Science and Technology*, 2003. 63(15): p. 2171-2185.

[36] Aziz, S.H. and M.P. Ansell, The effect of alkalization and fibre alignment on the mechanical and thermal properties of kenaf and hemp bast fibre composites: Part 1 – polyester resin matrix. Composites Science and Technology, 2004. 64(9): p. 1219-1230.

[37] Gojny, F.H., et al., Influence of nano-modification on the mechanical and electrical properties of conventional fibre-reinforced composites. *Composites Part A: Applied Science and Manufacturing*, 2005. 36(11): p. 1525-1535.

[38] Tzetzis, D. and P.J. Hogg, Bondline toughening of vacuum infused composite repairs. *Composites Part A: Applied Science and Manufacturing*, 2006. 37(9): p. 1239-1251.

[39] Dhakal, H.N., Z.Y. Zhang, and M.O.W. Richardson, Effect of water absorption on the mechanical properties of hemp fibre reinforced unsaturated polyester composites. *Composites Science and Technology*, 2007. 67(7–8): p. 1674-1683.

[40] Lomov, S.V., et al., Full-field strain measurements for validation of meso-FE analysis of textile composites. *Composites Part A: Applied Science and Manufacturing*, 2008. 39(8): p. 1218-1231.

[41] Sudarisman and I.J. Davies, The effect of processing parameters on the flexural properties of unidirectional carbon fibre-reinforced polymer (CFRP) composites. *Materials Science and Engineering*: A, 2008. 498(1-2): p. 65-68.

[42] Colin de Verdiere, M., et al., Evaluation of the mechanical and damage behaviour of tufted non crimped fabric composites using full field measurements. *Composites Science and Technology*, 2009. 69(2): p. 131-138.

[43] Liu, R.-m. and D.-k. Liang, Experimental study of carbon fiber reinforced plastic with embedded optical fibers. *Materials & Design*, 2010. 31(2): p. 994-998.

[44] Heß, H. and N. Himmel, Structurally stitched NCF CFRP laminates. Part 2: Finite element unit cell based prediction of in-plane strength. *Composites Science and Technology*, 2011. 71(5): p. 569-585.

[45] Vaughan, T.J. and C.T. McCarthy, A micromechanical study on the effect of intra-ply properties on transverse shear fracture in fibre reinforced composites. *Composites Part A: Applied Science and Manufacturing*, 2011. 42(9): p. 1217-1228.

[46] Demers, C.E., Tension–tension axial fatigue of E-glass fiber-reinforced polymeric composites: fatigue life diagram. Construction and Building Materials, 1998. 12(5): p. 303-310.

[47] Kalam, A., et al., Fatigue behaviour of oil palm fruit bunch fibre/epoxy and carbon fibre/epoxy composites. *Composite Structures*, 2005. 71(1): p. 34-44.

[48] Kawai, M., et al., High-temperature off-axis fatigue behaviour of unidirectional carbon-fibre-reinforced composites with different resin matrices. *Composites Science and Technology*, 2001. 61(9): p. 1285-1302.

[49] Aymerich, F., P. Priolo, and C.T. Sun, Static and fatigue behaviour of stitched graphite/epoxy composite laminates. *Composites Science and Technology*, 2003. 63(6): p. 907-917.

[50] Chambers, A.R., et al., The effect of voids on the flexural fatigue performance of unidirectional carbon fibre composites developed for wind turbine applications. *International Journal of Fatigue*, 2006. 28(10): p. 1389-1398.

[51] Ferreira, J.A.M., et al., Analysis of fatigue and damage in glass-fibre-reinforced polypropylene composite materials. *Composites Science and Technology*, 1999. 59(10): p. 1461-1467.

[52] Bertin, M., F. Touchard, and M.-C. Lafarie-Frenot, Experimental study of the stacking sequence effect on polymer/composite multi-layers submitted to thermomechanical cyclic loadings. *International Journal of Hydrogen Energy*, 2010. 35(20): p. 11397-11404.

[53] De Monte, M., E. Moosbrugger, and M. Quaresimin, Influence of temperature and thickness on the off-axis behaviour of short glass fibre reinforced polyamide 6.6 – cyclic loading. *Composites Part A: Applied Science and Manufacturing*, 2010. 41(10): p. 1368-1379.

[54] Ferreira, J.A.M., J.D.M. Costa, and P.N.B. Reis, Static and fatigue behaviour of glass-fibre-reinforced polypropylene composites. *Theoretical and Applied Fracture Mechanics*, 1999. 31(1): p. 67-74.

[55] Fuqiang, W. and Y. Weixing, A Model of the Fatigue Life Distribution of Composite Laminates Based on Their Static Strength Distribution. *Chinese Journal of Aeronautics*, 2008. 21(3): p. 241-246.
[56] Caprino, G. and A. D'Amore, Flexural fatigue behaviour of random continuous-fibre-reinforced thermoplastic composites. *Composites Science and Technology*, 1998. 58(6): p. 957-965.
[57] Epaarachchi, J.A., Effects of static–fatigue (tension) on the tension–tension fatigue life of glass fibre reinforced plastic composites. *Composite Structures*, 2006. 74(4): p. 419-425.
[58] Epaarachchi, J.A. and P.D. Clausen, An empirical model for fatigue behavior prediction of glass fibre-reinforced plastic composites for various stress ratios and test frequencies. *Composites Part A: Applied Science and Manufacturing*, 2003. 34(4): p. 313-326.
[59] Quaresimin, M., L. Susmel, and R. Talreja, Fatigue behaviour and life assessment of composite laminates under multiaxial loadings. *International Journal of Fatigue*, 2010. 32(1): p. 2-16.
[60] Flore, D. and K. Wegener, Modelling the mean stress effect on fatigue life of fibre reinforced plastics. *International Journal of Fatigue*, 2016. 82, Part 3: p. 689-699.

IOP Conf. Series: Materials Science and Engineering **388** (2018) 012014 doi:10.1088/1757-899X/388/1/012014

Pragmatic approach to cure profile enhancement for improved fatigue performance of thermoset matrix composites

Ulrich A. Mortensen, Tom L. Andersen, Lars P. Mikkelsen

DTU Wind Energy, Section of Composite Mechanics and Structures, Frederiksborgvej 399, 4000 Roskilde, Denmark

E-mail: ulmo@dtu.dk

Abstract. The current paper proposes a low cost scheme for enhancement of the fatigue life properties of fibre reinforced composites through minimization of residual stresses induced from the composite cure cycle. The enhancement scheme works on the presumptions that a low processing temperature at the event of resin gelation causes reduced residual stresses. The requirements for material characterization and numerical implementation of the scheme is low compared to optimization schemes available in the literature. The enhancement scheme is implemented and used to produce enhanced two stage cure cycles for a commercially available epoxy resin.

1. Introduction

Most of the fibre reinforced composites manufactured today are based on thermoset matrix resins which intrinsically builds up residual stresses in the final composite products. Recent studies have found that high levels of residual stresses impair the fatigue life properties of fibre reinforced composites. Hüther et al [1] showed that the fatigue properties of UD non-crimp fabric composites were impaired by cure cycles with higher amount of process induced strains. These findings were supported by Mikkelsen et al [2]. Mikkelsen et al measured the magnitude of process induced strain due to various temperature profiles in neat resin samples and used the same temperature profile when curing otherwise identical composite laminates reinforced with non-crimp glass fibre fabrics. Mikkelsen et al [2] found that the composite laminates cured with temperature profile that produced high magnitudes of process induced in the neat resin samples had impaired fatigue life properties compared composite laminates cured with temperature profiles that induced low amounts of strain in the neat resin samples. Warnet et. al [3] conducted bending experiments on cross-ply laminates and observed a decrease in experimental energy release rate of bending specimens when the residual stresses where increased. Experimental observations on transverse cracking in cross-ply laminates were also made recently by Zobeiry et al [4]. In their work laminates were produced with two different cure cycles that produced different levels of residual stresses in the cross-ply laminate. It was shown that higher levels of residual stresses caused transverse matrix cracks to form at lower levels of applied strains during mechanical testing. Asp et al [5] came up with a possible reason for the residual stresses impair the strength of composites. Asp et al showed that cavitation-induced cracks could occur before matrix yielding in polymers where significant hydrostatic stresses existed.

Cure profiles of composites can be designed to minimize the residual stresses in the final laminate. A key observations was made by Prasatya et al [6] who concluded that a direct relation existed between the temperature of the resin material at the point of gelation and the resulting residual stresses in the final composite material. The gelation point were described as the point where the resin material changes from liquid to solid phase. The observations made by Prasatya is supported by a recent study made by Mortensen et al [7] in which an optical Fibre Bragg Grating (FBG), made from silica, was embedded in a neat resin sample during cure. Mortensen et al found that tempeature at the point of gelation resulted in lower strains induced to the FBG fibre. Other researches have made conclusions that implicitly supports the idea of that low temperatures of gelation. Genidy et al [8] concluded that curing resins at a low temperature for a long time increased effective thermal expansion [when heating from the first to second temeperature stage of a two stage cure, red], because it counteracts the effect of chemical and cool down shrinkage. White and Hahn [9] conducted experiments on unsymmetrical laminates to find an optimized cure profile for lower residual stresses and concluded that lowering the initial cure temperature and holding it for longer time reduced the residual stresses while maintaining the mechanical properties. They noted that laminates with degrees of cures less than 0.95 had inferior properties to the full cured laminates, but that laminates with a higher than 0.95 degree of cure performed as well as laminates cured with the manufacturer recommended cure cycle. White and Hahn [9] also noted that high magnitudes of cool down rates reduced the residual stresses build up in the composite laminates. Gopal et al [10] focused on the cooling and heating rates of two stage cures, and claimed that for a two stage cure cycle it was optimal to go from the first temperature hold to the second temperature hold as fast as possible. Shah et al [11] made a complex constraint optimization of cure cycle parameters to reduce residual stresses. The results for two stage cures showed that the temperature of the first hold was generally lowered when the cure was allowed to take up more time. While Shah et al did not point this out them selves, the tendency of their optimized cure cycles fit with the idea of minimizing the temperature of gelation in order to minimize residual stresses. The work of Shah et al [11] succeeded in producing an optimization scheme that can effectively lower the magnitude of residual stresses, but the sheer complexity of proposed method makes it unsuitable for industrial purposes. The method requires implementation of very sophisticated numerical methods and expensive experimental material characterization of the desired resin for manufacturing.

The following work will present a scheme for designing two stage cure cycles for thermoset resin based composites with minimized levels of residual stresses based on a 0-D numerical cure model that is suitable for thin laminates. The scheme requires relatively inexpensive material characterization and a relatively low effort in terms of implementation into a numerical code. The idea behind the scheme is to design two-stage cure cycles that minimizes the temperature of gelation as a method for minimizing the resulting residual stresses. The scheme will be subject to a constraint of maximum allowable manufacturing time, as well as an additional constraint of material quality based on the thermoset resin degree of cure.

2. Cure kinetic model and Process simulation

The degree of cure of a thermoset resin is often defined by the amount of reaction heat released pr. unit mass of the material H, relative to the ultimate amount of reaction heat pr. unit mass H_U [12] [13]. This definition is expressed in eq. (1). The rate of the resin cure follows from the same definition, thus being proportional to the rate of heat of reaction as prescribed in eq. (2). Dynamic (temperature ramp) DSC measurement are widely used to determine the ultimate heat of reaction H_U. Figure 1 shows three measurements performed on the resin used in this article where the crosshatched area for each of the heat flow curves represents the ultimate heat of reaction H_U for each of the measurements. Isothermal DSC scans, as shown in Figure 2, are used for non-linear regression to determine the reaction order model.

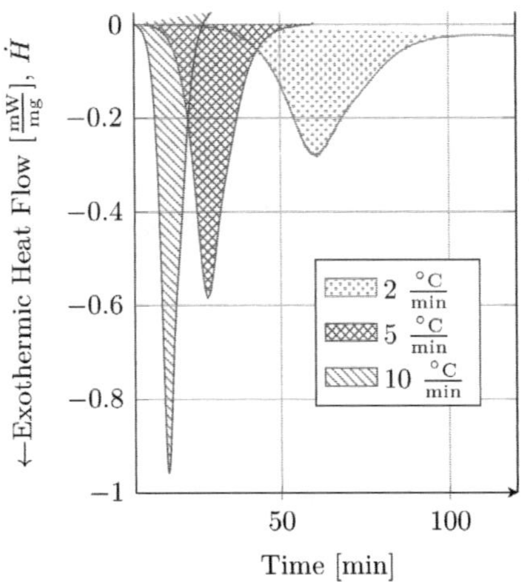

Figure 1: Dynamic DSC measurement of an epoxy resin with different heating rates. All measurements were started at -20 °C and ended at 300 °C.

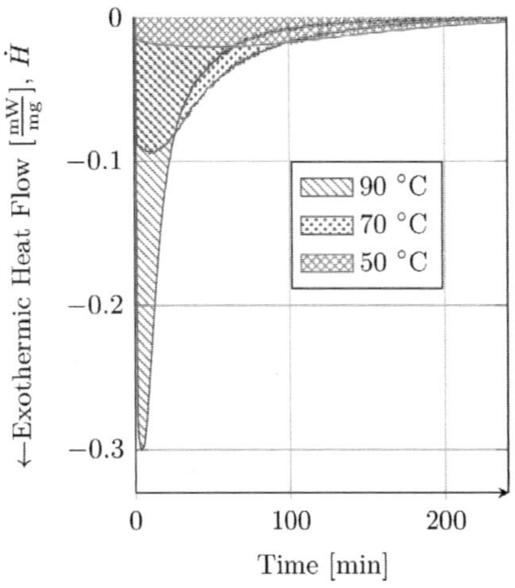

Figure 2: Isothermal DSC measurement of an epoxy resin. Each measurement were kept at the isothermal temperatures for 4 hours.

$$\alpha(t) = \frac{H(t)}{H_U}, \qquad \alpha \in [\,0\,;1\,] \tag{1}$$

$$\frac{\mathrm{d}\alpha}{\mathrm{d}t} = \frac{1}{H_U}\frac{\mathrm{d}H(t)}{\mathrm{d}t} \tag{2}$$

The general form of most phenomenological cure kinetic models can be described by eq. (3) where the term $f(\alpha)$ is the reaction model which is specific to the applied cure kinetics relations. The $K(T)$ term on the right hand side is the temperature dependence of the reaction rate, which, for all cure models described in the current paper, will be defined by by an Arrhenius expression as printed in eq. (4). The constant A $[-]$ is a pre-exponential constant, E $\left[\frac{\mathrm{J}}{\mathrm{mol}}\right]$ is the activation energy, $T\,[\mathrm{K}]$ is the material temperature and $R = 8.3145\left[\frac{\mathrm{J}}{\mathrm{K{\cdot}mol}}\right]$ is the ideal gas constant.

$$\frac{\mathrm{d}\alpha}{\mathrm{d}t} = K(T)\,f(\alpha) \tag{3}$$

$$K = A\exp\left(-\frac{E}{RT}\right) \tag{4}$$

$$f(\alpha) = \alpha^m(1-\alpha)^n \tag{5}$$

The cure kinetic model for thermoset resins that will be applied in the current work is classified as a phenomenological model. The expression for the reaction model $f(\alpha)$ is printed in eq. (5). The reaction order parameters n and m as well as the reaction rate constants of the Arrhenius expression A and E are determined using non-linear regression. The data input for the non-linear regression model is the degree of cure, computed using eq. (1), and the rate of cure, computed using eq. (2), as well as the corresponding temperature. Figure 3 shows the non-linear regression data fit to each of the isothermal DSC measurements. The degree of cure and rate of cure in Figure 3 is derived from the isothermal measurement data shown in Figure 2. The table in Figure 3 presents the parameters found for the cure kinetic model used in the regression, and these parameters along with the phenomenological cure model of eqs. (3)-(5) will be applied in section 5. A more detailed description of the process of obtaining the parameters for the cure kinetic model can also be found in Khoun et. al [14] or Yousefi et al. [12].

Having established a cure kinetic model, and the corresponding parameters for the model, simulation of the resin cure can be performed using numerical integration tools such as MatLab's ODE solver[1] or SciPy's odeint [2] library or other numerical tools. Solving the cure kinetic equations with a given initial value for the degree of cure will be denoted as a cure simulation in section 4 of this paper.

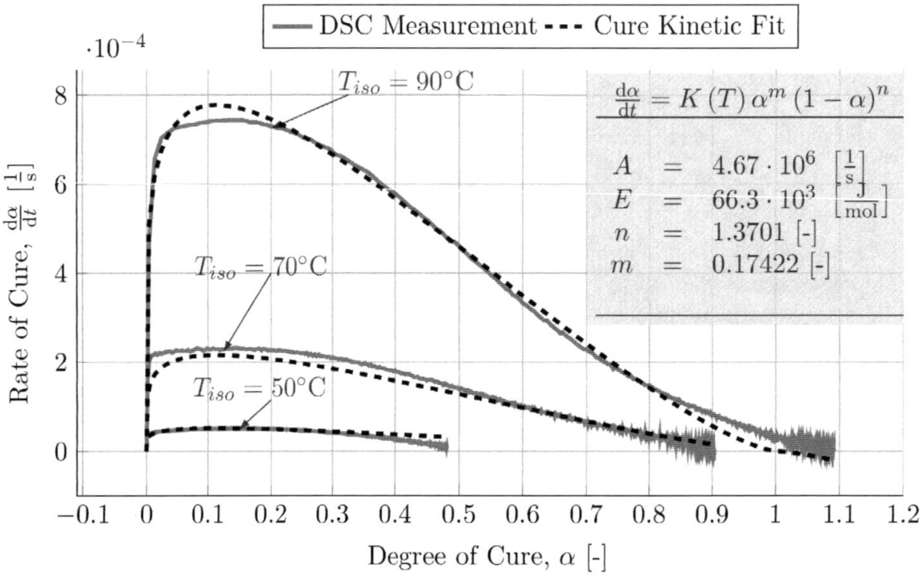

Figure 3: Fit of cure kinetic model (eq. (5) with input from the isothermal DSC measurements.

3. Gelation and Process Induceds Strain

While at least one study [4] have suggested that tool-part interaction can produce internal stresses when the thermoset matrix is in the liquid phase it is generally assumed in the literature that no significant stresses can airise prior to gelation (see e.g. [13] [14] [15] [16]). The residual stresses σ_{res} in a thermoset matrix composite is then dependent on the cure history *after* gelation and the mechanical constraints dictated by the method of manufacturing. The point of gelation is most often found using rheometry as this technique readily spans both the liquid and solid phase of polymeric materials and can be used to generate visco-elastic material models [14]. Khoun et al used rheometry to establish a degree of cure of gelation of $\alpha_{gel} = 0.7$ [14]. From the perspective of polymer chemistry the point of gelation can be defined as the time of critical conversion (degree of cure) where infinite three-dimensional polymer networks can exist [17]. This definition allows the gelation degree of cure to be computed based on stoichometry of the resin. O'Brien [13] did however note that the theoretically calculated degree of cure of gelation from the resin they investigated was found to be $\alpha_{gel} = 0.58$, significantly lower than the point they obtained using rheometry which was $\alpha_{gel} = 0.71$. Khoun et al [14] and O'Brien [13] both defined the gelation as the point of equality between the storage and loss moduli of

[1] `https://se.mathworks.com/help/matlab/numerical-integration-and-differential-equations.html`
[2] `https://docs.scipy.org/doc/scipy/reference/generated/scipy.integrate.odeint.html\#scipy.integrate.odeint`

their respective rheometry measurements.

3.1. Temperature of Gelation

In a recent study by Mortensen et. al [7] the total process induced strain was found to be closely related to the temperature of gelation T_{gel}. The study measured chemical and thermal strain using optical Fibre Bragg Gratings (FBG) on a neat epoxy sample that was unaffected by mechanical constraints. The phenomenon that is thought to govern the relation between temperature of gelation and total process induced strain can be illustrated by the the strain development during cures shown in Figures 4 and 5 for a one stage and a two stage cure respectively. It should be noted here that the event of gelation is assumed to occur at a specific degree of cure, α_{gel} the degree of cure of gelation, independently of the material temperature at the instant of the gelation. This means that the temperature of gelation T_{gel} is defined as the temperature of the material in the instant that the degree of cure of gelation α_{gel} is reached such that $\alpha(T, t_{gel}) = \alpha_{gel}$. The gelation of a resin will be defined as the point in time where the resin can obtain normal forces and are capable of inducing measurable deformations of small objects, such as a single glass fibre FBG. Furthermore, the total process induced strain at room temperature ε_{tot} is assumed to be the sum of the chemical (shrinkage) strain ε_{ch} and the thermally induced strain ε_{th} as given by eq. (6).

$$\varepsilon_{tot} = \varepsilon_{ch} + \varepsilon_{th} \tag{6}$$

The measurements shown in Figures 4 and 5 are based on the experiments conducted by Mortensen et al [7]. In the one stage cure shown in Figure 4 the gelation point are reached when the temperature is close to the isothermal cure temperature. The temperature after gelation is nearly constant such that no thermal strains are expected to develop during the temperature hold. The strain induced during the temperature hold (region A1 in Figure 4) are caused solely by chemical shrinkage of the resin. When the temperature hold ends and the material is cooled and compressive thermal stresses arise due to thermal contraction of the material. This means that using a one stage cure produces a total process induced strain that is the sum of the total amount of chemical shrinkage strain and the thermal strain of the temperature difference from the cure temperature to the operating temperature, which in most cases are at room temperature of the final manufactured part.

For the two stage cure shown in Figure 5 the point of gelation occurs at the end of the first temperature hold, thus causing a low temperature of gelation. After the material has reached gelation, and is capable of attaining normal stresses, the processing and material temperature are increased. In region A2 in Figure 5 the temperature is increased. The temperature increase causes the the rate of cure to increase, thus resulting in chemical strain that causes compression. The chemical strain is however counteracted by thermal expansion of the resin, resulting in a net positive increase in strain. In the A2 region in Figure 5 the thermal expansion and chemical shrinkage acts in opposite direction. After the peak, when the material temperature is constant, only chemical strain due to curing are acting on the FBG strain sensor, causing a slight decrease in strain. At the end of the second temperature stage the process induced strain is in tension before the material is cooled to room temperature where only thermal strains are induced to cause a final state of compression. The final level of total process induced strain is lower for the two stage cure due to the thermal expansion during the temperature ramp up that happens after gelation and counteracts the chemical shrinkage. The level of process induced strain before the final cool down shows an important difference between the one stage and two stage cures because it must be assumed that the final cool down to room temperature must cause similar amount of thermal stresses for both of the materials. In the on stage cure the strain before the final cool down equals the inevitably level of chemical strain, while for the two stage the strain is in the tension regime before the final cool down. It should further be remarked that a significant porting of the the thermally induced strain comes from the final cool down of the material to room temperature, as pointed out in region C2 of Figure 5, and as such the maximum temperature of the second temperature stage should be identical to the ultimate T_g^{ult} of the resin material such that full cure can be achieved with as low a second stage temperature as possible.

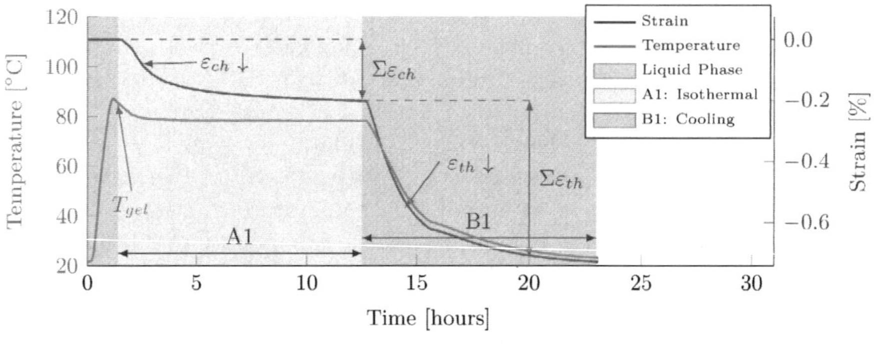

Figure 4: Strain and Temperature development over time for a one stage cure with 80°C curing temperature for 12 hours.

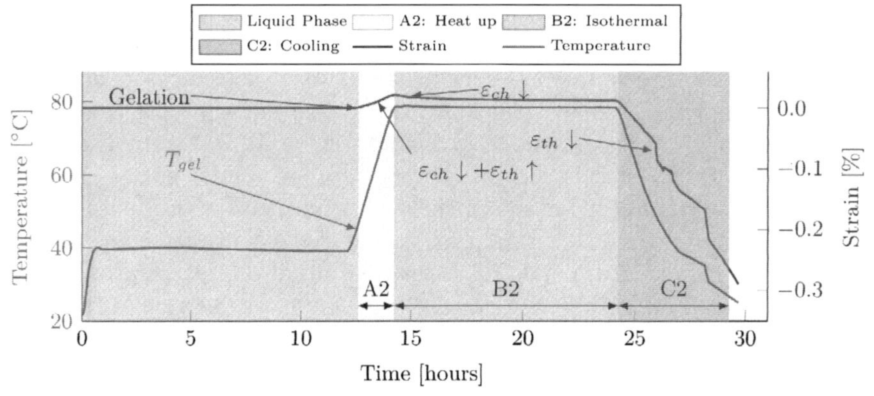

Figure 5: Strain and Temperature development over time for a two stage cure with 40°C for 12 hours followed by 80°C for 12 hours.

The experiments conducted by Mortensen et al [7] also included implicit measurement of the degree of cure using dielectric analysis. This allowed the degree of cure of gelation to be established as $\alpha_{gel} = 0.66 \pm 0.05$ for the epoxy resin. This value, $\alpha_{gel} = 0.66$, will be used in cure simulations in section 5.

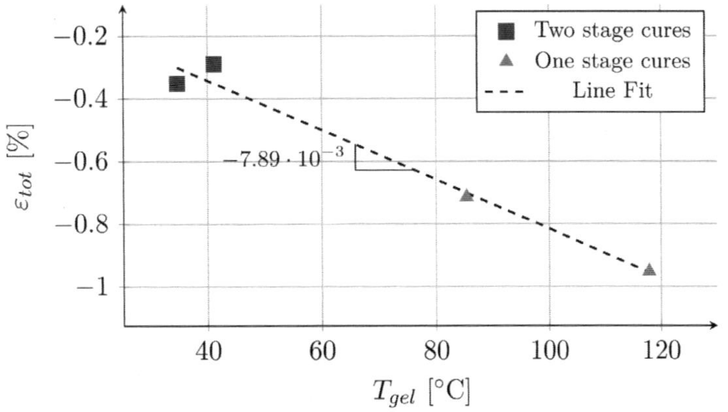

Figure 6: The total process induced strain at room temperature (25 °C) plotted against the temperature of gelation T_{gel}.

Figure 6 shows how the total process induced strain is related to the temperature of gelation for the epoxy used by Mortensen et al [7]. The two data points with low temperatures of gelation is a product of two stage cures, and the high temperature of gelation results from one stage cures. There is a clear tendency for the total process induced strain to be increasingly compressive (negative) for an increase in the temperature of gelation. This tendency constitutes an important observation when designing cure cycles with the aim of lowering the residual stresses: Even though the measured process induced strain in a neat epoxy sample are not directly proportional to the level of residual stresses in a manufactured composite, the clear negative correlation between temperature of gelation and process induced strain means that the residual stresses in a composite must also increase in magnitude for an increase in the temperature of gelation. The validity of this principle is also supported in the literature where Prasatya concluded that "*the results demonstrate a direct relationship between the temperature at which gelation occurs and the magnitude of isotropic residual stresses at room temperature*" [6].

4. Cure optimization

4.1. Problem formulation

Based on the works of Hüther et al [1], Warnet et al [3] and Mikkelsen et al [2] an increased level of residual stresses σ_{res} in thermoset composites leads to impaired fatigue life of the composites. An improved cure cycle for a composite must then minimize residual stresses in order to improve the fatigue life of a thermoset matrix composite. An overall goal of minimizing the residual stresses can be posed as a constraint minimization problem, as printed in eq. (7), where the objective function is the total process induced strain subjected constraints of maximum time t_{max}, maximum temperature T_{max} and minimum attained degree of cure α_{end}. The optimization problem does however come with the caveat of having an output variable (α_{end}) as non-linear constraint. Furthermore, having process induced stresses, or strains, as the objective function requires expensive experimental resin characterization in order to build a thermo-chemical viscoelastic material model.

$$\begin{aligned} \textbf{min} \quad & \sigma_{res}(T,t) \\ \textbf{subject to} \quad & \max(t) \leq t_{max} \\ & \max(T) \leq 1.2 \cdot T_g \\ & \alpha(T, t = t_{max}) \geq \alpha_{end} \end{aligned} \tag{7}$$

Instead of solving the non-linear constraint optimization problem posed in eq. (7) the observations from section 3 can be applied to simplify the problem. Specifically the consequence that minimizing the temperature of gelation T_{gel} also minimizes the total process induced strains, and hence the residual stresses, is applicable in formulating a more suitable minimization problem. The revised problem, for an assumed two stage cure, is given in eq. (8). It should be noted that the constraints are the same as given in eq. (7), but temperature of gelation T_{gel} can not readily be defined as an objective function.

$$\begin{aligned} \textbf{min} \quad & T_{gel} \\ \textbf{subject to} \quad & \max(t) \leq t_{max} \\ & \max(T) \leq 1.2 \cdot T_g \\ & \alpha(T, t = t_{max}) \geq \alpha_{end} \end{aligned} \tag{8}$$

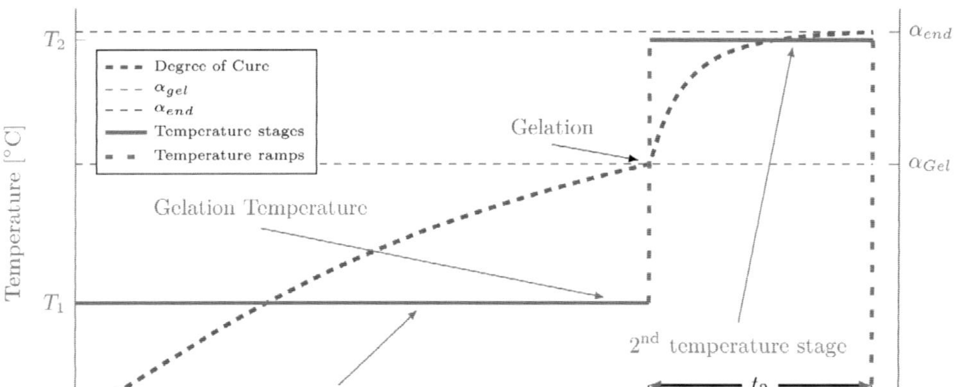

Figure 7: Principal sketch of a the parameters that define the optimal two stage cure.

4.2. Minimization of the Temperature of Gelation

The solution to eq. (8) will not be defined here in terms of classical optimization, but will instead apply some known quantities about the resin cure cycle aimed to be enhanced. Figure 7 shows schemtically the parameters that define a two stage cure cycle, excluding heating and cooling ramps to and from the temperature stages. The temperature of the first stage T_1 and second stage T_2 and the amount of time, t_1 and t_2 respectively, the temperatures are held at

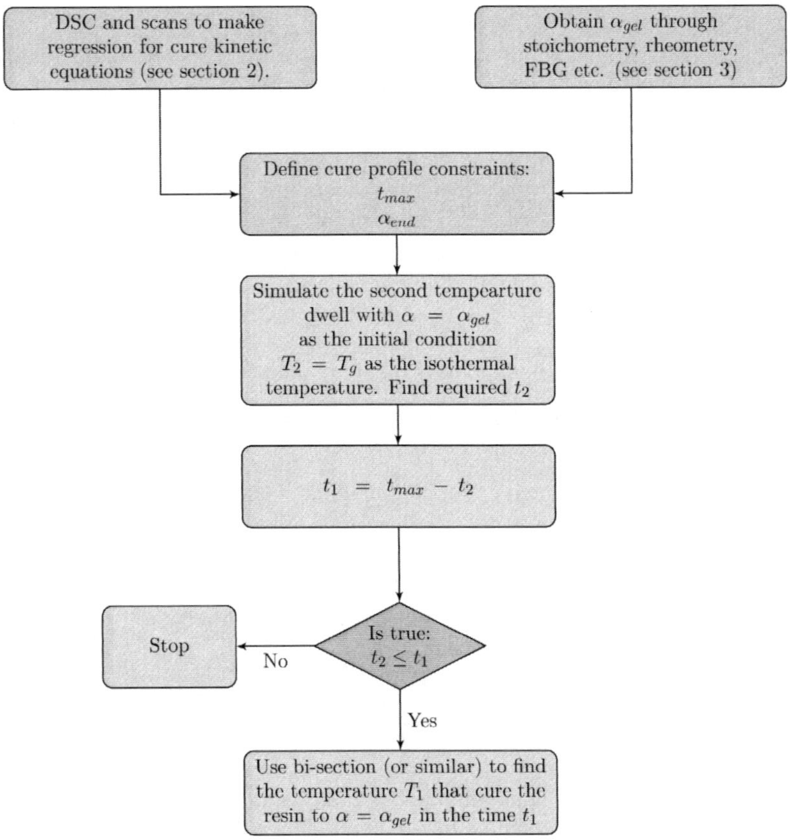

Figure 8: Flowchart of the minimization of the temperature of gelation. Blue boxes are experimental procedures and user input. Grey boxes are simulation process steps.

those stages. The following will assume that the degree of cure of gelation α_{gel} is a known resin quantity and that the temperature T_2 of the second stage is identical to the ultimate T_g of the resin material, henceforth denoted T_g^{ult}. The choice of $T_2 = T_g^{ult}$ for the second curing stage is based on the general need for the process temperature of the material to reach the temperature of ultimate glass transition temperature to ensure cure. In other words T_g^{ult} is the lower bound temperature the resin needs to reach in order to attain a full cure. The development of the degree of cure can be simulated by numerical integration of the cure kinetics equation from section 2. Using $\alpha = \alpha_{gel}$ as initial condition and $T = T_2 = T_g^{ult}$ the time (t_2) required for the resin degree of cure to reach the specified minimum degree of cure α_{end}, can be found. When t_2 is a known time the amount of time (t_1) available for the first temperature stage is known as the remaining time from the defined maximum available time t_{max}. The required solution to eq. (8) can then be reduced to finding the temperature T_1 of the first temperature stage that satisfies eq. (9).

$$\alpha(T_1, t_1) = \alpha_{gel} \tag{9}$$

The described procedure for minimizing the temperature of gelation T_{gel} is schematically outlined in Figure 8. An additional condition are adviced in the flowchart in Figure 8, namely the requirement that the time for cure during post-gelation t_2 should not exceed the time t_1 for cure during pre-gelation. The condition is based on practical experiments from simulations that shows the process to be ineffective if $t_2 > t_1$.

IOP Conf. Series: Materials Science and Engineering **388** (2018) 012014 doi:10.1088/1757-899X/388/1/012014

5. Results

The optimization process described in section 4 has been implemented in a MatLab routine. The routine follows the simulation steps (grey boxes) in Figure 8 and takes in α_{end} and t_{max} as constraint parameters. The routine furthermore takes a cure kinetic function and a set of parameters as arguments as well as a value for the gelation degree of cure. For the results shown in this section applies the cure kinetic model is described in section 2, and the specific model fit and parameters that are shown in Figure 3. Figure 9 shows the found cure cycles with three common time constraints of 8, 16 and 24 hours. All three of the shown cure cycles were subjected to the constraint of $\alpha_{end} \geq 0.99$. The decreasing of the initial temperature T_1 resulting for increase of the maximum allowable time should come as no surprise; the entire optimization flow was designed to do exactly this.

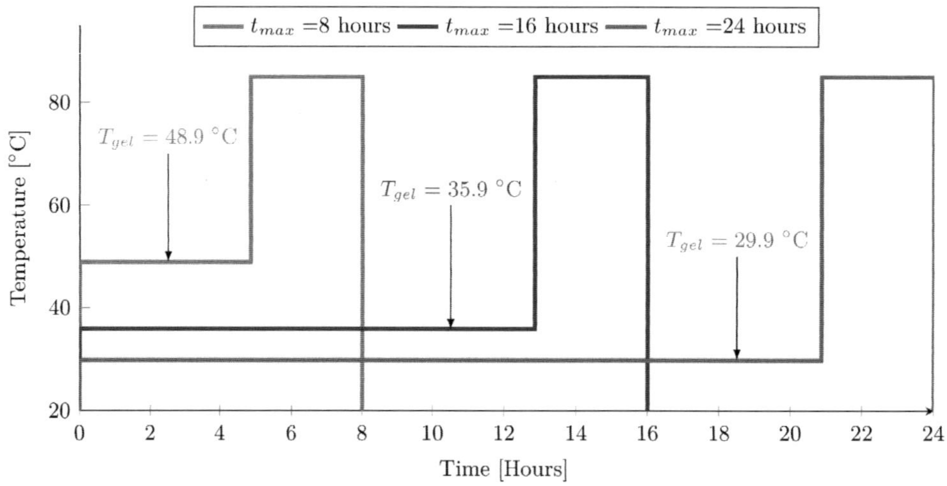

Figure 9: Cures with minimzed temperature of gel

Given that the temperature of gelation found using the optimization process is a direct product of the time and curing constraints, t_{max} and α_{end} respectively, a small investigation into the effect of the constraints were conducted. Figure 10 shows the temperature of gelation, found using the optimization process, plotted against the time constraint t_{max}. The Figure shows curves for various levels of the constraint for the minimal attained degree of cure α_{end}. Each of the curves in Figure 10 starts at the value of t_{max} that allows at least half of the cure time to be spent at the low curing temperature t_1. This essentially means that the first point of each of the curves are the lowest t_{max} value where the $t_2 \leq t_1$ requirement is fulfilled. The resulting curves shows that having a tight requirement of the minimum attained degree of cure makes the optimization less effective, and essentially means that longer time is needed to reach a degree of cure that satisfies the constraint. Figure 10 also shows that significant decreases in the temperature of gelation can be achieved by increasing the the maximum allowed time for the curing of a composite laminate. Especially cure cycles constrained to take less than 5 hours could see significant lowering of the temperature of gelation by allowing additional time for curing.

6. Discussion

The implementation of the proposed cure profile optimization have been implemented and tested on a commercially available low temperature epoxy. The applied degree of cure of gelation in the results was $\alpha_{gel} = 0.66$. This was the mean value report in the experiments by Mortensen

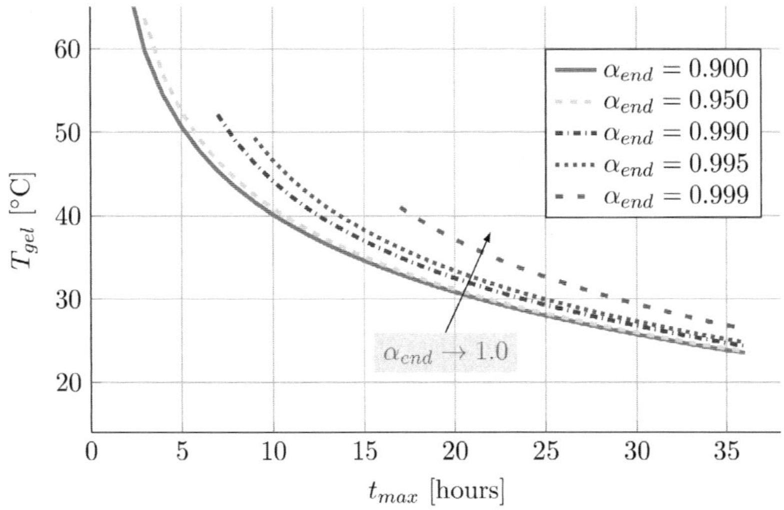

Figure 10: The obtained value for the temperature of gelation plotted as an output function of the time constraint t_{max}. Various curves for the minimum attained degree of cure α_{end} are shown.

et al, who also reported an uncertainty of ± 0.05. It may be argued that since the optimization scheme will only function if the α_{gel} is reached at the low temperature stage the proper α_{gel} input value to the optimization would be $\alpha_{gel} = 0.66 + 0.05 = 0.71$. In general it should be suggested that the α_{gel} input should be the upper bound value of experimental observations of α_{gel}. Common values of α_{gel} for epoxies are 0.6 - 0.7 [9] [13]. Other types of thermoset resins may have significantly different values for α_{gel}. Polyesters are reported to have gelation degree of cures as low as $\alpha_{gel} = 0.1 - 0.3$, which may also mean that the prospect of changing the temperature of gelation of a cure cycle for other types of thermoset resins are very different than the prospect for epoxy resins.

The requirement for the minimum acceptable degree of cure α_{end} after completing the curing cycle may also affect the outcome of the optimization process. It is noted here that White and Hahn [9] showed that epoxies cured to less than 0.95 degree of cure suffered from impaired mechanical properties. It is thus recommend that $\alpha_{end} \geq 0.95$ for epoxies.

The optimization process outlined in this paper is based on a 0-D curing model that do not take into account the thermal conduction problems of laminates with finite thickness. For thin plates with thickness's less than 3 mm the through thickness thermal gradients can be assumed to be low enough that the cure model will provide sufficiently accurate results. For thick laminates the cure kinetic model would have to be implemented into a 1-D thermo-chemical model in order to account for temperature and cure gradients through the laminate thickness. The procedure would have to satisfy eq. (9) at all points through the thickness of the laminate. The procedure would also have to avoid high exothermic reactions during the first temperature stage. Low exothermic peaks are however an intrinsic property of low temperature cures as the low cure rate guarantees a low exothermic heat release rate. The optimization process also excludes the required temperature ramp both from the room temperature to the first temperature stage T_1 as well as the temperature ramp from the first to the second temperature stage with temperature T_2. In order to implement a temperature cycle in a practical application these temperature ramps can of course not be neglected. The temperature ramps can be included either by using a fixed heating rate and deducting the time needed to heat from room temperature to T_2, which will correspond to the total time needed for heating. More sophisticated approaches would include additional iterative procedures and will not be further detailed here.

7. Conclusion

A scheme for optimizing thermoset resin cure cycles has been presented. The scheme takes advantage the idea that lowering the temperature of gelation for a thermoset resin will result in lower magnitudes of residual stresses in composite laminates. The outlined optimization scheme requires a low degree of experimental characterization because it does not use on residual stresses or strains as an objective function, which would require costly resin characterization experiments in order create a cure dependent thermo-viscoelastic material model. Instead the optimization relies on minimizing the temperature of gelation of the thermoset resin used in composite manufacturing. The use of the temperature of gelation also means that the implementation of the scheme in a numerical code is greatly simplified compared to the required numerical implementation of cure dependent visco-elastic material models in a classical constraint optimization scheme.

Acknowledgements

The authors gratefully acknowledges the financial support from Civilingeniør Frederik Leth Christiansen Almennyttige Fond.

References

[1] Hüther J and Brøndsted P 2016 *IOP Conference Series: Materials Science and Engineering*
[2] Mikkelsen L, Hüther J, Pereira G and Brøndsted P 2018 *Composites Science and Technology* Submitted pre-print
[3] Warnet L, Akkerman R and Reed P 2003 *Fracture of Polymers, Composites and Adhesives II* **32** 465 – 476 ISSN 1566-1369 URL `http://www.sciencedirect.com/science/article/pii/S1566136903801176`
[4] Zobeiry N, Forghani A, Li C, Gordnian K, Thorpe R, Vaziri R, Fernlund G and Poursartip A 2016 *Philosophical Transactions of the Royal Society of London A: Mathematical, Physical and Engineering Sciences* **374** ISSN 1364-503X (*Preprint* `http://rsta.royalsocietypublishing.org/content/374/2071/20150278.full.pdf`) URL `http://rsta.royalsocietypublishing.org/content/374/2071/20150278`
[5] Asp L, Berglund L and Talreja R 1996 *Composites Science and Technology* **56** 1089 – 1097 ISSN 0266-3538 URL `http://www.sciencedirect.com/science/article/pii/0266353896000747`
[6] Prasatya P, McKenna G B and Simon S L 2013 *Journal of Composite Materials* **35** 826–848 URL `https://doi.org/10.1177/a037322`
[7] Mortensen U A, Andersen T L, Christensen J and Maduro M A M 2018 *Proceedings of the 18th European Conference on Composite Materials*
[8] Genidy M S, Madhukar M S and Russell J D 1999 *Journal of Reinforced Plastics and Composites* **18** 1304–1321 URL `https://doi.org/10.1177/073168449901801403`
[9] White S and Hahn H 1993 *Journal of Composite Materials* **27** 1352–1378 URL `https://doi.org/10.1177/002199839302701402`
[10] Gopal A K, Adali S and Verijenko V E 2000 *Composite Structures* **48** 99 – 106 ISSN 0263-8223 URL `http://www.sciencedirect.com/science/article/pii/S026382239900080X`
[11] Shah P, Halls V, Zheng J and Batra R 2018 *Journal of Composite Materials* **52** 773–792 URL `https://doi.org/10.1177/0021998317714317`
[12] Yousefi A, Lafleur P G and Gauvin R 1997 *Polymer Composites* **18** 157–168 URL `https://onlinelibrary.wiley.com/doi/abs/10.1002/pc.10270`
[13] O'Brien D J and White S R 2003 *Polymer Engineering & Science* **43** 863–874 URL `https://onlinelibrary.wiley.com/doi/abs/10.1002/pen.10071`
[14] Khoun L, Centea T and Hubert P 2010 *Journal of Composite Materials* **44** 1397–1415 (*Preprint* `https://doi.org/10.1177/0021998309353960`) URL `https://doi.org/10.1177/0021998309353960`
[15] Bogetti T A and John W Gillespie J 1992 *Journal of Composite Materials* **26** 626–660 (*Preprint* `https://doi.org/10.1177/002199839202600502`) URL `https://doi.org/10.1177/002199839202600502`
[16] Wisnom M, Gigliotti M, Ersoy N, Campbell M and Potter K 2006 *Composites Part A: Applied Science and Manufacturing* **37** 522 – 529 ISSN 1359-835X internal Stresses in Polymer Composites URL `http://www.sciencedirect.com/science/article/pii/S1359835X0500179X`
[17] Flory P J 1953 *Principles of Polymer Chemistry* (Ithaca, New York, USA: Cornell University Press) ISBN 0-8014-0134-8

Investigation on multiaxial strength reduction for multi-directional laminates under variable amplitude loading

M Möller[1], J Blaurock[1], G Ziegmann[2] and A Esderts[3]

[1] TH Köln, Institute for Automotive Engineering, Betzdorfer Straße 2, 50679 Cologne, Germany, https://www.th-koeln.de/

[2] TU Clausthal, Institute of Polymer Materials and Plastics Engineering, Agricolastr. 6, 38678 Clausthal-Zellerfeld, Germany, https://www.puk.tu-clausthal.de/

[3] TU Clausthal, Institute for Plant Engineering and Fatigue Analysis, Leibnizstr. 32, 38678 Clausthal-Zellerfeld, Germany, https://www.imab.tu-clausthal.de/

E-mail: marc.moeller@th-koeln.de, +49 2261-8196-6326

Abstract. Due to their excellent ratio of high strength and stiffness to low density, the use of composite materials in lightweight solutions is growing in widely varying industry sectors. Most of the lightweight structures are loaded cyclically and therefore there is the need for optimized computation models concerning the fatigue of composite materials. The present study mainly focuses on the fatigue life estimation of continuous fibre reinforced plastics under variable amplitude loads. The approach is based on iterative ply-wise calculation of multiaxial stress states using classical laminate theory and a subsequent analysis of the material effort with the use of the failure criteria by Puck. A number of suitable linear and nonlinear residual strength models are examined for an in-depth approach on the multiaxial strength reduction at the ply-level. The aim of this study is to calculate fatigue life of multi-directional laminates with input data from uni-directional plies only. The model is finally validated for a multi-directional laminate under variable amplitude loads with arbitrary pulsating and alternating stress ratios using experimental data from the well documented composite fatigue data base OptiDAT.

1. Introduction

Particularly in the era of lightweight manufacturing, resource conservation and sustainability, and therewith the steadily increasing demand for more efficient utilization of lightweight materials, deepening the understanding of the behavior of composite materials is becoming increasingly important. Whilst in operation, most structural parts are subjected to cyclic loads, which especially requires further research of the fatigue material behavior of composites. Therefore, the current research focuses on the calculation of fatigue life and residual strengths of multi-directional laminates via ply-wise structural analysis. The present paper mainly deals with variable amplitude loads in terms of two-block loading considering different residual strength models. Figure 1 shows a simplified routine for the lifetime estimation of composites using a ply-wise fatigue analysis based on a modified classical laminate theory (CLT). On this occasion, only data of uni-directional plies is used for fatigue life and residual strength estimation of multi-directional laminates. As shown in Figure 1 the current stiffness and strength of each ply is calculated within every cycle of the actual block load. The status of the laminate is updated and the current stress state within each ply is then calculated subsequently via CLT. The current

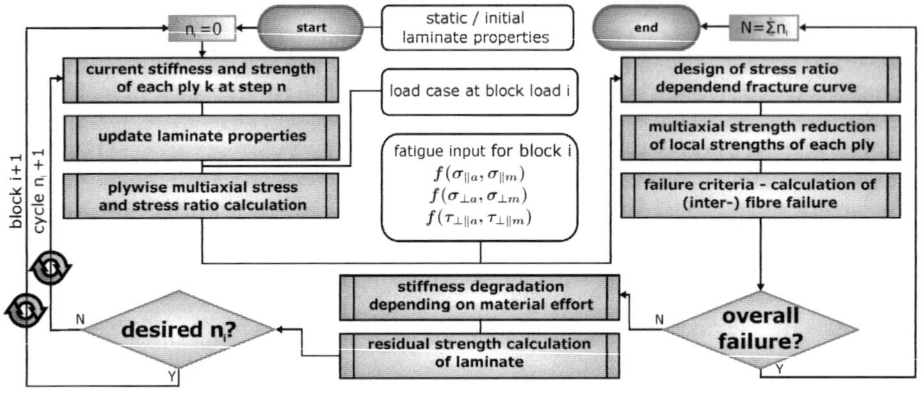

Figure 1. Routine for lifetime estimation of composites under variable amplitude loading

ply-related fracture curves are then designed with the use of S-N curves for the present stress ratios and a multiaxial strength reduction of all five in-plane strengths is carried out. After that, the failure conditions are checked on the ply-level and changes to the ply-related stiffness values are made. The procedure is repeated until failure occurred in every single ply.

2. Models

2.1. Failure Criteria

In the present paper, the failure criteria by Puck [2] is used for calculations of fibre and inter-fibre failure. To do justice to the increase of strengths due to the in-situ effect of the plies within the laminate, the equations for the stress expose factors under transverse tension (Mode A) and transverse compression (Mode B or Mode C) are modified as follows

$$f_{e,IFFA}(n) = \sqrt{\left(\frac{\tau_{\perp\parallel}(n)}{X_t(n)}\right)^2 + \left(1 - p^+_{\perp\parallel}\frac{Y'_t(n)}{S'_{\perp\parallel}(n)}\right)^2 \left(\frac{\sigma_\perp(n)}{Y'_t(n)}\right)^2} + p^+_{\perp\parallel}\frac{\sigma_\perp(n)}{S'_{\perp\parallel}(n)} \tag{1}$$

$$f_{e,IFFB}(n) = \frac{1}{S'_{\perp\parallel}(n)}\left(\sqrt{\tau^2_{\perp\parallel}(n) + \left(p^-_{\perp\parallel}\sigma_\perp(n)\right)^2} + p^-_{\perp\parallel}\sigma_\perp(n)\right) \tag{2}$$

$$f_{e,IFFC}(n) = \left[\left(\frac{\tau_{\perp\parallel}(n)}{2\left(1 + p^-_{\perp\perp}\right)S'_{\perp\parallel}(n)}\right)^2 + \left(\frac{\sigma_\perp(n)}{Y_c(n)}\right)^2\right]\frac{Y_c(n)}{(-\sigma_\perp(n))} \tag{3}$$

where $\sigma_\perp$ and $\tau_{\perp\parallel}$ are the transverse and in-plane shear stress, X_t is the parallel tension strength, Y_c is the transverse compression strength, $\{p^+_{\perp\parallel}, p^-_{\perp\parallel}, p^+_{\perp\perp}$ and $p^-_{\perp\perp}\}$ are the inclines of the failure envelope and Y'_t and $S'_{\perp\parallel}$ are the increased transverse tensile and in-plane shear strength for embedded plies by Wang and Karihaloo [3] according to the following equation

$$Y'_t = Y_t\Big[1 + \mathcal{A}f_t(\Delta\Theta)\Big], \quad S'_{\perp\parallel} = S_{\perp\parallel}\Big[1 + \mathcal{C}f_t(\Delta\Theta)\Big] \tag{4}$$

where $\mathcal{A}$ and $\mathcal{C}$ [3] are material parameters, which take into account the number of plies of a given thickness in the embedded lamina. The functions $f_t(\Delta\Theta)$ and $f_s(\Delta\Theta)$ are given by the orientation of the neighboring plies as follows

$$min\left[\frac{\sin^2(x)}{1+\sin^2(x)}, \frac{\sin^2(y)}{1+\sin^2(y)}\right] = \begin{cases} f_t(\Delta\Theta) \text{ with } x = \Delta\Theta_a & \text{and } y = \Delta\Theta_b \\ f_s(\Delta\Theta) \text{ with } x = 2\Delta\Theta_a & \text{and } y = 2\Delta\Theta_b \end{cases} \tag{5}$$

2.2. Strength Reduction

Different existing residual strength models are brought together to one procedure for the application within the computational routine. For the repetitive computations, the residual strength in every cycle is defined by

$$S_{r,n} = S_{r,n-1} - \left\{\left[(S_{st} - \sigma_{max,i})\left(1 - \left(\frac{n_{i-1}}{N_i}\right)^{\alpha_i}\right)^{\beta_i}\right] - \left[(S_{st} - \sigma_{max,i})\left(1 - \left(\frac{n_i}{N_i}\right)^{\alpha_i}\right)^{\beta_i}\right]\right\} \tag{6}$$

where the residual strengths $S_{r,n}$ and $S_{r,n-1}$ at the nth and the n-1th step as well as the static strength S_{st} represent the five local strengths in terms of the parallel tensile and compression strengths (X_t, X_c), the transverse tensile and compression strengths (Y_t, Y_c) or the in-plane shear strength ($S_{\perp\parallel}$). Equation 6 refers to different residual strength models with varying combinations of material parameters α_i and β_i. With α_i and β_i being both zero, there won't be any strength degradation of the addressed strength. For $\alpha_i = 1$ and $\beta_i = 1$ the model refers to the linear residual strength model by Broutman and Sahu [4]. As studies on the fatigue life under spectrum loads by Passipoularidis and Philippidis [5, 6] have shown, the linear model is generally the most efficient choice for conservative predictions and without the need of any residual strength data. For arbitrary values of α_i with $\beta_i = 1$ the model represents the nonlinear one-parameter model used by [7, 8]. The strength degradation is then simulated with either a steep loss of strength at the beginning or at the end of the lifetime. For all other values of α_i and β_i the model refers to the "normalized residual strength model (NRSM)" by Stojkovic et al. [9]. By using the NRSM, the typical initial loss of strength as well as the sudden decrease in strength at the end of life can be modeled very well. Equation 6 is applicable for variable amplitude loads, as shown for two exemplary two-block loads in Figure 5. Two degradation

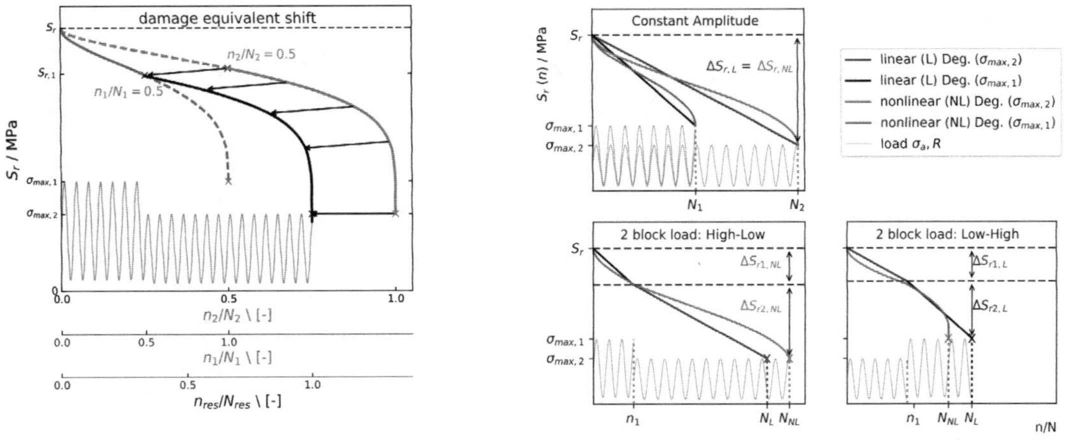

Figure 2. Life fraction equivalent shift of residual strength behaviors under block loads

behavior at $\sigma_{max,1}$ and $\sigma_{max,2}$ in High-Low and Low-High loads are illustrated exemplarily. If the maximum stress is changed, the new maximum stress is taken into consideration and the degradation starts at the equivalent life fraction of the new curve. The shape of the degradation curve stays the same while the start point of the second curve is moved to the equivalent life fraction of the first curve, as shown in Figure 2 for an exemplary High-Low block load.

2.3. Stiffness Degradation

In the current work Puck's model for the stiffness degradation of transverse and shear values due to inter-fibre fracture [10] is applied. The model is used for stiffness degradation in static failure analysis in combination with Puck's failure criteria, but has already been examined for the modeling under fatigue loads by Adden and Horst [11].

$$\begin{bmatrix} E_{\perp}(n) \\ G_{\perp\parallel}(n) \\ \nu_{\perp\parallel} \\ \nu_{\parallel\perp}(n) \end{bmatrix} = \begin{bmatrix} E_{\perp}^{0}\left(\dfrac{1-\eta_{r,\perp}}{1+c_{\perp}(f_{e,IFF}(n)-1)^{\xi_{\perp}}}+\eta_{r,\perp}\right) \\ G_{\perp\parallel}^{0}\left(\dfrac{1-\eta_{r,\perp\parallel}}{1+c_{\perp\parallel}(f_{e,IFF}(n)-1)^{\xi_{\perp\parallel}}}+\eta_{r,\perp\parallel}\right) \\ const. \\ \nu_{\perp\parallel}\cdot(E_{\perp}(n)/E_{\parallel}(n)) \end{bmatrix} \tag{7}$$

The following parameters are used for stiffness degradation under inter-fibre failure mode A, B and C for all further investigations:

- $\eta_{\perp_A} = 0.3$ and $\eta_{\perp_{B,C}} = 0.5$
- $\eta_{\perp\parallel_{A,B,C}} = 0.25$, $\xi_{\perp_{A,B,C}} = 1.31$, $c_{\perp_{A,B,C}} = 5.34$ and $\xi_{\perp\parallel_{A,B,C}} = 1.5$, $c_{\perp\parallel_{A,B,C}} = 0.7$

3. Experimental data

Fatigue data from experimental tests is taken from the OptiDAT[1] database [12]. All of the examined materials are glass fibre reinforced plastics (GFRP). The uni-directional material consists of non-woven unidirectional glass rovings with a minor amount of off-axis reinforcement, made of polyester (PES) yarn, and the bi-axial material consists of non-woven glass rovings in 2 layers (±45°). The dry fibers are are infused with the epoxy system "Prime 20" by vacuum assisted resin transfer molding and the system is post-cured at $80°C$ for 4 hours [13].

Mat.#1-#3 The basic material configurations are the uni-directional laminate $[0]_5$ ("UD2" [12]), transverse tested uni-directional laminate with stacking sequence $[0]_7$ ("UD3" [12]) and biaxial laminate with stacking sequence $[\pm 45]_5$ ("MD3" [12]).

Mat.#4 The multi-directional laminate with stacking sequence $[(\pm 45/0)_4/\pm 45]$ ("MD2" [12]) is used for the validation of the simulation method.

Only input data for static and fatigue tests of uni-directional (Mat.#1 and Mat.#2) and bi-axial (Mat.#3) specimen is used as input data for the simulations. Table 1 shows the static input values from static tests of flat specimen, which consist of fibre volume contents between 51,94 % and 53,73 %. Figure 3 examplarily shows the fit of equation 6 to experimental residual

[1] The OPTIMAT BLADES project is a European research project, supported by the European Union. Project Coordinators: Wind Turbine Materials and Constructions (WMC, Netherlands) and Energy Reasearch Centre of Netherlands (ECN - Netherlands). Database Contributors: Center for Renewable Energy Sources (CRES - Greece), German Aerospace Center (DLR - Germany), Rutherford Appleton Laboratory (RAL - United Kingdoms), National Research Laboratory (RISOE - Denmark), University of Patras (UP - Greece), Technical Research Center (VTT - Finland), Free University Brussels (VUB - Belgium) and Wind Turbine Materials and Constructions (WMC - Netherlands).

	$E_{\parallel}$ GPa	$E_{\perp}$ GPa	$G_{\perp\parallel}$ GPa	$\nu_{\perp\parallel}$	$R_{\parallel}^{+}$ MPa	$R_{\parallel}^{-}$ MPa	$R_{\perp}^{+}$ MPa	$R_{\perp}^{-}$ MPa	$R_{\perp\parallel}$ MPa
mean value	38,43	14,07	4,23	0,2893	810,56	469,96	55,88	164,95	56,1

Table 1. Calculated static values for experimental data from [12]

strength data of Mat.#1 and the estimation of $\alpha_{\parallel}^{+}, \beta_{\parallel}^{+}$ for a description of the fibre parallel residual strength behavior under tension-tension load. Equation 6 is fitted to residual strength

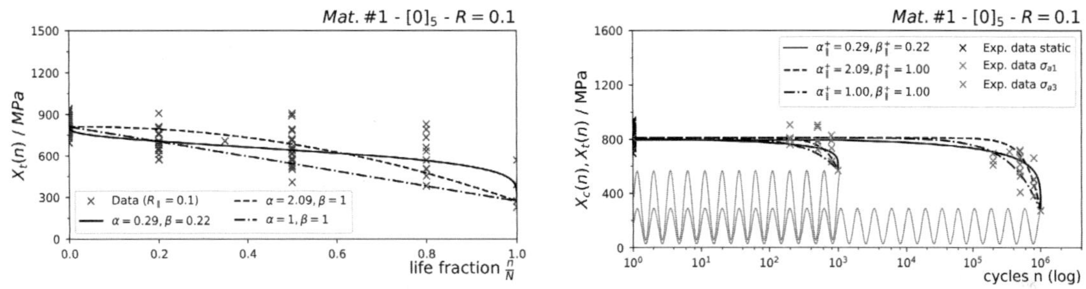

Figure 3. Estimation of $\alpha_{\parallel}^{+}, \beta_{\parallel}^{+}$ for $R_{\parallel}$= 0.1

data of all five local strengths X_t, X_c, Y_t, Y_c or $S_{\perp\parallel}$ at the ply-level for a deeper investigation on multiaxial fatigue. Table 3 shows the values for α_i and β_i concerning each strength strength for tension-tension and tension-compression loads.

	ratio	$\alpha_{\parallel}^{+}$, $\beta_{\parallel}^{+}$	$\alpha_{\perp}^{+}$, $\beta_{\perp}^{+}$	$\alpha_{\perp\parallel}$,$\beta_{\perp\parallel}$	$\alpha_{\parallel}^{-}$, $\beta_{\parallel}^{-}$	$\alpha_{\perp}^{-}$, $\beta_{\perp}^{-}$
conf. #1	*all*	1.00 , 1.00	1.00 , 1.00	1.00 , 1.00	0.00 , 0.00	0.00 , 0.00
conf. #2	0.1	0.29 , 0.22	1.02 , 0.66	0.70 , 0.37	0.00 , 0.00	0.00 , 0.00
	−1	0.44 , 0.29	0.32 , 0.24	0.70 , 0.37	0.00 , 0.00	0.00 , 0.00

Table 2. Coefficents for residual strength simulation with experimental data from [12]

4. Results

The results for predicted fatigue life and residual strength of the multi-directional laminate Mat.#4 and related experimental data for constant amplitude loads are shown in Figure 4. The Figure illustrates the predictions for a low and a high amplitude load regarding the further considered block load simulations. As can be seen, the fatigue life of the multi-directional laminate is estimated very accurately for constant amplitude loads. There a growing deviations between predicted and experimental fatigue life at higher number of endured cycles for R=0.1, but the estimations are on the safe side. In the initial phase, the use of the nonlinear model leads to slightly lower residual strength values than the linear model, because of the initial loss in strength. While there are nearly no differences between the estimations until approximately 30-40% lifetime, there are rapidly growing deviations between the two models in the late stage, particularly in the region of 80% lifetime. There is no big difference in the predicted fatigue life, but earlier strength degradation leads to slightly earlier failure in the ±45°-plies and, consequently, to earlier stress redistribution into the 0°-plies. Taking into account the

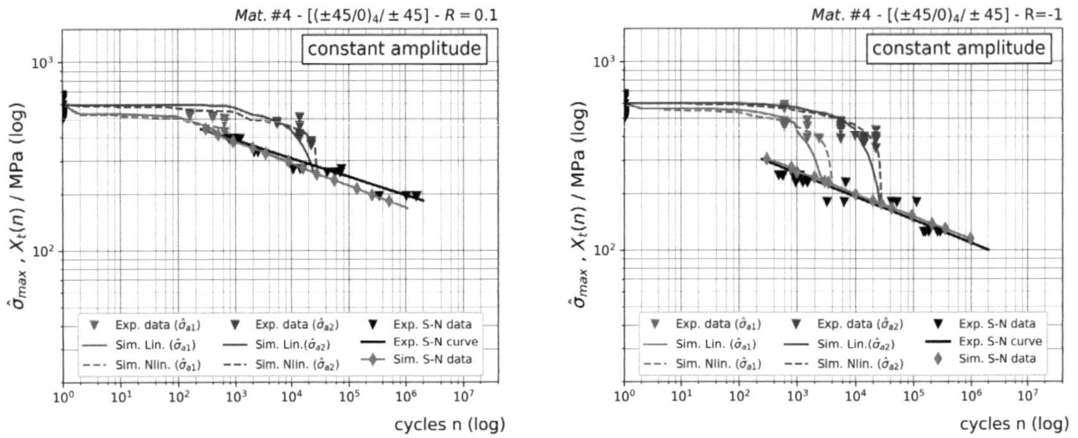

Figure 4. Predicted S-N curves and residual strength for constant amplitude loads

logarithmic scale, the differences in residual strength estimation between the use of linear and nonlinear model are significant and the nonlinear model leads to considerably better results. On the other hand, an enormous amount of data is needed for a holistic approach on multiaxial strength degradation with the nonlinear model. Nonetheless, it might also be beneficial to only use the nonlinear model for fibre parallel strength reduction and reduce the amount of data needed. Figure 5 shows the predicted life and residual strength for a High-Low and a Low-High load compared to experimental data. In the right pictures, the x-axis is divided into two parts, for a better view on the differences at the late stages. As shown in the box-plots, the interquartile range is very small at Low-High loads compared to the box-plot in High-Low loads and would therefore be difficult to examine within the plot. The four lower pictures of Figure 5

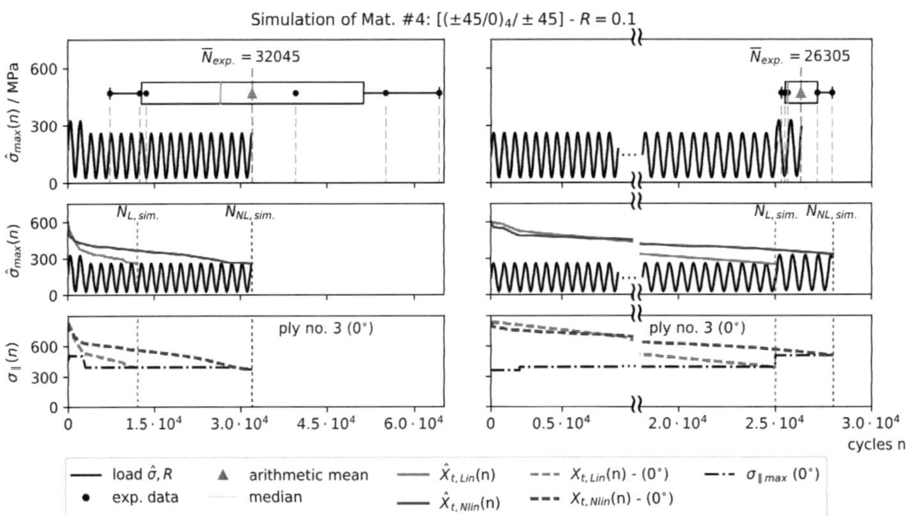

Figure 5. Predicted fatigue life and strength for High-Low and Low-High block loads at R=0.1

show, that the laminate failure is of course accompanied by the failure of the 0° plies. The use of the nonlinear model leads to more precise and less conservative predictions of fatigue life for the High-Low block load, when compared to the simulations with the linear model. Concerning the Low-High tension-tension block loads, both models lead to a very early failure after the start of the second block load, which matches the experimental observations. Figure 6 shows the predictions for a High-Low and a Low-High block load with a alternating stress ratio R=-1. Similar to the findings in Figure 5, the predictions of fatigue life are improved with the use of

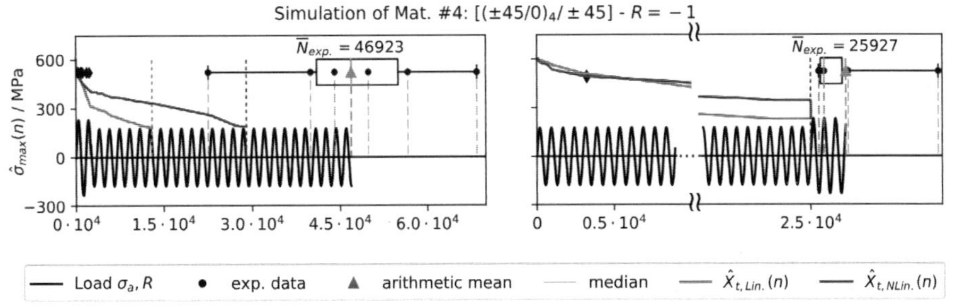

Figure 6. Predicted fatigue life and strength for High-Low and Low-High block loads at R=-1

the nonlinear model. While the use of both models still yield very conservative predictions, both predictions for fatigue life yield large differences from the arithmetic mean and are outside of the interquartile range for the experimental data. Anyway, there is quite a few experimental data with prematurely failed specimen in the first block load in Figure 6, which have not been taken into consideration for statistical evaluations and box-plot design. Predictions for the Low-High block load are also very similar to the results in tension-tension loads. Both models lead to conservative predictions with early failure in the first cycle after the start of the second block load. As shown in Figure 7, the simulation results for block loads with mixed stress ratios R=0.1 and R=-1 are largely a projection of those seen in Figure 5 and 6. The high tension-tension load

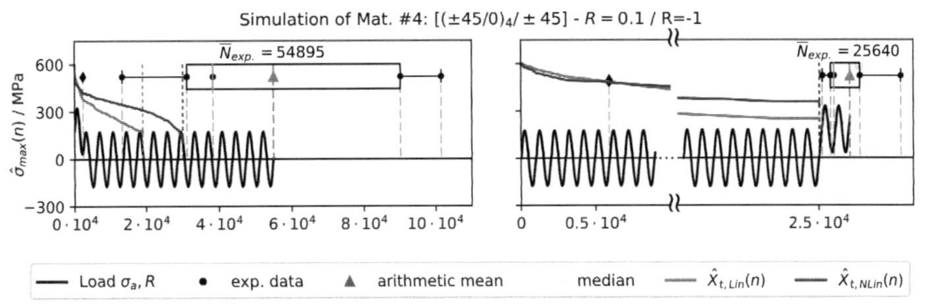

Figure 7. Fatigue life and strength for High-Low and Low-High block loads at R=0.1 and R=-1

followed by the low tension-compression load leads to the same conservative predictions as seen before. Still the nonlinear model is improving the predictions and yields less large deviations from the arithmetic mean on the safe-side. Similar to the results before, the start of the high tension-tension load after the initial low tension-compression load leads to a sudden failure in the first cycle of the second load block. Compared to the experimental data, the simulations are on the safe side, although the standard deviations of the experimental data are very small in that case. For a better comparison of the results, the model error function used by Post et al. [14] is used to compare the predicted fatigue life to the mean life from experimental data as follows

$$\text{Model Error} = M_e = log\left(\frac{N_{model}}{N_{experiment}}\right) \tag{8}$$

The model errors for the investigated block loads are illustrated in Table 3.

	R=0.1			R=-1			R=0.1 / R= -1		
	$\overline{N}_{exp}$	Lin .	Nonlin.	$\overline{N}_{exp}$	Lin .	Nonlin.	$\overline{N}_{exp}$	Lin.	Nonlin.
H-L	32045	-0.44	0.00	46923	-0.55	-0.21	54895	-0.46	-0.26
L-H	26305	-0.02	0.02	25927	-0.01	-0.01	25640	-0.01	-0.01

Table 3. Coefficents for residual strength simulation with experimental data from [12]

5. Conclusions

In summary, the use of the nonlinear model improves the predictions of residual strength in constant amplitude loads and the predictions of fatigue life in High-Low block loads. There is nearly no difference in the simulations results for the Low-High block loads, since the failure almost always occurs right at the beginning of the second cycle.

6. Acknowledgments

The authors wish to thank the contributors to the OptiDAT database [12] for the possibility to validate their investigations with the extensive amount of provided fatigue data.

References

[1] T.P. Philippidis and V.A. Passipoularidis: *Residual strength after fatigue in composites: Theory vs. experiment.* International Journal of Fatigue, Vol. 29, pp. 2104-2116, 2007.

[2] A. Puck and H. Schürmann: *Failure analysis of FRP laminates by means of physically based phenomenological models.* Composites Science and Technology, Vol. 62, pp. 1633-1662, 2002.

[3] J. Wang and B.L. Karihaloo: *Optimum in situ strength design of composite laminates. Part I: in situ strength parameters* Journal of Composite Materials, pp.1314–1337, 1996.

[4] L. Broutman and S. Sahu: *A new theory to predict cumulative fatigue damage in fiberglass reinforced plastics.* Composite Materials: Testing and Design (2nd Conference), ASTM STP497, Corten HT (1972) 170188.

[5] V.A. Passipoularidis and T.P. Philippidis: *A study of factors affecting life prediction of composites under spectrum loading.* International Journal of Fatigue, vol. 31, no. 3, pp. 408 - 417, 2009.

[6] V.A. Passipoularidis and T.P. Philippidis and P. Brondsted: *Fatigue life prediction in composites using progressive damage modelling under block and spectrum loading.* International Journal of Fatigue, vol. 33, no. 2, pp. 132 - 144, 2011.

[7] J.R. Schaff and B.D. Davidson: *Life prediction methodology for composite structures. part i - constant amplitude and two stress level fatigue.* Journal of Composite Materials, pp. 170188, 1997.

[8] K. Reifsnider and W.W. Stinchcomb: *A critical-element model of the residual strength and life of fatigue-loaded composite coupons.* Composite Materials: Fatigue and Fracture K.L., ASTM STP-907, American Society for Testing and Materials, Philadelphia, PA (1986), pp. 298-313, 1986.

[9] N. Stojkovic and F. Radomir and H. Pasternak: *Mathematical model for the prediction of strength degradation of composites subjected to constant amplitude fatigue.* International Journal of Fatigue, pp. 478487, 2017.

[10] A. Puck and M. Mannigel: *Physically based non-linear stress-strain relations for the inter-fibre fracture analysis of FRP laminates.* Composites Science and Technology, vol. 67, pp. 1955 - 1964, 2007.

[11] S. Adden and P. Horst: *Stiffness degradation under fatigue in multiaxially loaded non-crimped-fabrics.* International Journal of Fatigue, pp. 108–122, vol. 32, 2010.

[12] R. Nijssen: *Optidat - fatigue of wind turbine materials database.* Knowledge Centre WMC, Available at $https://wmc.eu/optimatblades_optidat.php$.

[13] T. K. Jacobsen: *Material Specification, Reference Material (OPTIMAT) - Glass-epoxy material.* Optimat Blades, Material Specification, Doc. No: 10004, revision 1, May 7th, 2002.

[14] N.L. Post, S.W. Case, J.J. Lesko: *Modeling the variable amplitude fatigue of composite materials: A review and evaluation of the state of the art for spectrum loading.* International Journal of Fatigue, vol. 30, no. 12, pp. 2064 - 2086, 2008.

IOP Conf. Series: Materials Science and Engineering **388** (2018) 012016 doi:10.1088/1757-899X/388/1/012016

Tension-compression fatigue behavior of 3D woven composites

Yao Siyuan and Chen Xiuhua

School of Aeronautics and Astronautics, Shanghai Jiao Tong University, 800 Dongchuan Road, 2100240, Shanghai, P.R.China

E-mail: chenxiuhua@sjtu.edu.cn

Abstract. In order to investigate the fatigue behavior of three-dimensional (3D) woven composites under tension-compression cyclic loading, fatigue tests which had a stress ratio of R=-1 were conducted. Under different loading levels, weft and warp fatigue tests were conducted. The hysteresis loops and curves of residual stiffness ratio via fatigue life were obtained. It is implicated that the fatigued damage progress of 3D woven composites under tension-compression cyclic loading contains three stages mainly, in which the failure of matrix, transverse crack propagation inside the yarns and the fracture of yarns occur in sequence. The crush and debonding of matrix, the avulsion of yarns which are perpendicular to the loading direction and the fracture of yarn which are along the loading direction are major failure modes. Semi-empirical models are obtained to describe the fatigue process and forecast the fatigue life of 3D woven composites. The mathematical equations fit the experimental results well.

1. Introduction

3D woven composites are currently being used as a class of aerospace material which combine lightweight with high damage tolerance. The high damage resistance of 3D woven composites is derived from through-thickness structural fibers, which resist large-scale delamination crack growth [1]. These enhancements motivate much of the present research on the mechanical properties of 3D woven composites.

Jin L et al [2] investigated the three-point bending fatigue damage behavior of 3D angle-interlock woven composites. The resin cracks, the yarns cracks and the resin-yarns interface debonding are the three modes of fatigue damage for the 3D woven composites. Dai S et al [3] presents a study on the open-hole quasi-static tensile and tension-tension fatigue behavior of an orthogonal and an angle-interlock 3D woven composites. The residual fatigue strength was found to be similar to the quasi-static tensile strength in both weaves. The surface crack initiation and progression during fatigue loading was identified using thermoelastic stress analysis which revealed that the orthogonal weave had larger surface damage area than the angle-interlock weave. Yu B et al [4] conducted a detailed investigation of failure mechanisms for angle-interlocked (AI) and modified layer-to-layer (MLL) 3D woven composites under tension-tension fatigue loading using surface optical microscopy, cross-sectional SEM imaging, and non-destructive X-ray computed tomography. A higher crack density was found in the AI composite than the MLL composite. Transverse cracking initiates in the fiber rich regions of weft yarns rather than the resin rich regions. Delamination in the failed MLL specimen were more extensive than the AI specimen. Using time-lapse X-ray computed tomography (CT), Yu B et al [5] also found the fatigue damage of 3D woven composites distributed regularly throughout the composite according to the repeating unit, even at large fractions of the total life. Roirand Q et al [6] modelled the failure mechanisms of 3D woven composites on 3D periodic cells by finite element calculations. A stochastic approach was applied to a simple representative elementary periodic cell. It was shown that this stochastic approach allowed more realistic failure simulations avoiding the idealized symmetry due to the deterministic modelling. Wilkinson M P and Ruggles-Wrenn M B [7] studied the tension-tension fatigue behavior of two polymer matrix composites (PMCs) at elevated temperature, which are the non-crimp 3D orthogonal weave composite (3D PMC) and the laminated composite (2D PMC). The 3D PMC offered improved delamination resistance.

Previous studies on fatigue properties of 3D woven composite materials mainly focus on three point bending fatigue and tension-tension fatigue. In order to investigate the tension-compression fatigue behavior of 3D woven composites, both weft tension-compression fatigue tests (WETCF) and warp tensions-compression fatigue tests (WATCF) were conducted. The fatigue damage was described based on the experimental phenomenon and results. Semi-empirical models are obtained to describe the fatigue process and forecast the fatigue life of 3D woven composites. The experimental results verified the validity of the models.

2. Experimental

2.1. Material

The unit cell of the material is shown in figure. 1, consisting of weft, warp and binder carbon fiber yarns and resin matrix. Each periodical unit cell contains four bundles of weft yarns in the direction of length. One binder yarn spans three weft yarns, weaving two layers of weft yarns together. The warp layer is located between the two layers of weft yarns and is arranged with binder yarns. The thickness of a specimen is determined by the number of layers. The specimens used in the fatigue tests contains 7 layers of weft yarns and 6 layers of warp yarns.

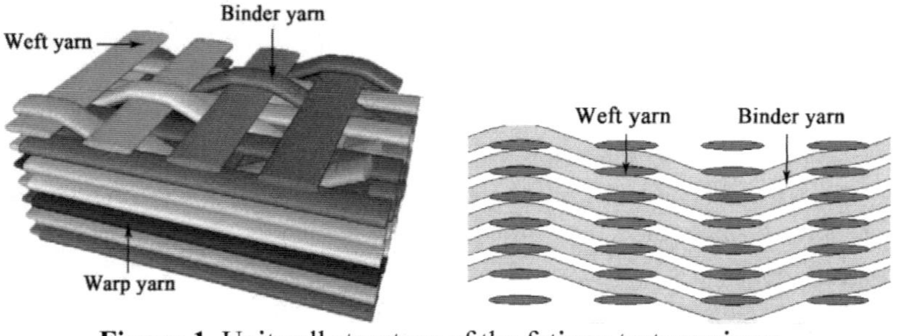

Figure 1. Unit cell structure of the fatigue test specimen.

The tension and compression strength values of the material are shown in table 1, which are obtained from static tests.

Table 1. Ultimate strength of 3D woven composites.

Load Direction	Tension strength/MPa	Compression strength/MPa
Weft	687	370
Warp	855	487

2.2. Tension-compression fatigue tests

Fatigue tests along weft and warp direction which had a stress ratio of R=-1 were conducted. Both weft and warp fatigue tests contain four specimens under different stress level. The stress level of tension-compression fatigue tests is defined as the percentage of peak stress over compression strength. The arrangement of stress level is shown in table 2.

Table 2. Stress level arrangement of fatigue tests.

Direction	Stress level/%			
Weft	47.7	39.8	31.8	23.9
Warp	48.3	42.3	36.2	30.2

The equipment and installation are shown in figure 2. The test machine recorded the number of cyclic loading, the loading force and the displacement of hydraulic chuck. the extension meter records the shape variables that occur during the test process.All tests are conducted under load control, of which the loading frequency is 15 Hz. The tests are completed when the fracture of specimen occurs or the specimen cannot carry the test load.

Figure 2. Apparatus and installation fatigue tests

3. Results and discussions

3.1. Fatigue fracture analysis

Table 3 shows the fatigue life of tension-compression fatigue specimen for 3D woven composites.

Table 3. Fatigue life a of tension-compression fatigue specimen.

WETCF			WATCF		
No.	Stress level/%	Fatigue life	No.	Stress level/%	Fatigue life
S1	47.7	11048	S1	48.3	46155
S2	39.8	132672	S2	42.3	100008
S3	31.8	895942	S3	36.2	481172
S4	23.9	7133147	S4	30.2	2173790

The failure region of specimens are observed under an optical microscope, as shown in figure 3. The crush and debonding of matrix, the avulsion of warp yarns and weft yarns and the fracture of weft yarns are major failure modes of weft tension-compression fatigue specimens. The fracture is nearly neat because of the flat shape of weft yarns. The cross sections of warp and binder yarns are all distorted and cracks distribute on the surface randomly. Specimens under high stress level tend to appear large distortion of warp and binder yarns, while the surface of specimens under low stress level show more fracture of weft yarns and matrix failure.

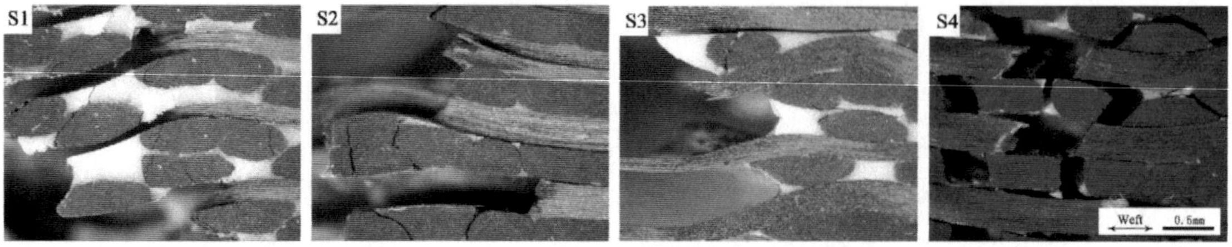

(a)WETCF

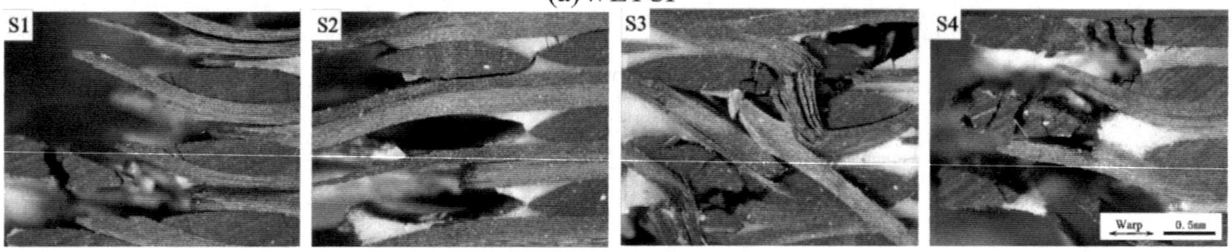

(b)WATCF

Figure 3. Micro pictures of the fracture region.

For warp tension-compression fatigue tests, the cross sections of warp yarns are round and thick, which makes it a longer period of crack propagation. At the end of crack propagation, the remaining warp yarns are not enough to bear the load. Thus irregular breakage and separation of yarns will occur. Kink and avulsion of binder yarns can be observed because of compression load and the concentrated stress given by weft yarns. Specimen under low cyclic load accumulates many crack propagation path. The weft yarn in the fracture path is eventually perforated by multiple cracks.

3.2. Stiffness reduction mechanism

Curves of residual stiffness ratio via normalized fatigue life are obtained, in which the residual stiffness ratio is defined by residual structure stiffness over initial structure stiffness. As shown in figure 4, it is implicated that the fatigued damage progress of 3D woven composites under tension-compression cyclic loading contains three stages mainly, in which the failure of matrix, transverse crack propagation inside the yarns and the fracture of yarns occur in sequence.

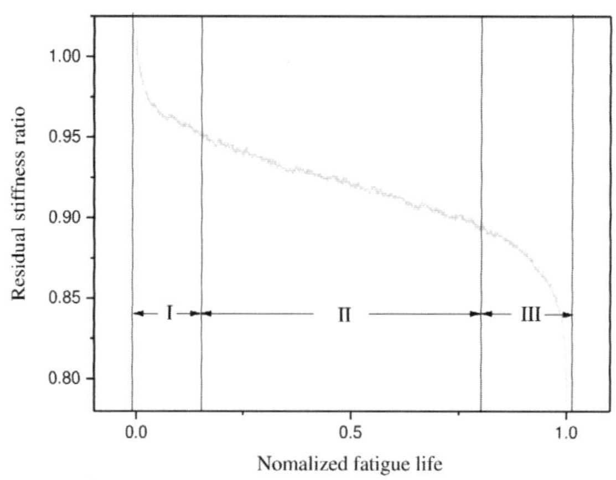

Figure 4. The three stages of tension-compression fatigue damage process.

Figure 5 shows all the specimen's curves of residual stiffness ratio via normalized fatigue life.

Higher stress level results in a longer period of stage II, while lower stress level causes longer stage I and III. Under lower stress amplitude, the strain of matrix is smaller. Thus it is slower for matrix to become failure, which means a longer period of stage I. In the meantime, lower stress makes the fracture of fiber yarns more stochastic instead of deterministic, so stage III tends to be a longer period.

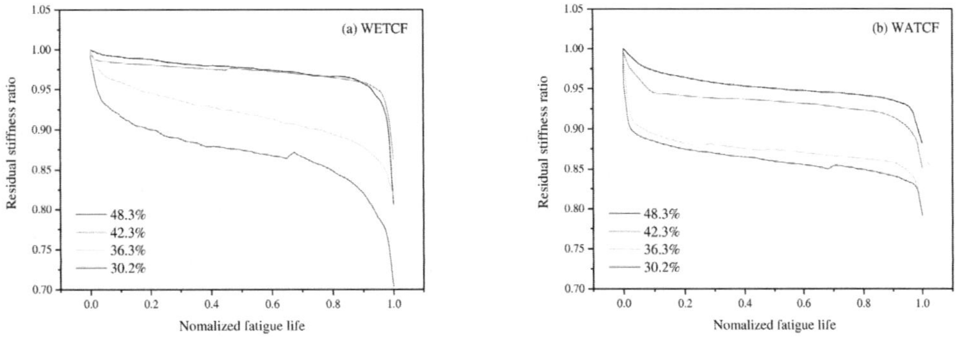

Figure 5. Residual stiffness ratio of tension-compression fatigue specimens.

3.3. Semi-empirical model

A semi-empirical models is investigated to fulfil the needs of engineering and design, as presented in Eq. (1).

$$\frac{E}{E_0}=1-\left(1-\frac{E_f}{E_0}\right)\left[q\left(\frac{n}{N_f}\right)^{m_1}+(1-q)\left(\frac{n}{N_f}\right)^{m_2}\right] \tag{1}$$

where E is the elastic modulus of the material during fatigue tests and E_0 is the elastic modulus of the material before the fatigue tests. E_f is the final modulus of the specimen when the test is completed. In reality, it would probably be zero or a value which is determined by the engineering practice. n is a certain cycle during the fatigue test and N_f is the final fatigue life. In addition, q , m_1 and m_2 are constants determined by the type of material and loading. These constants can be obtained through the least squares fitting method. Eq. (1) is similar to the equation in reference [8].

The mathematical equation describes the relation between the residual stiffness ration and normalized fatigue life. It can be applied to predicate the fatigue life of a serving 3D woven composite structure under tension-compression cyclic load, of which the stress ratio is R=-1. Table 3 shows the parameters of WETCF and WATCF under certain stress level.

Table 4. Fitting parameters of the semi-empirical model.

WETCF				WATCF			
No.	m_1	m_2	q	No.	m_1	m_2	q
S1	0.221445	342.7864	0.268961	S1	0.317528	21.0989	0.422061
S2	0.405357	118.5152	0.386728	S2	1.139266	138.5319	0.374316
S3	0.213681	36.32672	0.172382	S3	0.428574	66.56646	0.222634
S4	0.14014	120.9351	0.634219	S4	0.897774	123.3586	0.155348

Now that the parameters of WETCF under the stress level 31.8% have been obtained, fatigue life can be predicted for specimen under the same stress level. Another two WETCF specimen, S3-1 and S3-2, were tested under the stress level 31.8%, of which the fatigue life are 701931 and 471422, separately. Based on the experimental data, the E/E_0 of S3-1 at cycle 200000 is 0.937348 and that of S3-2 is 0.934345. Given E_f = 0.5, thus the calculated fatigue lives are 890868 and 715819. The errors are 26.9% and 51.8%, which are acceptable.

4. Conclusions

The following conclusions can be drawn from the study.

1) The fracture of fibers, the avulsion and kink of yarns and the crush and debonding of matrix are major failure modes of tension-compression fatigue specimens.
2) The fatigue damage progress of 3D woven composites under tension-compression cyclic loading contains three stages mainly. During stage I, the stiffness of specimen declines rapidly and then slows down before stage II, which is dominated by matrix failure. During stage II, yarn cracks propagate perpendicular to the loading direction, causing the steady decline of structure stiffness. The fracture of yarns occurs during stage III, which results in the sharp decline of the stiffness.
3) Stress level effects a lot on the fatigue process of 3D woven composites under tension-compression fatigue. Higher stress level results in a longer period of stage II and lower stress level results in longer periods of stage I and III.
4) Semi-empirical models are obtained to describe the fatigue process and forecast the fatigue life of 3D woven composites. The mathematical equations fit the experimental results well.

References

[1] Stegschuster G et al 2016 *Composites Part A: Applied Science and Manufacturing* **84** 308-315.
[2] Jin L et al 2013 *International Journal of Damage Mechanics* **22(1)** 3-16.
[3] Dai S et al 2015 *Composite Structures* **131** 765-774.
[4] Yu B et al 2015 *Composites Part A* **77** 37-49.
[5] Yu B et al 2016 *Composites Part A* **82** 279-290.
[6] Roirand Q et al 2017 *Continuum Mechanics & Thermodynamics* **29(5)** 1081-1092.
[7] Wilkinson M P and Ruggles-Wrenn M B 2017 *Applied Composite Materials* **2017** 1-20.
[8] Song J et al 2017 *Composite Structure* **166** 77-86.

IOP Conf. Series: Materials Science and Engineering **388** (2018) 012017 doi:10.1088/1757-899X/388/1/012017

Prediction of fatigue limit for unidirectional carbon fibre/epoxy composites

Bent F. Sørensen and Stergios Goutianos

The Technical University of Denmark, Department of Wind Energy, Section of Composites Mechanics and Structures, Risø Campus, Frederiksborgvej 399, DK-4000 Roskilde, Denmark

E-mail: bsqr@dtu.dk

Abstract. A micromechanics model is used for the prediction of the fatigue limit of unidirectional carbon fibre /epoxy composite materials. The model is based on the hypothesis that failure of a fibre will result in fibre/matrix debonding of the broken fibre. The associated debond crack tip stress fields will raise the stress in the neighbour fibres as the debond crack tips move along the broken fibre and can thus cause failure of the neighbouring fibres. The fatigue limit is defined from the maximum applied cyclic stress that does not induce failure of any neighbour fibres. Effects of microscale mechanical properties are investigated. The model predicts that the fatigue limit, expressed in terms of stress, increases with fibre volume fraction until 50-60 %, whereafter the fatigue limit decreases with increasing fibre volume fraction. With other parameters held fixed, the fatigue limit increases with increasing interfacial frictional sliding shear stress and with decreasing interfacial fracture energy.

1. Introduction

Fatigue - material degradation under cyclic loading - is a phenomenon that greatly influences the design limits of composite materials for load-carrying structures. Examples are wind turbine rotor blades that are designed to be in operation for 20-25 year, during which they will be subjected to cycling varying loads as the blades rotate. During one rotation, each blade will go from a low aerodynamical load (in the bottom position in front or behind the tower) to a high aerodynamical load (at the top position) as well as will undergo a complete change in the direction of the gravitation force. With a rotation time of about 5 seconds per rotation, 25 years of constant use corresponds to more than 150 million rotations. For such high number of load cycles, the use of traditional S-N data may not be attractive, since it is very time consuming and thus expensive to conduct cyclic experiments and establish lifetimes up to 150 million cycles (or higher). Instead, it may be more appropriate to focus on which conditions the material might possess a fatigue limits, i.e., whether a maximum load (or strain) value exists, below which the material would sustain an infinite number of load cycles.

Frequently, the run-out of 1 million cycles has been used as an engineering fatigue limit. Over the past few years, more experiments have been conducted far beyond 1 million cycles. Mandell *et al.* [1] loaded specimen cyclically up to 10^{10} cycles. They found that fatigue failure did happen after 1 million cycles at strain values below the engineering fatigue limit corresponding to run-out at 1 million cycles. These finding do not support the hypothesis of the existence of a true fatigue limit. However, neither do they rule out the possibility that a true endurance fatigue limit can exist.

One hypothesis could be to claim that a true fatigue limit can exist if the material does not develop any damage at all during cyclic loading. While such a hypothesis might be correct, is could also come out being very conservative. For instance, Gamstedt and Talreja [2] investigated the microstructural damage state and found that isolated fibre failures were always presented in specimens having been subjected to cyclic loading. This could lead to the discouraging conclusion that a fatigue limit may never exist. Another hypothesis, therefore, could be that a fatigue limit can exist if the situation of isolated fibre failures is stable in the sense that the failure of one isolated fibre does not lead to failure of neighbor fibres. It is therefore of interest to investigate the influence of one broken fibre on the nearest neighbour fibres. The effect of a broken fibre on neighbour fibres was investigated by use of Raman spectroscopy [3]. It was found that the stress in neighbour fibres was increased at the position of the debond crack tip of a broken fibre. This suggests that the debond crack tip stress field of the broken fibre plays an important role in the occurrence of fibre failure in neighbour fibres. An analytical micromechanical model that accounts for this has recently been developed [4]. The model enables the prediction of a fatigue limit as a a function of microscale parameters. In the present paper we will use this model for the prediction of a fatigue limit for unidirectional carbon fibre composites using available litterature data for AS4 carbon fibre in epoxy matrix.

2. Model summary

2.1. Interface degradation laws

In this section we will briefly describe the main assumptions behind the analytical model. The situation we analyse is a broken fibre surrounded by intact fibres. It is not of importance exactly at which load level the fibre has broken. It is assumed that debonding and frictional sliding occur along a part of the fibre/matrix interfaces in connection with the breakage of the fibre. The mechanics of the fibre/matrix interface is described in terms of an interfacial fracture energy (debond energy), G_c^i, and a frictional sliding shear stress, τ_s. When subjected to cyclic loading, the crack tip can advance in accordance with a Paris-Erdogan law. A threshold value is assumed to exist for the crack tip stress intensity range $(\Delta K)_{th}$; if the crack tip stress intensity range, ΔK_{tip}, is below this value, no crack tip advancement will occur under cyclic loading. Furthermore, it is assumed that during repeated forward and reverse slip, the interfacial frictional sliding shear stress decreases to a lower value denoted τ_s^c due to changed along the fibre/matrix interface e.g. by wear of asperities or plastic deformation. We take the assumption that the fibers do not degrade during cyclic loading. The fatigue model thus assumes that the fatigue mechanism is primarily driven by interface properties.

2.2. First loading

In the first load cycle, following fibre failure, the debond length will increase until the maximum applied strain value, $\bar{\varepsilon}_{max}$. At the maximum applied strain, the debond length, ℓ_d, is given by [4]

$$\frac{\ell_d}{r} = \frac{E_f}{2\tau_s}\left(\bar{\varepsilon}_{max} - \frac{\bar{\sigma}_i}{E_c}\right), \tag{1}$$

where r is the fibre radius, E_f is the Young's modulus of the fibres and E_c is the Young's modulus of the composite given by the rules of mixture

$$E_c = V_f E_f + (1 - V_f)E_m, \tag{2}$$

where E_m is the Young's modulus of the matrix and V_f is the volume fraction of fibres in the composite. The parameter $\bar{\sigma}_i$ denotes the a so-called initiation stress given by

$$\frac{\bar{\sigma}_i}{E_c} = \frac{(1-V_f)E_m}{E_c}\Delta\varepsilon^T + 2\sqrt{\frac{(1-V_f)E_m}{E_c}\left(\frac{G_c^i}{E_f r}\right)}, \tag{3}$$

where $\Delta\varepsilon^T$ is the so-called mismatch strain, i.e, the difference in the in-elastic strains of the fibre and matrix occurring during processing of the composite. As an example, in case $\Delta\varepsilon^T$ is due to differences in thermal expansion coefficients and the cool-down from processing, $\Delta\varepsilon^T$ is given by $\Delta\varepsilon^T = (\alpha_f - \alpha_m)(T - T_0)$, where α_f and α_m are the thermal expansion coefficients of fibre and matrix, respectively, T is the present temperature and T_0 is the processing temperature.

2.3. After many load cycles

We now consider the situation after so many load cycles that the interfacial sliding shear stress has decreased to τ_s^c along the part of the interface experiencing cyclic reverse and forward slip. It turns out [4] that cyclic slippage along the fibre/matrix interface occur along only a part of the debonded fibre; the region near the debond crack tip will experience sticking friction with τ_s. As a result, the crack tip stress intensity factor will remain at the same value as it was at the maximum applied stress for the very first load cycle. The stress intensity range ΔK_{tip} is thus zero, i.e. below the threshold value, $(\Delta K)_{th}$ [4]. Consequently, the debond crack tip remains stationary. The length of the debond crack under cyclic loading is predicted to be:

$$\frac{\ell_d^c}{r} = \frac{E_f}{2\tau_s}\left\{\frac{\bar{\varepsilon}_{max}}{2}\left[(1+R) + (1-R)\frac{\tau_s}{\tau_s^c}\right] - \frac{\bar{\sigma}_i}{E_c}\right\}, \tag{4}$$

where r is the fibre radius, E_f is the Young's modulus of fibre, $\bar{\varepsilon}_{max}$ is the maximum applied strain, R is the R-ratio, i.e. the ratio between the minimum and maximum applied strain, $R = \bar{\varepsilon}_{min} \,/\, \bar{\varepsilon}_{max}$.

2.4. Stress from debond crack tip

Next, the stress induced to the nearest neighbour fibres from the debond crack tip stress field is found as [4]

$$\hat{\sigma}_f^K = 0.435\frac{E_f}{E_m}\sqrt{\frac{G_c^i \bar{E}^*}{d^*}}, \tag{5}$$

where d^* is the fibre spacing given by (assuming perfect hexagonal fibre packing)

$$\frac{d^*}{r} = \sqrt{\frac{2\pi}{\sqrt{3}V_f}} - 2, \tag{6}$$

and $\bar{E}^*$ is given as

$$\frac{1}{\bar{E}^*} = \frac{1}{2}\left(\frac{1}{\bar{E}_f} + \frac{1}{\bar{E}_m}\right). \tag{7}$$

2.5. Prediction of fatigue limit

Finally, the fibre strength variation is described in terms of a Weibull distribution, with σ_0 being the characteristic strength, L_0 the reference length and m being the Weibull modulus. As mentioned, it is assumed that the fibres do not undergo any degradation due to the cyclic loading.

When the debond crack tip propagates along the interface of the broken fibre, its crack tip stress field will add an additional stress to its nearest fibres. More precisely, the debond crack tip stress field will influence the nearest 6 fibres (assuming a hexagonal fibre packing arrangement) as the singular crack tip stress field moves along the broken fibre, from the locus of fibre breakage, $z = 0$ to the final positions of the debond crack tip, $z = \pm\ell_d^c$. Using the Weibull model with the condition that the probability equal to 1/6 (corresponding to failure of one of the six neighbour fibres), the condition for the maximum strain level becomes:

$$A_1\left\{\frac{\tau_s L_0}{E_f r}\left[\left(\frac{\sigma_f^+(\bar{\varepsilon}_{fl})+\hat{\sigma}_f^K}{\sigma_0}\right)^m+5\left(\frac{\sigma_f^+(\bar{\varepsilon}_{fl})}{\sigma_0}\right)^m\right]^{-1} ln\left(\frac{5}{6}\right)+\frac{\bar{\sigma}_i}{E_c}\right\}-\bar{\varepsilon}_{fl}=0 \tag{8}$$

where the parameter A_1 depends on both the decrease in the interfacial frictional shear stress and the R-ratio:

$$A_1 = \frac{2\tau_s^c}{(1+R)\,\tau_s^c+(1-R)\,\tau_s}. \tag{9}$$

Generally, A_1 lies in the range of 0 to 1. Decreasing τ_s^c/τ_s and decreasing R lead to a decrease in A_1. The stress $\sigma_f^+(\bar{\varepsilon}_{fl})$ is the stress in fibres remote from the fibre break at the strain level corresponding to the fatigue limit. $\sigma_f^+(\bar{\varepsilon}_{fl})$ is given by

$$\frac{\sigma_f^+(\bar{\varepsilon}_{fl})}{E_f} = -\Delta\varepsilon^T\frac{(1-V_f)\,E_m}{E_c}+\bar{\varepsilon}_{fl}. \tag{10}$$

Furthermore, the parameter $\hat{\sigma}_i$ in Eq. (8) is given by Eq. (3), and the parameter $\hat{\sigma}_f^K$ is given by Eq. (5).

Eq. (8) is not easy to solve explicitly for $\bar{\varepsilon}_{fl}$ - recall that the parameter σ_f^+ depends on $\bar{\varepsilon}_{fl}$. However, Eq. (8) can be solved by numerical methods. In the remainder of this paper, the fatigue limit $\bar{\varepsilon}_{fl}$ is found by solving Eq. (8) numerically using a Newton-Raphson method and a convergence criterion based on the difference between two successive iteration values of $\bar{\varepsilon}_{fl}$ being less than 10^{-9}.

3. Results: Predictions of fatigue limit for carbon/epoxy composites

3.1. Material parameters

In the following we present predictions of the fatigue limit for an AS4 carbon fibre/epoxy composites. The following material parameters were used: E_f = 231 GPa, r = 3.55 μm, σ_0 = 4680 MPa, L_0 = 20 mm, m = 8.86 (data from Jimenes and Pellegrini [5]) and E_m = 4 GPa. Values for $\Delta\varepsilon^T$ =0.00407 (estimated from thermal expansion mismatch and cool down from processing temperature). In addition, values for τ_s = 128 $\pm$ 30 MPa and G_c = 6 J/m^2 were estimated from the data for debond length as a function of applied strain given by Kim and Nairn [6] using the procedure described elsewhere [7]. However, in the following we will investigate the effects of τ_s, τ_s^c, G_c^i and the parameter A_1 on the predicted fatigue limit, $\bar{\varepsilon}_{fl}$.

3.2. Model predictions

Fig. 1 shows the predicted fatigue limit as a function of fibre volume fraction and the parameter A_1 for τ_s = 128 MPa. The fatigue limit, expressed in terms of strain, $\bar{\varepsilon}_{fl}$, is seen to decrease with increasing V_f and to increase with increasing A_1.

Next, the fatigue limit expressed in terms of stress, $\bar{\sigma}_{fl}$, is calculated using (2) and is shown in Fig.2. Starting from a low value of V_f, $\bar{\sigma}_{fl}$ is seen to increase with V_f attains a maximum value and decrease with V_f. For A_1 = 1.0, the maximum fatigue limit (1830 MPa) is predicted for V_f= 0.54, while for A_1 = 0.25, the maximum value of $\bar{\sigma}_{fl}$ is 1327 MPa is attain at V_f = 0.47.

Fig. 3 shows $\bar{\varepsilon}_{fl}$ as a function of the interfacial fracture energy, G_c^i for various values of A_1. Initially, $\bar{\varepsilon}_{fl}$ decreases from a value around 0.020 (weakly dependent on A_1) for vanishing G_c^i,

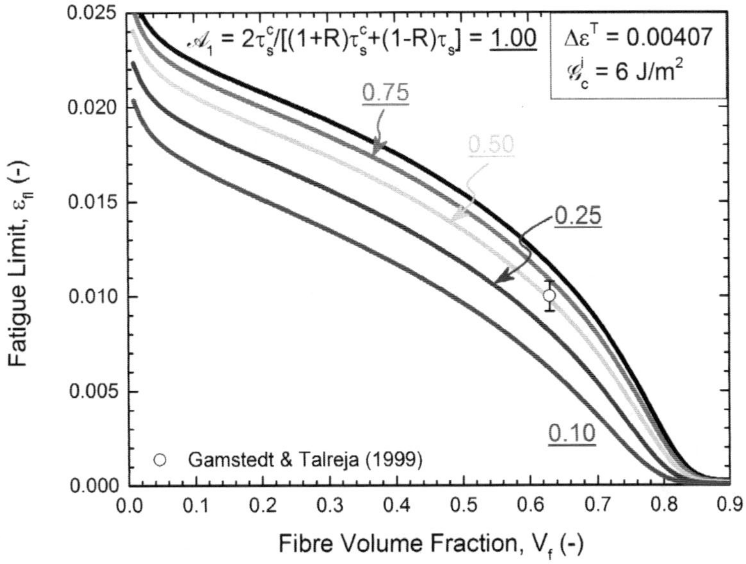

Figure 1. The predicted fatigue limit (in terms of strain) for unidirectional carbon fibre/epoxy composites is shown as a function of fibre volume fraction for various values of A_1 ($\tau_s = 128$ MPa). The fatigue limit found experimentally by Gamstedt and Talreja [2] is included as a point.

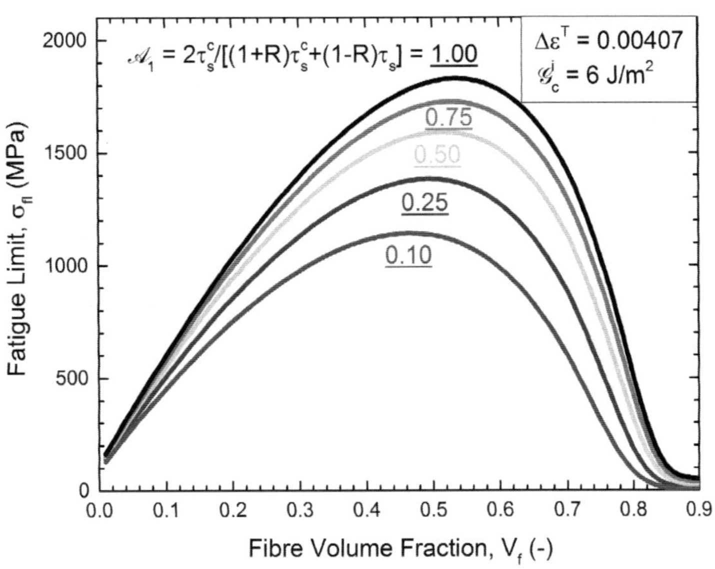

Figure 2. The predicted fatigue limit (in terms of stress) for unidirectional carbon fibre/epoxy composites is shown as a function of fibre volume fraction for various values of A_1 ($\tau_s = 128$ MPa).

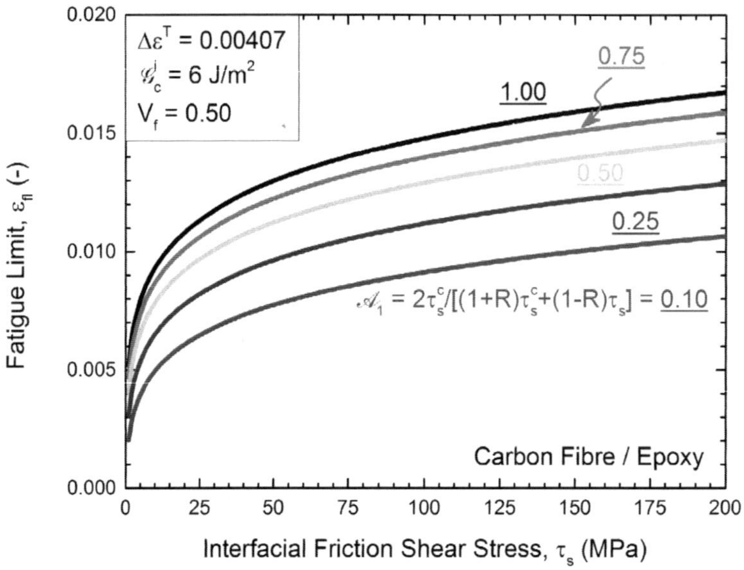

Figure 3. The predicted fatigue limit (in terms of strain) for unidirectional carbon fibre/epoxy composites is shown as a function of frictional shear stress for various values of A_1.

to a minimum value around (occurring at $G_c^i \approx 50$ J/m^2), and then slowly increases (almost linearly) with G_c^i reaching 0.0012 to 0.0046 (depending on the value of A_1) for $G_c^i = 250$ J/m^2.

The predicted fatigue limit in terms of strain, $\bar{\varepsilon}_{fl}$, is shown as a function of the initial interfacial frictional shear stress, τ_s, for various values of A_1 in Fig. 4. $\bar{\varepsilon}_{fl}$ increases monotonically with increasing τ_s; the increase is particularly strong for small values of τ_s. For large value of τ_s, say larger than 100 MPa, the increase in $\bar{\varepsilon}_{fl}$ is more moderate.

4. Discussion

4.1. Determination of τ_s^c, G_c^i and τ_s^c

In the following we sketch how the model can be applied for the prediction of the fatigue limit. First, interface properties must be determined from experiments. For instance, ℓ_d can be obtained from in-situ observations of composites or single filament specimens subjected to monotonic loading. Then, τ_s and G_c^i can be calculated by the use of micromechanical models, e.g. Sørensen [7]. Next, and ℓ_d^c can be measured from cyclic experiments as the maximum debond length that is reached after many load cycles. The analysis for determination of τ_s^c is as follows: We subtract (1) from (4) and rewrite:

$$\frac{\tau_s}{\tau_s^c} = 1 + 4\frac{\tau_s}{E_f\,\bar{\varepsilon}_{max}\,(1-R)}\left(\frac{\ell_d^c - \ell_d}{r}\right). \tag{11}$$

Eq. (11) thus enables the calculation of τ_s^c.

4.2. Comparison with experiments

The fatigue limit of 1.0% ± 0.08% (run-out for 10^6 cycles, $R = 0.1$) for a unidirectional AS4 carbon fibre/epoxy composite is superimposed as a point in Fig. 1. The data point lies comfortably within the range predicted by the model, just slightly above the curve for A_1=0.50,

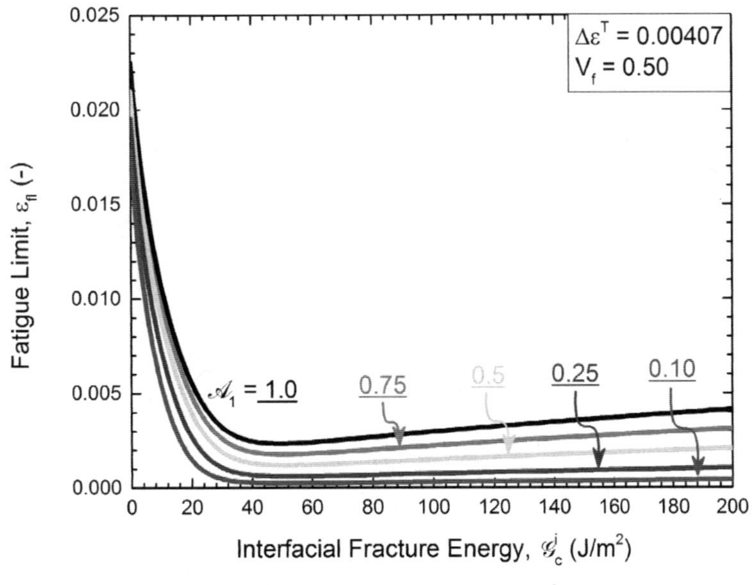

Figure 4. The predicted fatigue limit (in terms of strain) for unidirectional carbon fibre/epoxy composites is shown as a function of interfacial fracture energy for various values of A_1 (τ_s = 128 MPa).

lending confidence in the model predictions. However, this may be somewhat fortuitous since the curves for the predicted fatigue limit depends strongly on the chosen value of G_c^i, as indicated in Fig. 4.

We proceed to test the model predictions further. First, Gamstedt and Talreja [2] found that the debond length after millions of cycles was approximate 23 μm. Next, from the data from the monotonic single fibre fragmentation tests of Kim and Nairn [6], ℓ_d/r was around 1.5 for a strain value of 1.3-1.5 %. Extrapolating to a strain value of 1% gives ℓ_d/r =0.5 or ℓ_d =1.78 μm. Next, knowing ℓ_d and ℓ_d^c, we use (11) to calculate $\tau_s^c = 51.5$ MPa. Finally, these values, together with $R = 0.1$ are inserted into (9) to give $A_1 = 0.60$. This value is within the error bars of the experimental point from Gamstedt and Talreja [2], see Fig. 1. Recall that the prediction is made using data from two un-related studies, this finding is encouraging.

It can be noted that a similar comparison between model predictions and experimental data for fatigue limit of glass-fibre composites also gives a good agreement.

4.3. Raising the fatigue limit

One of the main reasons for the development of micromechanical models is that they can give model predictions to guide the development of future materials. The parameter study included in this paper suggests that the fatigue limit of unidirectional carbon fibre/epoxy composites can be raised by microstructural optimization. More specifically, the model predictions suggest that the fatigue limit (expressed in strain) can be raised by:

- decreasing V_f
- preventing a decrease (or minimizing the decrease) in the τ_s^c/τ_s ratio (preventing a decrease in A_1)

- increasing τ_s
- decreasing G_c^i, preferable below a few J/m^2.

Decreasing V_f may not be of great practical use, since a certain minimum value of V_f is often chosen in order to obtain a certain stiffness value. It is therefore also relevant to consider the fatigue limit in terms of stress. Then the situation is a bit more complicated since the maximum value of $\bar{\sigma}_{fl}$ as a function of V_f depends on a number of additional parameters, as can be seen from Fig. 2. A more detailed parameter study might be needed to determine the optimum set of interface parameters for a given material system.

Microstrutural optimization also offers a number of experimental challenges. First, improved experimental methods and data analysis approaches may be needed for characterisation of the interface properties obtained under a given set of processing parameters (see [7] for a more detailed discussion). Next, when the interfacial parameters of a given composite material have been well characterized, the next step will be to alter the processing conditions such that the optimal microscale parameters (e.g. optimal values for G_c^i, τ_s and τ_s^c) can be obtained. The results of the present study suggest that the interface fracture energy should be as low as possible whereas the friction should be as high as possible. It is recommended that more weight should be put in the design of fibre topology (e.g. roughness) and less on chemistry - the goal of the chemistry should be to decrease the fibre/matrix debond energy.

In a more broader sense, the model results suggest that the main issue in maximizing the fatigue limit is in controlling parameters that lowers $\hat{\sigma}_f^K$, the stress on neighbour fibres from the debond crack tip stress fields of the broken fibre.

Finally, it is noted that a decrease in the R-ratio will give a lower value of $\bar{\varepsilon}_{fl}$ since the parameter A_1 decreases with decreasing R, see eq. (9).

5. Concluding remarks

A micromechanical model was used for the prediction of the fatigue limit of unidirectional fibre composites. The effect of interface properties were investigated for a unidirectional AS4 carbon fibre/epoxy composite. The model predicts that the fatigue limit, expressed in terms of strain, decreases with increasing fibre volume fraction, decreases with decreasing τ_s^c/τ_s ratio, and increases with increasing interfacial frictional sliding shear stress. When expressed in terms of stresses, the maximization of the fatigue limit is more complicated. With other parameters held fixed, $\bar{\sigma}_{fl}$, attains a maximum at intermediate values of V_f. A further increase in the fibre volume fraction will then lead to a lower fatigue limit.

References

[1] Mandell J F, Samborsky D D, Wang L and Wahl N K 2003 New fatigue data for wind turbine blade materials *J. Solar Energy* **125** 506-14

[2] Gamstedt E K and Talreja R 1999 Fatigue damage mechanisms in unidirectional carbon-fibre-reinforced plastics *J. Mater. Sci.*, **34** 2235

[3] van den Heuvel P W J, Peijs T and Young R J 1998 Failure phenomena in multiple-fibre model composites - 3. A Raman spectroscopic study on the influence of interfacial debonding on the stress concentrations *Comp. Sci. Techn.* **58** 933-44

[4] Sørensen B F and Goutianos S 2018 Micromechanical model for prediction of the fatigue limit for unidirectional fibre composites *Mechn. Mater.* submitted

[5] Jimenez F L and Pellegrino S 2013 Failure of carbon fibers at a crease in fiber-reinforced silicone sheet *J. Appl. Mechn.* **80** 011020-1-8

[6] Kim B W and Nairn J A 2002 Experimental verification of the effect of friction and residual stress on the analysis of interfacial debonding and toughness in single fiber composites *J. Mater. Sci.* **37** 3965-72

[7] Sørensen, B F 2017 Micromechanical model of the single fiber fragmentation test *Mechn. Mater.* **104** 38-48

Stiffening effect of fatigue and creep loading in unidirectional flax fibre/epoxy composites

Yosuke Ueki[1], Hans Lilholt[2], and Bo Madsen[2]

[1] Reliability Science Research Department, Center for Technology Innovation – Mechanical Engineering, Research & Development Group, Hitachi, Ltd., Hitachinaka, Japan
[2] Section of Composite Materials, Department of Wind Energy, Technical University of Denmark, Risø Campus, Roskilde, Denmark

Corresponding author: Bo Madsen, e-mail: boma@dtu.dk

Abstract. A study related to the fatigue behaviour of natural fibre composites was conducted to expand their industrial applications. Unidirectional flax/epoxy composites were fabricated, and their fatigue and creep behaviour was investigated by applying a range of different testing conditions. Stiffness was found to increase during fatigue loading, and this was accompanied by accumulation of residual strain. By arranging the results in a stiffness-strain diagram, a linear trend was established with a positive slope representing the stiffening effect. A similar linear trend was obtained for creep loading, however, with a larger stiffening effect. For fatigue loading, the stiffening effect was found to be changed by the applied cycle frequency. The combined findings in the study suggest that stiffness change in the composites is reflecting a balance between stiffening due to residual strain, and softening due to fatigue damaging.

1. Introduction

Replacing synthetic fibres, such as glass fibres and carbon fibres, with plant-based fibres, widely denoted as natural fibres, is an effective way to realize materials carbon neutrality because plants absorb CO_2 during their growth. In addition, natural fibres are known to have mechanical properties comparable to glass fibres. Therefore, a large number of research studies related to natural fibres and their composites have been performed to evaluate materials characteristics, and to maximize materials performance.

Recent industrial use of natural fibre composites is mostly limited to non-structural components, such as body panels of automobiles made from non-woven random mats of natural fibres. One of the reasons why industrial applications have been limited so far is the lack of knowledge related to the long-term reliability of natural fibre composites. Especially fatigue behaviour should be clarified to expand their application to load-bearing structural components. A common observation in studies of fatigue behaviour of natural fibre composites is stiffness increase during fatigue loading [1,2], while conventional composites based on glass and carbon fibres are known to show stiffness decrease, which typically is assumed to be due to damage progression [3-5]. Liang et al. [2] presented that a natural fibre composite based on a cross-ply flax fabric showed continuous increase of stiffness during almost all of the fatigue life, while a glass fibre composites with an identical fibre configuration exhibited significant decrease of stiffness. Baley [6] reported that cyclic tensile loading of single natural fibres also showed stiffness increase. Nevertheless, the underlying mechanisms for this stiffness increasing behaviour remain unclear [1].

In the present study, it is hypothesised that the stiffness change observed in natural fibre composites during fatigue loading is the result of two counteracting mechanisms taking place: stiffening (i.e. stiffness increase) and softening (i.e. stiffness decrease). This is investigated by performing tension-tension fatigue tests of unidirectional natural fibre composites, and by focusing on the stiffness change and the concurrent change of residual strain. In addition, creep tests were performed to examine whether constant loading, i.e. a fatigue test without cyclic loading, induces a stiffening effect in the same way as fatigue loading does. This approach was used to separate the effects of the cyclic and constant loading components in fatigue loading.

2. Materials and methods

2.1. Fabrication of composite laminates and test specimens

Unidirectional natural fibre composites were fabricated from a flax yarn (Smeraldo, Nm 1/9.7, Linificio e Canapificio Nazionale SpA, Italy) and an epoxy resin (Araldite LY1564SP, epoxy; Araldite 3486, hardener; Huntsman, USA). The density of the flax fibres and the cured resin was measured to be 1.59 and 1.15 g/cm^3, respectively. The same type of flax yarn has been used in a previous study of natural fibre composites, but with focus on static tensile properties [7].

For fabrication of unidirectional flax/epoxy composite laminates, 14 layers of flax yarn were wound on to an aluminium plate (470 mm x 400 mm) using a custom-made winding machine. The misalignment of yarns was calculated to be negligible ($\approx$ 0.1°). The plate with the wound yarn was packed in a vacuum bag. After a vacuum infusion process with the epoxy resin, the entire aluminium plate with the impregnated yarn was placed in a thermo-chamber for the curing process, 5 hours at 50°C. After demoulding, the composite laminates were post-cured at 80°C for 4.5 hours. The density, volume fraction of fibres, and porosity of the four fabricated laminates were measured to be 1.28 ± 0.02 g/cm^3, 31.4 ± 0.2%, and 0.3 ± 0.3% (mean ± standard deviation), respectively. The laminates were cut into (i) butterfly-shaped test specimens (length 410 mm, R = 900 mm, gauge section 60 x 25 mm), a type of specimen geometry specially designed for tensile fatigue testing of composites [8], and (ii) rectangular test specimens (length 180 mm, gauge section 100 x 25 mm). Tapered glass fibre composite tabs were mounted on both types of test specimens.

2.2. Static tensile tests

To establish the testing conditions for the fatigue and creep tests, static tensile tests were initially performed with the rectangular specimens using a hydraulic testing machine (Instron, UK). The displacement rate was 2 mm/min. Strain was measured by two extensometers (50 mm gauge length) attached to each side of the specimens.

2.3. Fatigue tests and creep tests

Fatigue tests were performed using an identical setup as used for the static tensile test. The applied fatigue testing mode was tension-tension, as indicated in Figure 1 (left), and with a constant stress ratio of R = 0.1. Different levels of stress amplitudes and frequencies were examined, as presented in Table 1. Although almost all fatigue tests were performed on the butterfly-shaped specimens, a few tests that were focusing only on the stiffening effect were done on the rectangular specimens. During the fatigue tests, the stress-strain hysteresis loops were recorded intermittently, and stiffness was determined as the slope of a linear regression line fitted to the loops (see Figure 4).

Creep tests were performed using an electronic testing machine (Instron, UK). Figure 1 (right) shows the general time course of the creep tests. For each test, two values of stiffness were determined from the slopes of linear regression lines fitted to the two loading phases (1st and 2nd ramp). Two stress levels (σ_c = 100 and 150 MPa) and several holding times (t_h = 0 to 1800 sec) were examined, as presented in

Table 2. The magnitude of the two stress levels were selected to be within the range of mean stress levels applied in the fatigue tests.

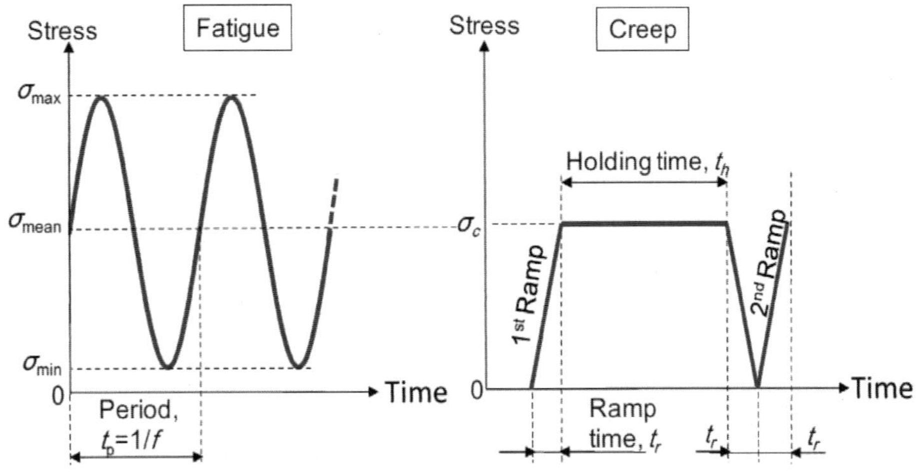

Figure 1. Time course for fatigue tests (left) and creep tests (right).

Table 1. Applied conditions for fatigue tests.

Specimen shape	Max. stress, σ_{max} [MPa]	Mean stress, σ_{mean} [MPa]	Stress ratio, $R = \sigma_{min}/\sigma_{max}$	Frequency [Hz]
Butterfly	110	61	0.1	5
Butterfly	129	71	0.1	2
Butterfly	159	87	0.1	2
Butterfly	180	99	0.1	2
Butterfly	203	112	0.1	2
Butterfly	220	121	0.1	2
Butterfly	240	132	0.1	2
Rectangular	220	121	0.1	2
Rectangular	220	121	0.1	1
Rectangular	220	121	0.1	0.5
Rectangular	220	121	0.1	0.25

Table 2. Applied conditions for creep tests.

Stress level, σ_c [MPa]	Holding time, t_h [sec]	Ramp time, t_r [sec]
100	0	1
100	1	1
100	5	1
100	30	1
100	300	1
100	1800	1
100	0	0.3
100	0	0.5
150	0	1
150	1	1
150	5	1
150	30	1
150	300	1
150	1800	1
150	0	0.3
150	0	0.5

3. Results and discussion

3.1. Tensile properties

Based on the measured static tensile stress-strain curves of the unidirectional flax/epoxy composites, the stiffness and the ultimate tensile stress (UTS) were determined to be 18.5 ± 0.5 GPa and 282 ± 14 MPa (mean ± standard deviation, n = 5), respectively. These properties are comparable to previously reported properties of unidirectional natural fibre composites with similar fibre contents of about 30 % [7, 9]. The stress-strain curves showed transition regions (or yielding regions) in the strain range 0.2 - 0.4%. This specific characteristic has been reported in several studies both for flax fibres and for flax fibre composites [6, 10], though the underlying mechanism is still unclear. In the present study, the applied strain levels in the fatigue and creep tests are all above 0.5%, and therefore, they exceed the transition region of the composites.

3.2. Fatigue behaviour

Figure 2 shows the typical fatigue failure mode of the composites. At first, a localized damage region appeared at the specimen edge within the gauge section. After another several hundreds of cycles, a splitting crack propagated from the initial damage region to the start of the tab, presumably due to the increase of shear stresses along the fibres. Finally, the decrease of the load carrying area induced bulk failure at the start of the tab. It can be noted that due to the stochastic nature of the fatigue failure mode, it is inevitable that the recorded maximum number of cycles at failure (N) will show some degree of scatter, even for identical materials and test conditions.

Based on the results of the fatigue tests, the obtained S-N diagram is shown in Figure 3, together with the results of the static tensile tests. Generally, S-N data points of composites tested with a constant stress ratio (i.e. R-value) are known to be well fitted using an empirical power law relation, $S = C{\cdot}N^{-1/m}$, where S is maximum stress, and, C and m are fitting coefficients. As shown in Figure 3, the coefficient m, which determines the slope of the S-N curve, is 9.7 for the unidirectional flax/epoxy composites. This is similar to the m-value of about 10 seen for unidirectional glass/epoxy composites [11].

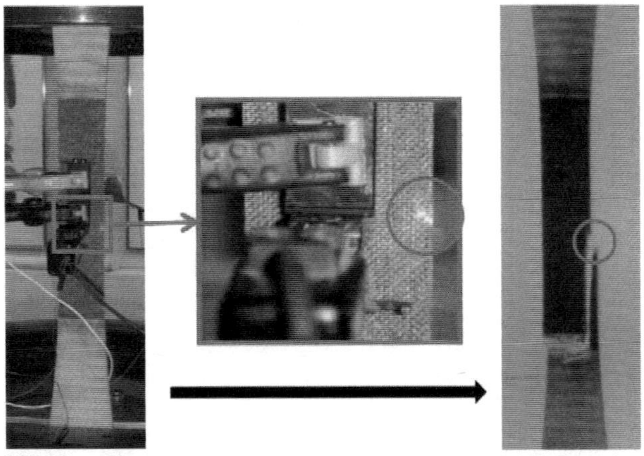

Figure 2. Typical fatigue failure mode observed for the unidirectional flax/epoxy composites. The red circles indicate the position of the visible initial damage region.

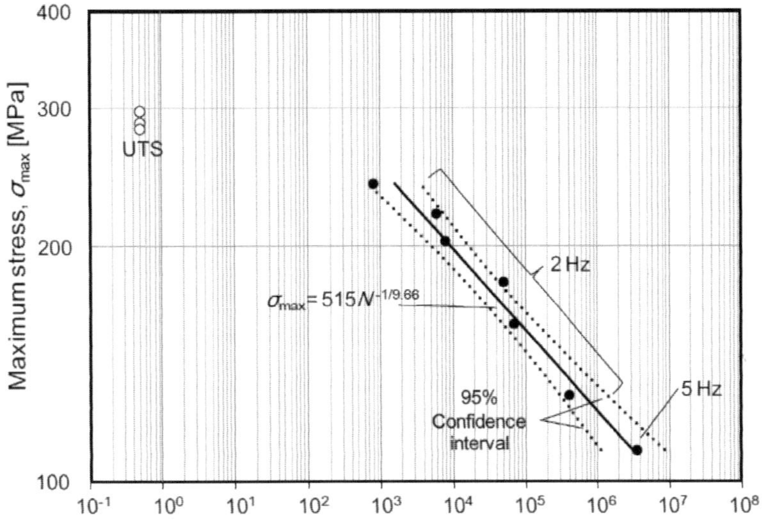

Figure 3. The logS - logN fatigue diagram of the unidirectional flax/epoxy composites, together with the results of the static tensile tests.

A special fatigue behaviour observed for unidirectional flax/epoxy composites, not seen for conventional types of composites, is the change in stiffness during fatigue testing. Figure 4 shows examples of the recorded stress-strain hysteresis loops during two fatigue tests with mean stress levels of 61 and 121 MPa (corresponding to maximum stress levels of 110 and 220 MPa). For both stress levels, the slope of the loops at halfway of fatigue life is increased compared with the slope at the start of the fatigue test. For the high mean stress level of 121 MPa, this trend is continued towards the end of fatigue life, whereas for the low mean stress level of 61 MPa, the slope is decreased at the end of fatigue life. For both stress levels, the position of the loops is consistently shifted to the right during the fatigue testing. The starting strain value for each loop is denoted *residual strain*. The two examples in Figure 4 show that the composites become stiffer during a major part of their fatigue life, while at the same time, the residual strain is increased.

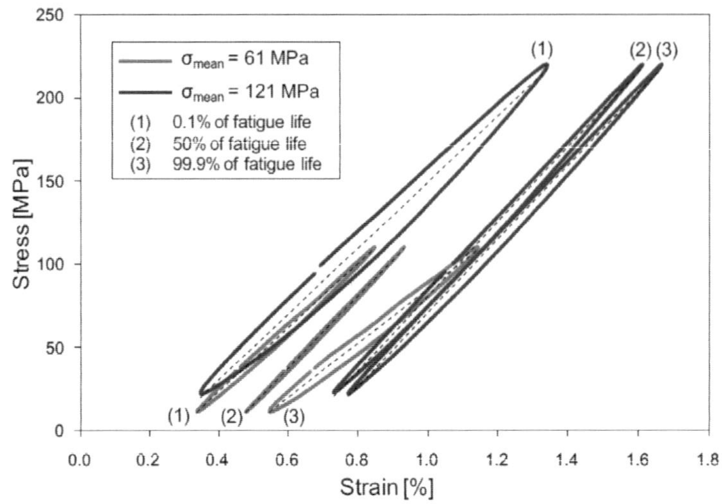

Figure 4. Examples of hysteresis loops recorded during two fatigue tests with mean stress levels of 61 and 121 MPa. Dotted lines show linear regression lines of each loop to determine stiffness.

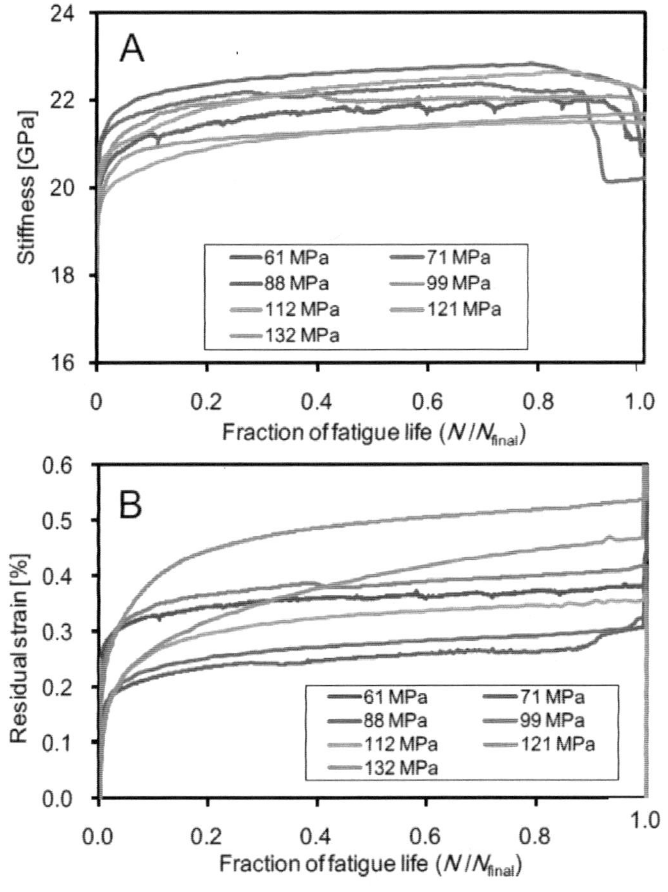

Figure 5. Stiffness (A) and residual strain (B) as a function of fraction of fatigue life. Results are shown for fatigue tests with different mean stress levels.

For the performed fatigue tests with the different mean stress levels, Figure 5 shows the determined stiffness and residual strain as a function of the fraction of fatigue life (N/N_{final}). For both stiffness and residual strain, it is difficult to see a common trend with respect to the mean stress level. However, as shown in Figure 6, if the relative change in stiffness ($(E-E_0)/E_0 = \Delta E/E_0$, where E and E_0 are the current and the initial stiffness, respectively) is plotted as a function of the residual strain, the curves for the different mean stress levels share a common linear trend. Such a diagram is hereafter referred to as a *stiffness-strain diagram*. The established correlation shown in Figure 6 suggests that stiffening of the composites is induced during accumulation of residual strain. The stiffening effect is represented by the slope of the linear trend line.

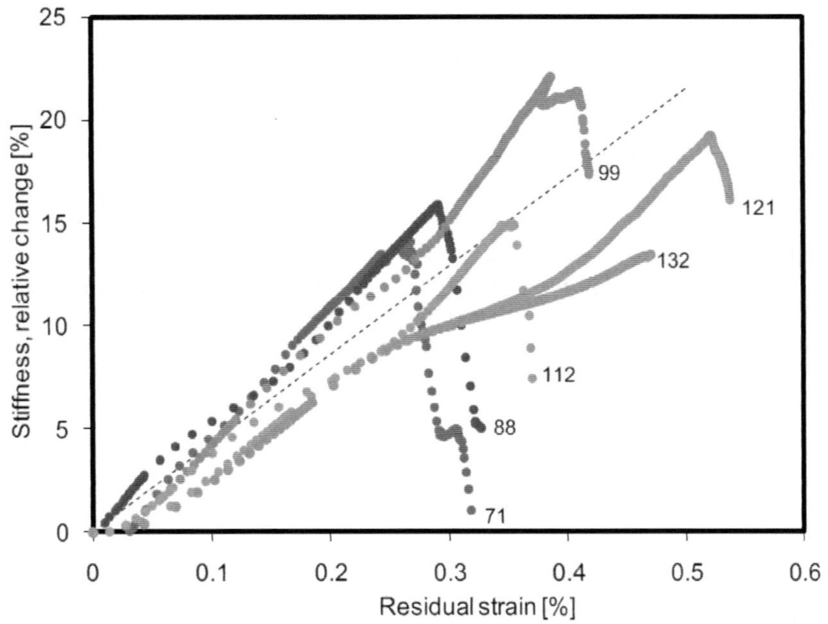

Figure 6. Stiffness-strain diagram for fatigue. Relationship between relative change in stiffness and residual strain. Numbers at the curves indicate the applied mean stress levels. Dotted black line is a linear regression line for all stress levels.

3.3. Creep behaviour

In the results of the fatigue tests presented above, a stiffening effect was clearly observed. If this stiffening effect is induced by the residual strain, a similar effect should expectedly be observed in creep tests, which correspond to fatigue tests with no stress amplitude, i.e. the applied creep stress can be assumed equivalent to the mean fatigue stress (see Figure 1). In Figure 7, an example of results of a creep test with an applied stress of 150 MPa is shown. As can be observed in Figure 7A, the onset strain value is increased above zero for the 2nd ramp, which shows that residual strain is generated. In addition, as can be observed in Figure 7B, the slope of the 2nd ramp is steeper than the slope of the 1st ramp. This is indicating a stiffening of the composites as an effect of the constant loading. Analogous to the fatigue test, the relative change in stiffness ($(E_2-E_1)/E_1 = \Delta E/E_1$, where E_1 and E_2 are stiffnesses obtained from the 1st and the 2nd ramps, respectively) can be calculated.

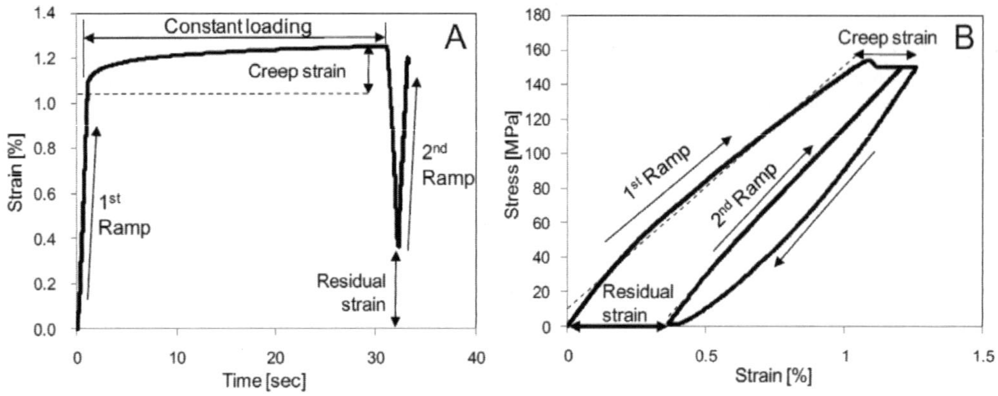

Figure 7. An example of results of a creep test with a stress level of 150 MPa, and a holding time of 30 sec. A: Strain-time, B: Stress-strain.

For the applied two stress levels in the creep tests (100 and 150 MPa), Figure 8 shows the determined relative change in stiffness and the residual strain as a function of the holding time. It is clearly indicated that the effect of holding time on stiffness and residual strain is similar. Both properties are increased rapidly in the range of short holding times, and the rate of increase is reduced in the range of long holding times. In addition, as expected, the increase of the stress level leads to an increase of both the stiffness and the residual strain.

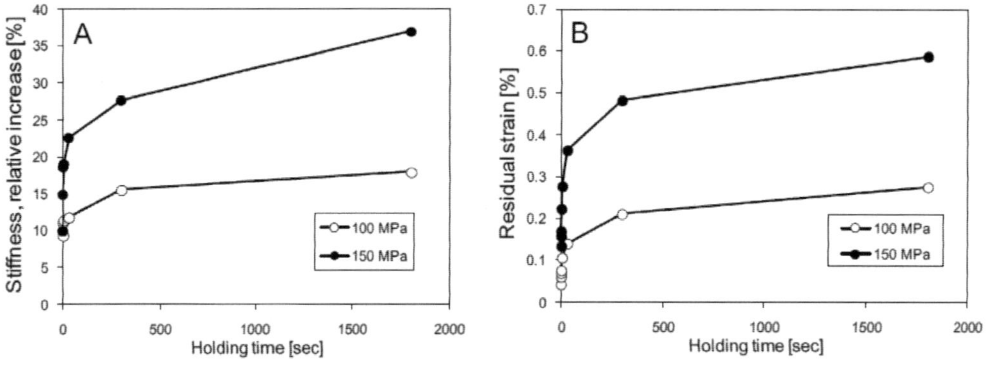

Figure 8. Relative change in stiffness (A) and residual strain (B) as a function of the holding time in the creep tests.

Based on the results from the creep tests, Figure 9 shows the stiffness-strain diagram, i.e. the relative change in stiffness as a function of residual strain. The results from the two stress levels share a common linear trend, as was also observed for the fatigue tests. Thus, it is clearly demonstrated that the stiffening effect does not require the repeated loading-unloading cycles of the fatigue tests.

3.4. Fatigue and creep behaviour

Figure 10 is a co-plot of the results of the fatigue tests and the creep tests in the same stiffness-strain diagram. As can be observed, the slope of the linear trend line is clearly lower for the fatigue tests, than for the creep tests. This indicates the existence of a concurrent mechanism in the fatigue test that is counteracting the stiffening effect. This mechanism is likely to be the damage induced by the repeated loading in the fatigue tests, which is the typically stated reason for the softening seen for conventional composites, such as glass fibre composites [3, 4]. In the case of the flax/epoxy composites in the present study, it is indicated that stiffening proceeds simultaneously with softening, and this explains why the fatigue tests result in a lower slope of the linear trend line compared to the creep tests. Thus, the slope in the stiffness-strain diagram is reflecting the balance between stiffening related to residual strain, and softening related to fatigue damaging.

Next, the balance between stiffening and softening will be studied by changing the applied loading frequency in the fatigue test.

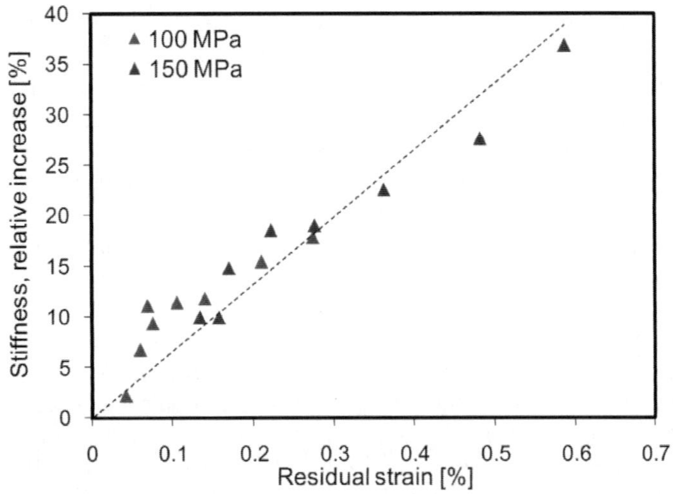

Figure 9. Stiffness-strain diagram for creep. Relationship between relative change in stiffness and residual strain. Dotted black line is a linear regression line for all data points.

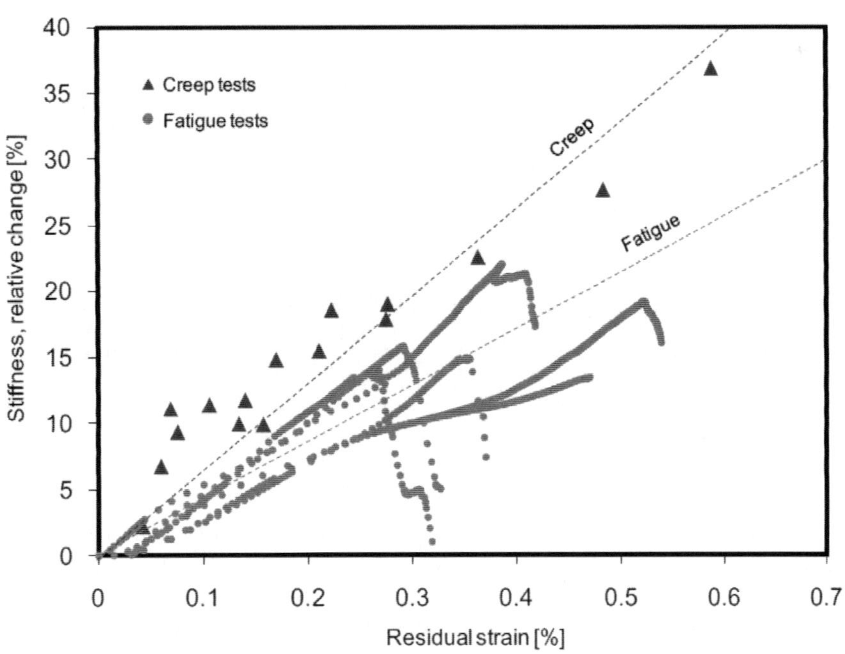

Figure 10. Stiffness-strain diagram for fatigue and creep.

3.5. *Effect of loading frequency on fatigue behaviour*

Additional fatigue tests were performed with a mean stress level of 121 MPa, and with loading frequencies of 0.25, 0.5, 1 and 2 Hz (see Table 1). In these tests, the constant stress component is the same, which means that the induced residual strain per unit of time can be assumed to be the same. The number of cycles per unit of time is however different between the tests, and this is assumed to lead to a difference in the amount of induced damage. Therefore, the balance between stiffening and softening is expected to be affected by the loading frequency, e.g. a lower frequency is expected to lead to a larger resulting stiffening effect due to less damaging. Figure 11 presents the results of these additional fatigue tests. As shown in Figure 11A, the relative change in stiffness proceeds faster for the lower frequencies, when the results are plotted as function of the number of fatigue cycles. Furthermore, as can be seen in the stiffness-strain diagram in Figure 11B, the slope of the curves is clearly steeper for the lower frequencies. Thus, as expected, the stiffening effect is getting larger when the cycle frequency is reduced. For a given residual strain induced by the constant stress component, the induced damage is lower for the fatigue tests with the lower frequency.

In the study by Shah [12] of comparable unidirectional natural fibre composites, fatigue testing was performed using a loading frequency of 10 Hz, and a negligible change of the stiffness was found. Qualitatively speaking, the results by Shah [12] are supporting the findings in the present study, i.e. an increase in loading frequency is leading to a decrease in the stiffening effect (Fig 11B).

Figure 11A shows that the specimens tested by the lower frequencies exhibit shorter fatigue life, although the fatigue damage per cycle should be same, i.e. it should be independent of the loading frequency. Interestingly, it can be observed that the final level of the relative change in stiffness is almost the same (about 16 %) for all frequencies. It is therefore indicated that the failure criterion for unidirectional natural fibre composites is not only governed by the number of loading cycles, i.e. by the magnitude of accumulated damage, but also the magnitude of the stiffening must be taken into account.

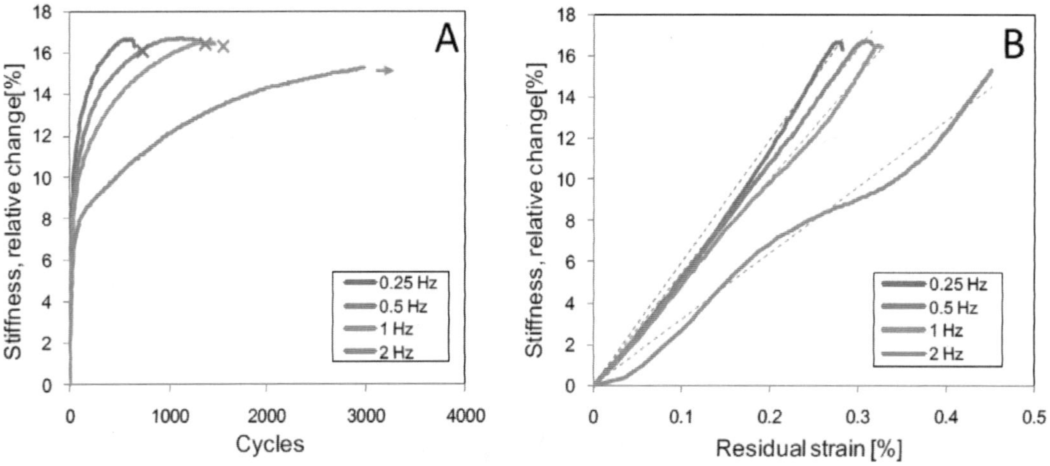

Figure 11. Results of fatigue tests performed with a mean stress level of 121 MPa and with various loading frequencies. Broken lines in B are linear regression lines for each curve.

3.6. *Underlying mechanisms for stiffening and softening*

In the present study, mechanical tests have been performed to quantify and analyse the mechanical response of unidirectional natural fibre composites to fatigue and creep loading. This has been done, however, without knowing the exact nature of the underlying mechanisms. In preliminary investigations by the authors, X-ray microtomography were used to observe the microstructure of the unidirectional flax fibre/epoxy specimens before and after testing, however, no clear differences were found (data not shown). In the case of natural fibres, Kohler and Spatz [13] reported a lowering of the cellulose microfibril angle during the initial phase of loading until the yielding region, where after the angle remained almost constant. Upon unloading, a residual deformation of the fibres remained. In the case of natural fibre yarns, as used in the present study, it must be considered that the twisting angle of the fibres in the yarn becomes lowered during fatigue and creep tests. These changes on the two structural levels of fibres and yarns support a stiffening effect in the composites.

4. Conclusions

In the present study, fatigue and creep tests under variable conditions were performed on unidirectional flax/epoxy composites. The results were analysed, and the following findings were obtained.

a) Increase of stiffness was correlated to increase of residual strain in both fatigue and creep tests. By arranging the results in a stiffness-strain diagram, a linear trend was established with a positive slope representing the stiffening effect.
b) Creep tests induced a larger stiffening effect than fatigue tests.
c) Fatigue tests performed with lower frequencies induced a larger stiffening effect.

Based on these findings, it is suggested that stiffening proceeds simultaneously with softening in the composites. The resulting stiffness change is reflecting the balance between stiffening due to residual strain, and softening due to fatigue damaging. Further work is needed to elucidate the underlying mechanisms.

Acknowledgements

The authors are most grateful for technical help and guidance provided by Jonas Kreutzfeldt Heininge, Jacob Christensen, Christian H. Madsen, Christen Malte Markussen (fabrication of composites and specimens), Vagn Jensen, Jan Sjølin, Steffen Rasmussen (fatigue and creep testing), and Kristine Munk Jespersen (tomography).

References

[1] Mahboob Z, Bougherara H 2018 Fatigue of flax-epoxy and other plant fibre composites: Critical review and analysis *Composites Part A* **109** 440-462
[2] Liang S, Gning P B, Guillaumat L 2012 A comparative study of fatigue behaviour of flax/epoxy and glass/epoxy composites *Composites Science and Technology* **72** 535-543
[3] Liu B, Lessard L B 1994 Fatigue and damage-tolerance analysis of composite laminates: Stiffness loss, damage-modelling, and life prediction *Composites Science and Technology* **51** 43-51
[4] Zangenberg J, Brøndsted P, Gillespie J J 2014 Fatigue damage propagation in unidirectional glass fibre reinforced composites made of a non-crimp fabric *Journal of Composite Materials* **48** 2711-2727
[5] Shirazi A, Varvani-Farahani A 2010 A stiffness degradation based fatigue damage model for FRP composites of (0/θ) laminate systems *Applied Composite Materials* **17** 137-150
[6] Baley C 2002 Analysis of the flax fibres tensile behavior and analysis of the tensile stiffness increase *Composites Part A* **33** 939-948
[7] Mehmood S, Madsen B 2012 Properties and performance of flax yarn/thermoplastic polyester composites *Journal of Reinforced Plastic Composites* **31** 1746-1757
[8] Korkiakoski S, Brøndsted P, Sarlin E, Saarela O 2016 Influence of specimen type and reinforcement on measured tension–tension fatigue life of unidirectional GFRP laminates *International Journal of Fatigue* **85** 114-129
[9] Madsen B, Hoffmeyer P, Lilholt H 2007 Hemp yarn reinforced composites – II. Tensile properties *Composites Part A* **38** 2204-2215
[10] Abdullah A H, Khalina A, Ali A 2011 Effects of fibre volume fraction on unidirectional kenaf/epoxy composites: the transition region *Polymer-Plastic Technology and Engineering* **50** 1362–1366

[11] Nijssen R P L, Brøndsted P 2013 Fatigue as a design driver for composite wind turbine blades *Advances in wind turbine blade design and materials* Woodhead Publishing pp 175-209
[12] Shah D U 2016 Damage in biocomposites: Stiffness evolution of aligned plant fibre composites during monotonic and cyclic fatigue loading *Composites Part A* **83** 160-168
[13] Kohler L, Spatz H C 2002 Micromechanics of plant tissues beyond the linear-elastic range *Planta* **215** 33-40

Development of a novel testing device for fabric reinforced carbon fibre composites under cyclic biaxial load applications

Carl H. Wolf[1], Marco Böcker[1], Christian Düreth[2], Stephanie Ackermann[1], Sebastian Henkel[1], Mike Thieme[2], Robert Böhm[2], Ilja Koch[2], Maik Gude[2] and Horst Biermann[1]

[1]Technische Universität Bergakademie Freiberg, Institute of Materials Engineering, Gustav-Zeuner-Straße 5, 09599 Freiberg, Germany

[2]Technische Universität Dresden, Institute of Lightweight Engineering and Polymer Technology, Holbeinstraße 3, 01307 Dresden, Germany

E-mail: Carl.Wolf@ww.tu-freiberg.de

Abstract. In this paper, a novel testing device has been developed and adopted to a planar-biaxial servo hydraulic testing machine to examine the multiaxial fatigue behaviour of woven CFRP under superimposed interlaminar cyclic shear and static or cyclic through-thickness compression stresses.

1. Introduction

Woven-fabric composites gain an increasing usage in modern advanced applications of carbon fibre reinforced plastics (CFRP), because of their inherit advantages [1]. Generic applications are fan blades, bolted joint, drive shafts and furthermore, where multiaxial stress states are likely common. Characteristic load applications are in-plane tensile or out-of-plane shear stress superimposed with through-thickness-compressive stresses (σ_1/σ_3^-- or τ_{13}/σ_3^--state of stresses). Particular interlaminar shear stresses can cause delamination leading to degradation and fatigue of the composite material.

In order to initiate such failure critical stress-states, like τ_{13}/σ_3^--state of stress, in a carbon fibre reinforced thermoset, a novel specimen and testing device has been iteratively developed and adapted to a planar-biaxial servo hydraulic testing machine, to ensure a homogenous stress distribution [2]. In multiaxial quasi-static tests (for example by [3, 4, 5, 6]), it has been experimentally proven that through thickness compression increases the interlaminar shear strength. Many failure criteria proposed in literature can describe this phenomenological and empirical evidence either by a maximum failure strain criterion, such as NU-Theory [7], Sun [8] or Christensen et. al. [9], or by an invariant based failure criterion, such as Cuntze [10].

The influences of the interface waviness and nesting of fabric reinforcements on the mode I and II energy release rates have been investigated in [11, 12, 13], but there is a lack of a reliable understanding of the delamination propagation during fatigue load applications especially with superimposed through-thickness compression [1]. Currently, there are plenty of biaxial testing devices for cyclic analysis, which can be distinguished in tests using a single loading system, where the biaxial stress ratios are determined by the specimens' geometry or the loading fixture configuration, and tests using two or more independent loading systems. But apparently these testing methods rely to in-plane characteristics of CFRP, except for the hydraulic bulge test, but an inhomogenous stress gradient in thickness direction of the specimen makes it challenging to estimate the stress distribution [14].

Consequently the described testing method for τ_{13}/σ_3^--stress states has been adapted to cyclic load applications, to describe the influence of through-thickness compression on the delamination propagation of CFRP. Although the proposed stamp-geometry, to initiate the through-thickness compression stresses, has been designed to minimize premature failure in the contact area, it could be observed that premature failure occurs at the outer-notches, when τ_{13}/σ_3^--stresses exceed the ratio of 1:2[1]. Hence a wider contact surface has been chosen to assure the same stress conditions at the notchs' corners. Unfortunately in cyclic load applications the frictional and wear behaviour of the CFRP has to be emphasised and hence it is challenging to predict the shear forces in the shear plane.

In this paper a testing device and a method to assess the stresses in an experimental approach are presented. Furthermore step-wise load increase tests were carried out to determine the influence of through-thickness compression on the interlaminar shear fatigue behaviour.

2. Experimental details

2.1. Planar biaxial testing system

The tests were carried out on a servo hydraulic planar biaxial system Instron 8800 (cf. Figure 1). This system has four actors, which are orthogonally aligned (x- and y-axes). Each actor can reach forces up to 250 kN. In this setup, a high speed camera (mvBlueCOUGAR-XD) has been incorporated and aligned on the region of interest respectively the hole notch pattern for image acquisition (cf. Figure 2). In a post-scripting process the images have been assessed by the Digital Image Correlation (DIC) software ARAMIS v6.1.1-8 for strain measurements. The camera has a native resolution of 2048 × 2048 pixels, which corresponds to a spatial resolution of 11 × 11 µm per pixel on the focused area.

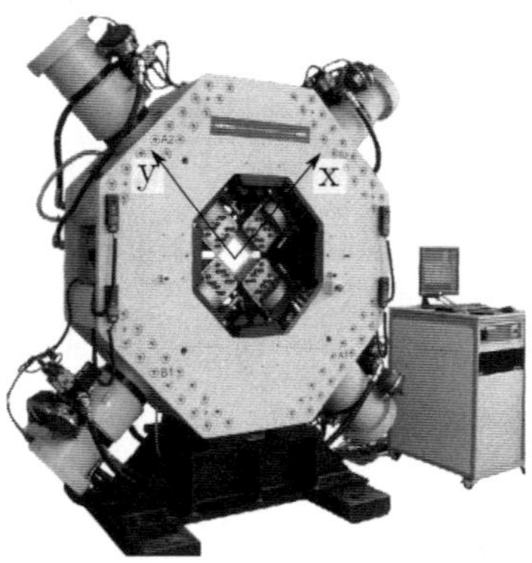

Figure 1. Servo hydraulic Planar biaxial testing system Instron 8800 with the axes x and y.

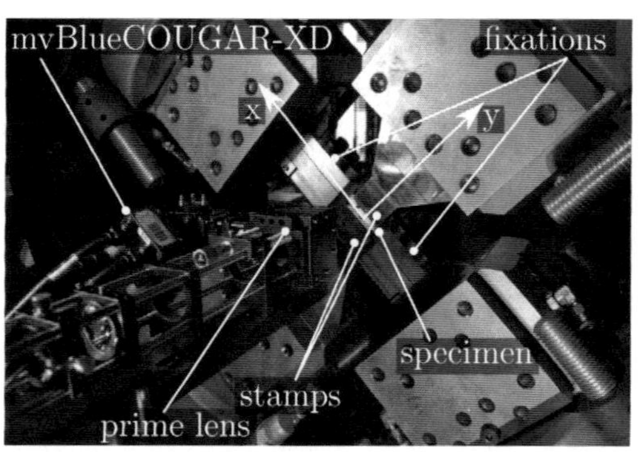

Figure 2. Illustration of the test setup with the high speed camera mvBlueCougar-XD.

[1] will be published on the European Solid Mechanics Conference 2.-6. July 2018 in Bologna („Experimental and numerical characterisation of the fracture behaviour of novel specimens for superimposed out-of-plane stresses").

Table 1. Selected material properties (a) elastic constants (b) strength of the orthotropic fabric reinforced polymer CF-EP - determined with standard test methods.

(a)

elastic constants		
$E_1 = 69.4\,\text{GPa}$	$G_{12} = 4.95\,\text{GPa}$	$\nu_{12} = 0.080$
$E_2 = 69.4\,\text{GPa}^*$	$G_{13} = 3.40\,\text{GPa}$	$\nu_{13} = 0.015$
$E_3 = 11.5\,\text{GPa}$	$G_{23} = 3.40\,\text{GPa}^*$	$\nu_{23} = 0.015^*$

*symmetry assumed

(b)

strength	
$R_1^+ = 911\,\text{MPa}$	$R_{12} = 185\,\text{MPa}$
$R_2^+ = 911\,\text{MPa}^*$	$R_{13} = 35\,\text{MPa}$
$R_3^- = 907\,\text{MPa}$	$R_{23} = 35\,\text{MPa}^*$

*symmetry assumed

2.2. Material and samples

The material used in this investigation is a carbon fibre reinforced thermoset, compounded from a textile plain weave fabric (ECC-style 450, areal density: $200\,\text{g/m}^2$), processed out of TENAX-E® HTMA 3K rovings, and an aerospace epoxy thermoset (RTM6) in a resin transfer moulding process. After consolidation of a 10 mm thick plate with an estimated fibre volume fraction of 60 % and a density of $1.52\,\text{g/cm}^3$, the specimens were cut out and the hole notch pattern has been added by drilling and milling.

According to the symmetrical rapport of a plain weave fabric and neglecting the influence of the weaving process, it can be assumed that the in-plane elastic and strength properties are identical in the material coordinate of the 1- and 2-axes. The through-thickness direction (material coordinate system: 3-axis) of this orthotropic material differs in its elastic and strength properties (cf. Table 1).

The geometry of the specimens, as illustrated in Figure 3, has been numerically and experimentally designed on the basis of the double lap shear [2, 15]. Due to the force redirection in the hole notch area, interlaminar shear stresses τ_{13} are initiated in the CFRP.

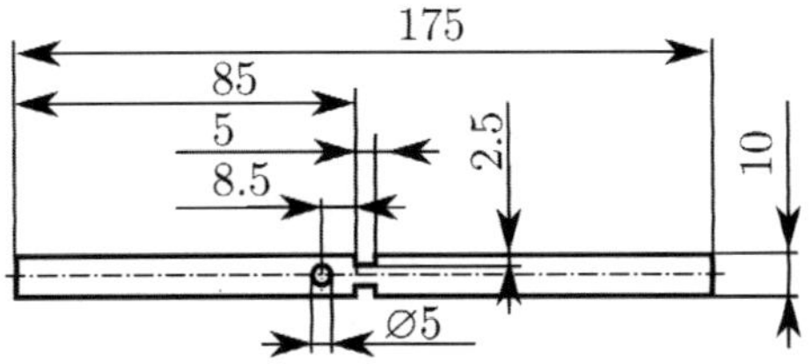

thickness: 10 mm

(a)

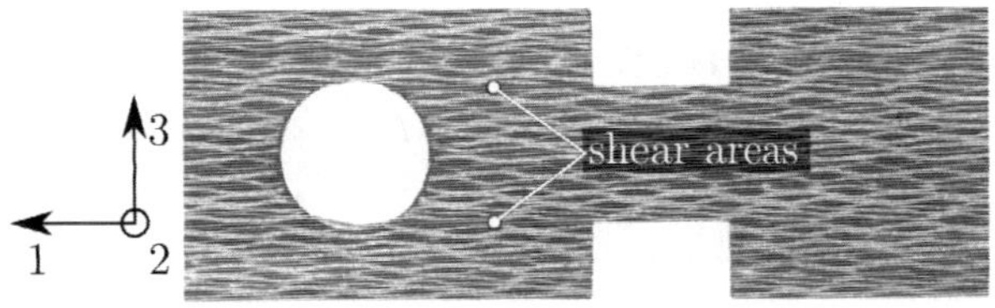

(b)

Figure 3. (a) Drawing of the specimen with hole-notch-combination. (b) Picture of the sample in the testing area with visible layers and two main shear areas (red).

2.3. Testing device

In comparison to the quasi-static test setup it has been decided for a fixation (cf. Figure 4) instead of a wedge clamping, because of the easier manufacturing process of the specimens and the low expected forces [2]. Additionally the noval test setup is able to apply either tensional or compressional forces in x- respectively 1-direction. To compensate any misalignment of the fixtures a dove tail for fixation is incorporated.

A further evolutionary step was to increase the stamps' contact surface to avoid premature failure at the outer notches, as mentioned before. Hence, equal stress states at the notches could be achieved. As in the former setup radii were added on the sides of the stamps to decrease

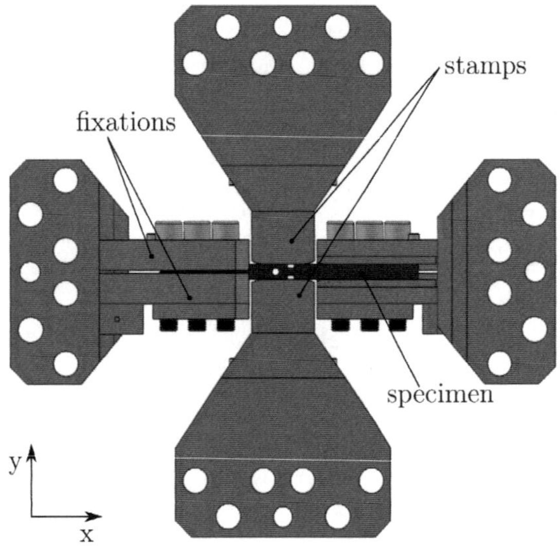

Figure 4. Illustration of the test setup with fixations and stamp alignment.

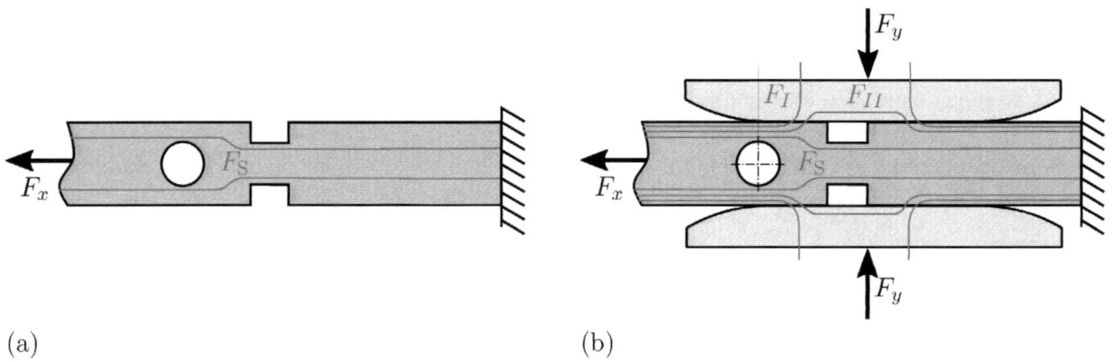

Figure 5. Illustration of the test setup and its schematic force fluxes for (a) uniaxial and (b) biaxial testing.

the Hertzian contact pressure and to avoid premature failure in that region. Both compression stamps and fixations were milled out of hardened 42CrMo4 steel without any surface treatment like diamond like coating as used in [2]. Nevertheless, on the one hand the exceptional higher force demand in y-direction compared to the solution proposed in [2] and on the other hand the previously indefinable effect on the stress distribution due to friction are challenging.

Without the use of the stamps in uniaxial shear testing, the axial force F_x is completely conducted over the shear planes, what results in a shear force F_S (cf. Figure 5 (a)). In biaxial testing, the axial force F_x is the sum of the shear force F_S and the friction forces F_I and F_{II} (cf. Figure 5 (b)). The force F_I results in a bending moment of the stamps, while F_{II} is reconducted through the stamps in the specimen. Without any reliable characterisation of the tribological system it is defying to estimate the shear stresses causing delamination by measuring the reaction forces. In this paper a second attempt is presented to obtain the forces by a strain-measuring method.

2.4. Sample testing

Varying tensile loads F_x and constant or cyclic pressure loads F_y were applied on the specimens. Also different phase shifts ϕ between F_x and F_y were examined (cf. Figure 6). The initial loads (cf. Table 2) were set to different levels in order to reduce testing time. After every 10^5 cycles, the loads were step-wise increased by 10 %. The first cycle of each level had a frequency that was defined by a constant force change rate of 2 kN/s (cf. Figure 7). All other cycles on that level were performed with a sinus shaped function and a frequency of 5 Hz.

Table 2. Initial test load parameters for the specimens and amount of load increasement (subscripts: »a« amplitude, »m« mean force).

Specimen	Initial load					Increase after 10^5 cycles			
	$F_{x,\text{ m}}$ [kN]	$F_{x,\text{ a}}$ [kN]	$F_{y,\text{ m}}$ [kN]	$F_{x,\text{ a}}$ [kN]	ϕ [°]	$\Delta F_{x,\text{ m}}$ [%]	$\Delta F_{x,\text{ a}}$ [%]	$\Delta F_{y,\text{ m}}$ [%]	$\Delta F_{y,\text{ a}}$ [%]
1	0.55	0.45	0	0	0	10	10	0	0
2	3.19	2.61	30	2.61	-90	10	0	0	0
3	0.83	0.68	10	0	0	10	10	0	0
4	2.18	1.78	30	0	0	10	10	0	0
5	2.92	2.39	2.92	2.39	180	10	10	10	10
6	3.19	2.61	3.19	2.61	90	10	10	10	10
7	3.19	2.61	3.19	2.61	-90	10	10	10	10
8	3.19	2.61	30	2.61	90	10	0	0	0

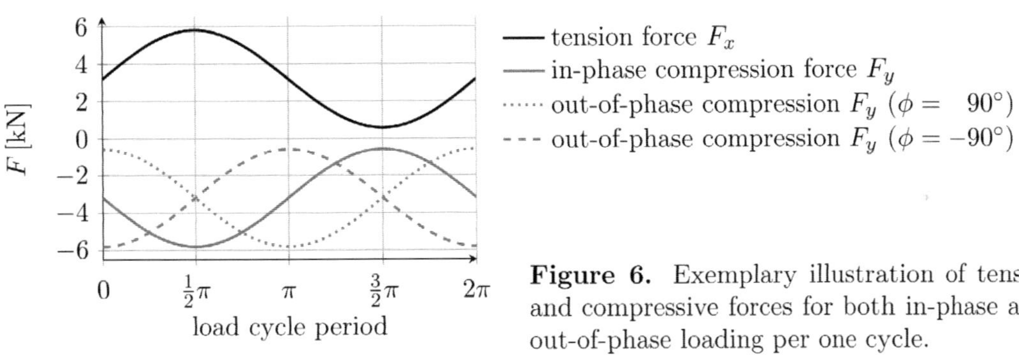

Figure 6. Exemplary illustration of tensile and compressive forces for both in-phase and out-of-phase loading per one cycle.

2.5. Measurement of strains and stresses in the shear areas

To evaluate the deformation and to calculate the stresses by a linear-elastic material model in the main shear areas on the surface, the acquired images were assessed by a DIC software. Before the test, one side of the sample was prepared with white paint and graphite spray to produce a unique randomly distributed gray-scale pattern with high contrast. The shear force F_S was calculated by the strain in 1-direction ε_1 in the area of the notches and the elastic behaviour of the material (cf. Figure 8). Also the average strain of the through-thickness compression ε_3 was measured between the notches and the hole by this method.

2.6. Simulation of the pressure distribution in the shear areas

For the understanding of the pressure distribution in the shear areas of the specimen by applied stamps, finite element analysis (FEA) has been used. For the specimens mesh hexahedron elements with a mean element length of 0.5 mm and a higher mesh density in the shear areas

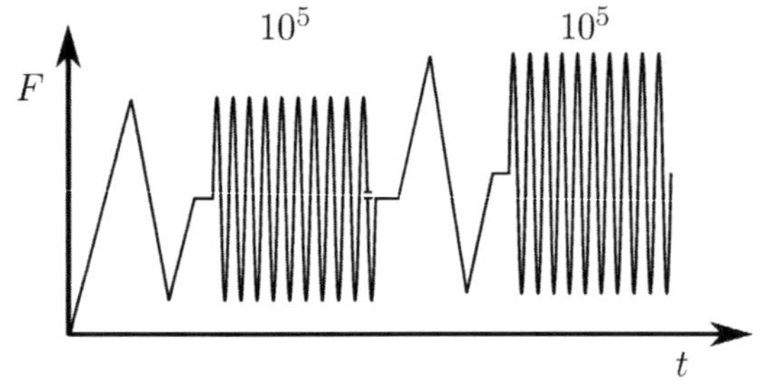

Figure 7. Schematic illustration of the sequence of the cyclic tests with step-wise load increase after 10^5 cycles.

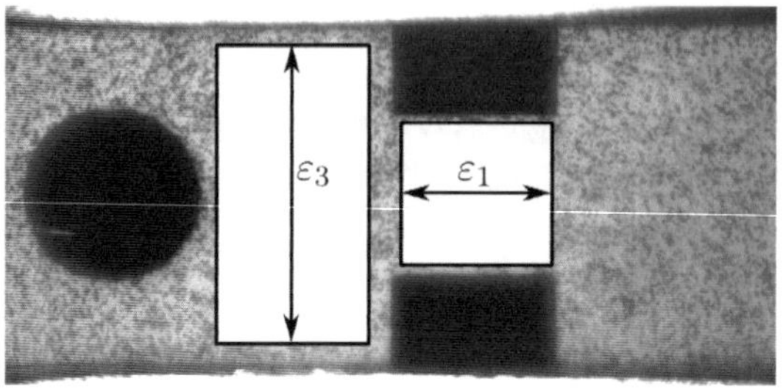

Figure 8. Measuring areas of the DIC for the through-thickness compression and the elongation resulting to shear stress.

down to 0.25 mm were used (cf. Figure 9 (a)). The two dimensional simulation was done using linear elastic orthotropic material behaviour.

3. Results and discussion

FEA of applied compressive forces F_y showed a much better pressure distribution in the shear planes with the stamps in comparison to [2]. In the area of the hole and the notches, the σ_3 stress was inhomogeneous. In some positions, e.g. at the hole and at the notches, even tensile stresses in 3-direction were detected (cf. Figure 9 (b)). Higher compressive forces showed a more homogeneous pressure distribution in the area between the hole and the notches.

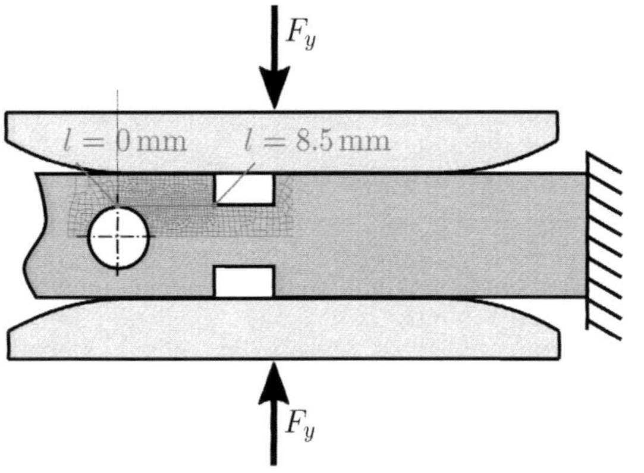

(a)

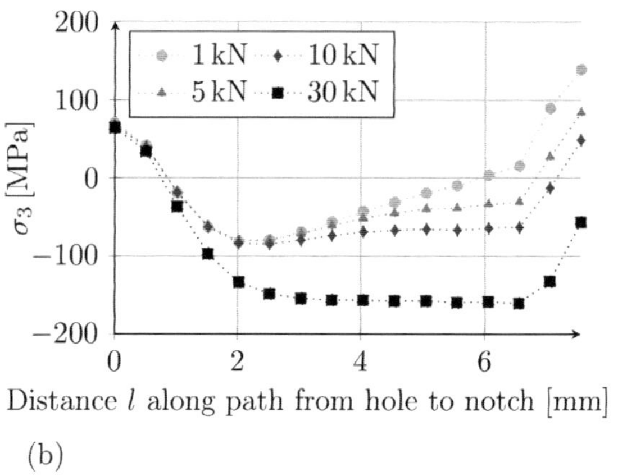

(b)

Figure 9. (a) FEA mesh in the shear area with the chosen line for pressure evaluation in 3-direction from hole ($l = 0$ mm) to notch ($l = 8.5$ mm, red line).
(b) Stress distribution in 3-direction σ_3 in the shear area along the red line of figure 9 (a) at different compressive forces F_y.

Because of the lower pressure or tensile stresses in 3-direction, crack initiation occurred in the area of the notches or of the hole (cf. Figure 10); the notch tip was not the position of crack initiation.

Fatigue crack growth was detected along the shear planes between the fabric layers (cf. Figure 11). The crack growth is connected with a reduction of the shear areas, what results in higher shear stresses and higher shear angles, if the same shear load F_S is applied. Furthermore, it was possible to determine the delamination propagation qualitatively by the strain-field assessment of ε_1 from DIC analysis.

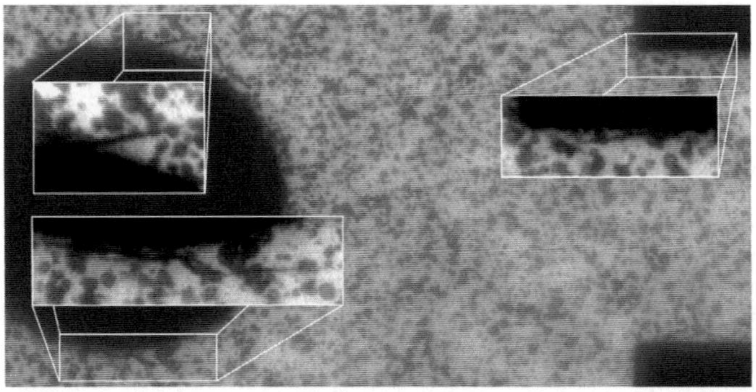

Figure 10. Crack initiation in the areas with low compressve stresses in 3-direction of specimen 3 at $N = 370\,000$ cycles.

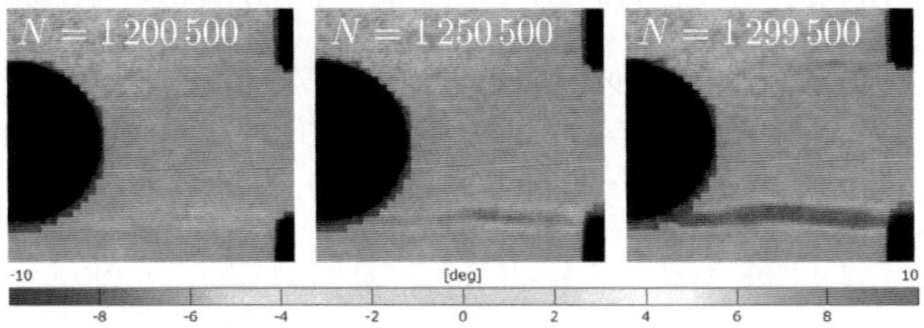

Figure 11. Shear angle at the last stage of specimen 8. The load on all pictures is the same. The absolute shear angle increases with decreasing shear area and increasing shear stress.

Unwanted delamination on the other side of the notches was prevented through the compression of the stamps. The friction between the sample and the fixations was high enough to apply tensile forces F_x on the sample up to 10 kN.

Only two samples failed directly after load increase. Again, failure occurred through shear on the shear planes. The maximum possible stresses estimated from DIC are shown in Figure 12 and are compared with static results. Here maximum cyclic delamination failure stress is below the

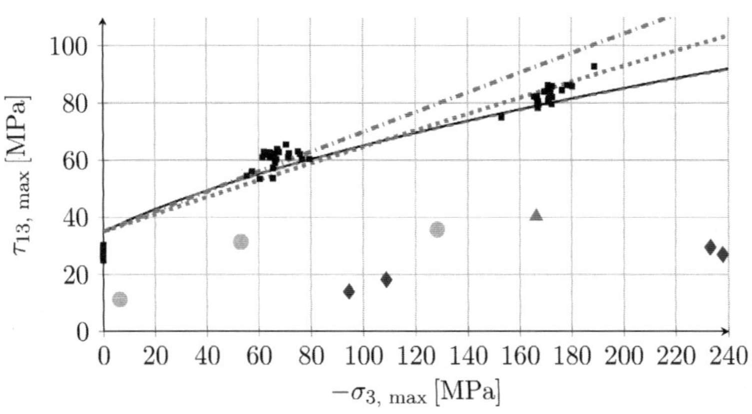

Figure 12. Cyclic failure level in comparison to static results and failure criteria by different authors.

static strength in dependence of the interlaminar pressure. Interlaminar pressure increases the shear failure level. Specimens with phase shifts of ±90° failed below all other specimens. Even if the layers in the shear areas were not connected anymore and were delaminated, the friction between the layers had to be overcome to notice a failure and a seperation of the specimen.

4. Conclusions

- Failure didn't occur directly after load increase in seven of nine specimens.
- The stresses in the shear areas could be estimated in a indirect way by obtaining the strain via DIC and a determination of the stresses with an linear elastic material behaviour.
- The maximum forces of shear fatigue failure were significantly lower than the static failure results.
- Phase shifts ϕ of $\pm 90°$ resulted in fatigue failure at even lower shear stress amplitude levels.
- Crack initiation occurred in the area of the notches and hole, because of the low pressure (or even tensile stress components) in 3-direction.
- Crack growth was detectable. The localisation of the crack tip on the surface under compression was possible with help of DIC.
- Through-thickness compression affects the delamination propagation and the fatigue behaviour positively.

Acknowledgements
The authors gratefully acknowledge the financial support of the German Research Foundation (DFG - GU 614/8-1). The authors also thank M. Sc. Andreas Seupel from the Institute of Mechanics and Fluid Dynamics of Technische Universität Bergakademie Freiberg for his support in FEA.

References
[1] Harris B 2003 *Fatigue in Composites: science and technology of the fatigue response of fibre-reinforced plastics* (Cambridge: Woodhead Publishing Limited)
[2] Böhm H, Schirner R, Gude M and Böhm R 2017 Experimental and numerical investigation of appropriate CFRP specimen designs for inducing in-plane tensile/through-thickness compressive and interlaminar shear/through-thickness compressive stress states *Proc. 8th Int. Conf. Compos. Test. Model Ident.*
[3] Gan K W, Hallet S R and Wisnom M R 2013 Measurement and modelling of interlaminar shear strength enhancement under moderate through-thickness compression *Compos. A* **49** 18-25
[4] Collings T A 1974 Transverse compressive behaviour of unidirectional carbon fibre reinforced plastics *Composites* **5** 108–16
[5] McManamy T J and Kanemoto G and Snook P 1991 Insulation irradiation test programme for the Compact Ignition Tokamak *Cryogenics* **31** 277-81
[6] DeTeresa S J, Freeman D C and Groves S E 2004 The Effects of Through-thickness Compression on the Interlaminar Shear Response of Laminated Fiber Composites *Journal of Composite Materials* **38** 681-97
[7] Daniel I M, Luo J J and Schubel P M 2008 Three-dimensional characterization of textile composites *Compos. B* **39** 13-19
[8] Sun C 2000 Strength analysis of unidirectional composites and laminates *Compr. compos. mat.* 641-66
[9] Christensen R M and DeTeresa S 2004 Delamination failure investigation for out-of-plane loading in laminates *J. Compos. Mat.* **38** 2231-38
[10] Cuntze R G and Freund A 2004 The predictive capability of failure mode concept-based strength criteria for multidirectional laminates *Compos. Sci. Tech.* **64** 343-77
[11] Hufenbach W, Horning A, Gude M, Böhm R and Zahneisen F 2013 Influence of the interface waviness on delamination characteristics and correlation of through-thickness tensile failure with mode I energy release rates in carbon fibre textile composites *Mat. Design* **50** 839-45
[12] Olave M, Vara I, Usabiage H, Aretxabaleta L, Lomov S V and Vandepitte D 2015 Nesting effect on the mode I fracture toughness of woven laminates *Compos. A* **74** 166-73
[13] Olave M, Vara I, Usabiage H, Aretxabaleta L, Lomov S V and Vandepitte D 2015 Nesting effect on the mode II fracture toughness of woven laminates *Compos. A* **74** 174-81
[14] Smits A, Ramault C, Makris A, Van Hemelrijck D V, Clarke A, Williamson C, Gower M, Shaw R, Mera R, Lamkanfi E and Van Paepegem W 2007 A review of biaxial test methods for composites *Experim. Anal. Nano Engin. Mat. Struct.* 933-34
[15] ASTM Standard 3528-96: Test Method for Strength Properties of Double Lap Shear Adhesive Joints by Tension Loading. West Conshohocken, PA: ASTM International, 2008